Molecular Biology Methods

Siddra Ijaz • Imran Ul Haq
Hayssam Mohamed Ali
Editors

Molecular Biology Methods

Springer

Editors
Siddra Ijaz
Centre of Agricultural Biochemistry and Biotechnology (CABB)
University of Agriculture
Faisalabad, Pakistan

Imran Ul Haq
Department of Plant Pathology
University of Agriculture
Faisalabad, Pakistan

Hayssam Mohamed Ali
Department of Botany and Microbiology
College of Science, King Saud University
Riyadh, Saudi Arabia

ISBN 978-981-92-2641-2 ISBN 978-981-92-2642-9 (eBook)
https://doi.org/10.1007/978-981-92-2642-9

This Springer imprint is published by the registered company Springer Nature Singapore Pte Ltd.
The registered company address is: 152 Beach Road, #21-01/04 Gateway East, Singapore 189721, Singapore

Preface

Almost every area of modern society is addressed by biotechnology. Biotechnology can therefore play a significant role in the development of national policies in the field of agricultural biotechnology. The Centre of Agricultural Biochemistry and Biotechnology (CABB) mission is to contribute to the country's economic development by supporting innovative development of technologies, world-class scientific research, and talented scientists. The contents of this professional book are meant to enhance the technical skills of Biotechnology graduates, to enable them to develop themselves as learned researchers capable of research activities relevant to recombinant DNA technology, plant tissue culture, plant transformation, forensic analysis of animals and plants, molecular breeding, analysis of transgenic plants, development of edible vaccines and therapeutics, protoplast fusion, metabolomics, and proteomics. This will not only allow them to get into the international arena, but they will be able to conduct excellent research on issues that have a stakeholder interest.

Biotechnology has become one of the most revolutionary and dynamic areas of modern science involving almost all aspects of human society such as agriculture, health, industry, environment, and food security. Agricultural biotechnology has become particularly important in the field of scientific innovation, since countries are increasingly depending on biotechnology advances to achieve sustainable development, economic growth, and an improvement in the quality of life. Highly trained professionals with modern technical knowledge and research skills are, therefore, essential for the development of effective national policies in agricultural biotechnology.

This professional book is written with these emerging needs in mind to improve the technical and practical ability of the Biotechnology graduate and young researcher.

This professional book is written with these evolving needs in mind to improve the technical and practical skills of the Biotechnology graduate and young researcher. The main aim of this book is to give students extensive knowledge and skill-based learning in different advanced fields of biotechnology which can help them to understand the theory and its application in real practice. Its vision is to nurture scientifically competent researchers to serve the scientific community nationally and internationally.

The contents of this book encompass a wide range of modern biotechnological techniques and applications, including recombinant DNA technology, plant tissue culture, plant transformation, molecular breeding, forensic analysis of plants and animals, analysis of transgenic plants, development of edible vaccines and therapeutics, protoplast fusion, proteomics, and metabolomics. Topics are well organized and contain conceptual and practical information which is essential for any research or industrial application in modern times.

This book also demonstrates the overarching goal of fostering scientific excellence, innovative technology development, and world-class research culture. It is hoped that the graduates will have advanced technical skills and research capacity which will enable them to compete with confidence at the international level and also will make a contribution towards solving real-world problems of farmers, industries, health sector, and other stakeholders. The knowledge and expertise acquired from this book will in the end help develop sustainable biotechnological solutions, which can contribute to national progress and economic well-being.

The aim of this book is to help students, researchers, academicians, and professionals deepen their knowledge and application of biotechnology. Special thanks and acknowledgments are given to the work of all contributors, researchers, and educators who have provided assistance in the development of this work.

Faisalabad, Pakistan Siddra Ijaz
Faisalabad, Pakistan Imran Ul Haq
Riyadh, Saudi Arabia Hayssam Mohamed Ali

Contents

Contributors

Amjad Abbas Department of Plant Pathology, University of Agriculture, Faisalabad, Pakistan

Hina Ali Ahmed Department of Zoology, Sardar Bahadur Khan Women University, Quetta, Pakistan

Muhammad Amjad Ali Department of Plant Pathology, University of Agriculture, Faisalabad, Pakistan

Saqib Ali Department of Computer Science, University of Agriculture, Faisalabad, Pakistan

Syed Zulfiqar Ali Department of Plant Pathology, Balochistan Agriculture College, Quetta, Pakistan

Mahrukh Babar Faculty of Pharmacy, IBADAT International University, Islamabad, Pakistan

Maria Babar Centre of Agricultural Biochemistry and Biotechnology (CABB), University of Agriculture, Faisalabad, Pakistan

Muhammad Ehetisham-Ul-Haq Oilseeds Research Institute, Ayub Agricultural Research Institute, Faisalabad, Pakistan

Zakia Habib Centre of Agricultural Biochemistry and Biotechnology (CABB), University of Agriculture, Faisalabad, Pakistan

Imran ul Haq Department of Plant Pathology, University of Agriculture, Faisalabad, Pakistan

Sheraz Hassan Centre of Agricultural Biochemistry and Biotechnology (CABB), University of Agriculture, Faisalabad, Pakistan

Siddra Ijaz Centre of Agricultural Biochemistry and Biotechnology (CABB), University of Agriculture, Faisalabad, Pakistan

Zobia Jabeen Department of Plant Pathology, Balochistan Agriculture College, Quetta, Pakistan

Muhammad Jabran State Key Laboratory for Biology of Plant Disease and Insect Pests, Institute of Plant Protection, Chinese Academy of Agricultural Sciences, Beijing, China

Muhammad Hassan Jahanzaib Centre of Agricultural Biochemistry and Biotechnology (CABB), University of Agriculture, Faisalabad, Pakistan

Aurang Zaib Jamali Department of Soil Science, Balochistan Agriculture College, Quetta, Pakistan

Ali Hassan Khan Horticulture Research Institute, National Agriculture Research Center, Islamabad, Pakistan

Nabeeha Aslam Khan Plant Pathology Section, PPRI, Ayub Agricultural Research Institute, Faisalabad, Pakistan

Hayssam Mohamed Ali Department of Botany and Microbiology, College of Science, King Saud University, Riyadh, Saudi Arabia

Bakhtawar Nasir The University of Faisalabad, Faisalabad, Pakistan

Ali Raza School of Environmental and Rural Science, University of New England, Armidale, NSW, Australia

Hafiza Arooj Razzaq Center of Agricultural Biochemistry and Biotechnology, University of Agriculture, Faisalabad, Pakistan

Muhammad Asif Sadsique University of Agriculture, Faisalabad, Pakistan

Sadaf Safder Center for Advanced Studies in Agriculture and Food Security, University of Agriculture, Faisalabad, Pakistan

Muhammad Huzaifa Tanveer Department of Plant Pathology, University of Agriculture, Faisalabad, Pakistan

Atta Ullah Department of Plant Breeding and Genetics, Balochistan Agriculture College, Quetta, Pakistan

Muhammad Usman Department of Plant Pathology, University of Agriculture, Faisalabad, Pakistan

Muhammad Waris Department of Plant Pathology, Balochistan Agriculture College, Quetta, Pakistan

Javeria Yasin Department of Plant Pathology, University of Agriculture, Faisalabad, Pakistan

Chapter 1
Fundamental Methods in Molecular Biology

Siddra Ijaz and Sheraz Hassan

1.1 Introduction

Various fields, such as medical diagnostics, molecular biology, biotechnology, and forensic science, require the fundamental process of DNA extraction from eukaryotic cells to advance their respective investigations. In this process, DNA is isolated from the cells by breaking them open, and then the isolated DNA is purified and precipitated using standard laboratory protocol. Over time, DNA extraction has improved a lot in producing high yields of quality DNA that can further be used in several applications like genetic cloning, polymerase chain reaction (PCR) amplification, analyzing gene expression, and sequencing. The main purpose of writing this review is to sum up the currently available literature on different approaches for DNA extraction and precipitation from eukaryotic cells with special emphasis on principal methodologies, major challenges, and key innovations in the field.

DNA and RNA are the information-carrying molecules that responsibly pass on genetic information from one generation to the other with high accuracy and precision. These biomolecules are also known as nucleic acids due to their nature, biochemical composition, and their relative position within the cell. In recent times, several techniques have been developed for the efficient analysis of these biomolecules, and the application of these novel approaches has revolutionized the whole world of science and technology. Forensic science, medical diagnostics, genetics, molecular biology, biotechnology, and biochemistry are among the most fortunate fields that use nucleic acid analysis techniques to carry out their investigations, which have led to the fundamental discoveries of the medical world. Nucleic acid analysis has a larger number of important applications, such as getting insights into evolutionary relationships, gene expression analysis, genetic mutation and

S. Ijaz (✉) · S. Hassan
Centre of Agricultural Biochemistry and Biotechnology (CABB), University of Agriculture, Faisalabad, Pakistan

S. Ijaz et al. (eds.), *Molecular Biology Methods*,
https://doi.org/10.1007/978-981-92-2642-9_1

abnormality detection, and gene function-related studies. This review provides a brief introduction to some of the nucleic acid analysis techniques, along with their prominent applications in various areas, limitations, and recent advancements.

Polymerase chain reaction (PCR), also known as molecular photocopying, is an extensively utilized technique in the fields of biochemistry, forensics, molecular biology, genetics, and biotechnology. It has upgraded the whole scenario of genetic cloning through the rapid multiplication of targeted nucleotide sequences. Kary Mullis invented this method in 1983 (Mullis et al. 1986), while it was patented two years later, in 1985. By using a small number of DNA samples, it can increase the copy number of the target region of DNA to millions or even billions in a few hours through repeated heating and cooling cycles. Owing to the sensitivity and efficiency of PCR, it is a common laboratory method in medical diagnostics and other related fields. With the advancement in technology and ever-changing demands of the medical world, and to boost the specificity, reliability, and application spectrum of PCR, several variants of PCR have been developed by slightly modifying the standard setup of PCR. This chapter outlines the basic working mechanism of PCR, along with some of its key variants, prominent applications, and implications in various research areas, including food science, agriculture, biotechnology, and biochemistry (Ijaz and Haq 2019).

Gel electrophoresis is a versatile technique commonly used in various research studies across genetics, molecular biology, forensics, biotechnology, medical diagnostics, biochemistry, and clinical pathology. The primary purpose of employing this technique is to achieve size-based resolution of fragments from different biomolecules, including DNA, RNA, and proteins. The results of this approach facilitate further analysis, such as Southern or Northern blotting, by utilizing the separated bands-containing gel for the detection of specific sequences of DNA or RNA through the process of probe hybridization. This review highlights the working principle and mechanism, various application-based gel types and compositions, and prominent application areas alongside the major pros and cons of the standard gel electrophoresis method. Agarose gel electrophoresis (AGE) is a technique commonly used for separating different single-stranded polynucleotide molecules (RNAs), double-stranded polynucleotide molecules (DNAs), and polypeptide molecules (proteins) based on their relative charges, structures, and sizes in an applied electric field. It is now an integral part of various areas of the life sciences, including diagnostics, molecular biology, genetic engineering, pharmacogenetics, biochemistry, and biotechnology. This technique has applications in determining the quality and purity of various DNA and protein samples (Ijaz and Haq 2019; Sambrook and Russell 2001). Similarly, DNA fragment analysis (required in forensics), analysis of gene expression level (imperative for drug designing), and sequencing (important for species profiling) use this approach routinely and regularly.

Polyacrylamide gel electrophoresis (PAGE) is a crucial technique for analyzing and quantifying various biomolecular samples, including proteins and nucleic acids (DNA or RNA), which are integral to advancing research studies in fields such as proteomics, molecular biology, forensics, and diagnostics. The separation of molecules is based on their relative charge, shape, and size, and it occurs along the vertical axis of the gel, with bands of molecular fragments arranged in increasing order of size from top to bottom. It is a technique that is employed for its unique and commendable features, such as high-resolution sample fragments, efficient separation and quantification of low-abundance molecules, and a flexible nature. This chapter provides an overview of this fundamental approach, with a special emphasis on its working principles, types, methodology, major areas of application, and technological innovations, alongside some of its major pros and cons.

In the case of the protein sample, sodium dodecyl sulfate-polyacrylamide gel electrophoresis (SDS-PAGE) is a preferred purification technique. Similarly, formaldehyde gel electrophoresis is a method of choice for the efficient resolution of the bands of RNA samples. This technique operates on the same generalized principle as the migration of charged fragments under the action of an applied electric field, with separation occurring based on their size-to-charge ratio. Formaldehyde can denature RNA in such a way that it does not allow RNA to adopt a secondary or tertiary structure. Thus, RNA fragments get separated in their denatured or linear form as required for many of their downstream applications, such as gene expression analysis, RNA size estimation, mRNA analysis, and quantification of RNA. This chapter deals with the general principle behind formaldehyde gel electrophoresis, the procedural steps involved, prominent applications, advantages, limitations, and useful innovations that effectively address the probable issues faced by the researcher during the process.

Protein isolation and analysis are the most widely used approaches in various protein-related research fields, including biochemistry, forensics, molecular biology, and diagnostics. The isolation and analysis are two distinct areas, as the isolation process involves the efficient extraction and recovery of different proteins from living cells and tissues using various tools and techniques. In contrast, the analysis of proteins refers to those methodologies that are employed for the efficient separation, quantification, visualization, and characterization of different protein samples. In this way, both of these processes are equally important for carrying out an in-depth study of fundamental aspects of proteins such as their structural conformation, functionality, and interactions with other molecules. These features of proteins help understand the basics of ongoing biological processes within the bodies of living organisms. Therefore, such methodologies are employed for the efficient isolation and analysis of protein samples, which do not disturb the structural and functional domains of proteins under study. This chapter provides an overview of some of the widely used techniques for protein isolation and analysis, with a special focus on the prominent applications, technical hurdles, and recent advancements that have occurred in these fields in recent times.

Electroelution is a combination of two fundamental molecular biology techniques that researchers routinely use for analyzing different biomolecules such as nucleic acids (DNA and RNA), proteins, and peptides. In electroelution, electro denotes the process of Electrophoresis that involves the resolution of biomolecular fragments (DNA, RNA, or proteins) under the influence of an electric field. In contrast, elution implies the process that involves the retrieval of electrophoretically resolved or separated bands of biomolecules from the gel matrix (Hames 1990). In the research world, electroelution is a method of choice for the efficient purification and recovery of separated fragments. Hence, it is a principal approach that facilitates the execution of several other biomedical analyses of importance, such as spectrometry, blotting, and expression analysis, among others. This chapter provides a brief but comprehensive overview of the working principle, important applications, advantages, recent innovations, and different drawbacks that must be addressed to enhance the efficiency and broaden the application spectrum of the electroelution process (Weber and Osborn 1969).

1.2 DNA Extraction and Precipitation from Eukaryotic Cells

The evolution of the DNA extraction process over time and the basic working mechanism of different components used in the procedure are:

1. Early Methods of DNA Extraction
In 1871, Friedrich Miescher became the first person in history to isolate nucleic acid (DNA) from eukaryotic white cells (pus) using a combination of alkaline and acidic solutions for subsequent extraction and precipitation of DNA. In this way, he led the foundation of a highly versatile process, which was not further improved until the 1960s. Then, in 1962, two scientists named Marmur and Doty came up with the idea of using detergents as they observed that detergents can easily break the cell membrane down to release its internal components. The method (they developed) was applied effectively to a diversity of eukaryotic cells. Hence, it marked the beginning of a new area of research, which led to further improvement in obtaining enough high-quality DNA with the help of easy and accessible methods.

2. Modern Methods of DNA Extraction
Innovations made the process of DNA extraction and precipitation procedure faster, easier, and more straightforward. In this process, the disintegration of the cell wall and cell membrane was considered one of the challenges. The availability of different detergents (i.e., SDS, Triton X-100, and CTAB) made it an easy task. CTAB, in particular, helped dissolve the resilient cell walls of plant cells (Porebski et al. 1997). Afterward, mechanical means of lysis (i.e., homogenization and grinding) made the process even better and efficient. Different approaches were devised to eliminate the impurities (i.e., RNA, proteins, and cell debris) that may interfere with the quality and purity of DNA. Similarly, DNA precipitation also improved since Miescher used an acidic solution for this purpose. The novelty, in this regard, was the introduction of alcohol (i.e., isopropanol) usage by replacing acidic solutions to safeguard the integrity of isolated DNA. In recent years, the development of automated DNA extraction kits has made the process faster, more reliable, and easier to manage.

3. General Principle of DNA Extraction

For the successful extraction of an ample quantity of high-quality DNA, eukaryotic cells have to pass through three different steps, including cell lysis (breaking open the cells to release their cellular components), removal of cellular debris (to eliminate other biomolecules and purify DNA), and DNA precipitation (to isolate DNA present in solution). Similarly, three distinct approaches are employed for this purpose: enzymatic (a mixture of cellulase, xylanase, and pectinase), organic (phenol-chloroform), and nonorganic (silica-based or magnetic bead-based). The type of eukaryotic cell (plant or animal), the expected percentage purity of the extracted DNA, and the possible application of the obtained DNA are the major determinants of the method of choice.

1.3 Process of DNA Isolation

Several intermediate stages are involved to get an isolated, purified, and ready-to-use DNA sample. These stages include:

1. Sample Collection

Like any other process, DNA extraction requires raw material to get started, which, in this case, is whole tissue or cultured cells. The sample should be chosen with great care, as it is an important determinant of the efficacy of the procedure. When it comes to the ease of DNA extraction, cultured cells are preferred over connective tissues, as the tissue is not an easy target for lysis.

2. Cell Lysis

The cell membrane of the eukaryotic cell is disintegrated in such a way that the intracellular material (i.e., DNA along with other biomolecules such as proteins, sugars, and lipids) comes out of it in the surrounding buffer solution. Improper handling may cause shearing of the extracted DNA, rendering the entire procedure useless. Lysis buffer contains salts like NaCl, which help reduce pressure on DNA, triggered by abrupt changes caused by membrane destabilization. Salt stabilizes DNA, while EDTA (ethylenediaminetetraacetic acid), an important component of buffer, sequesters metal ions such as Mg+2 to hinder the activity of DNases and prevent DNA degradation. Similarly, proteinase K removes contamination present in the form of protein to get purified DNA. However, it has a drawback, as it may remove essential components, such as DNA-associated proteins, making isolated DNA unproductive when used in specific studies.

Cell lysis can be accomplished by a variety of methods that include:

- *Mechanical lysis*: In this technique, cells having inflexible cell walls (i.e., plants and fungi) are frequently disturbed using various physical means, for example, sonication, grinding, homogenization, etc., to break open their cell walls and cell membranes (Möller et al. 1992).

- *Chemical lysis*: This technique employs various detergents such as SDS and Triton X-100 to disintegrate the cell membrane and maintain the integrity of the extracted DNA. For extracting DNA from animal cells, which are rich in histones and other proteins, the method of enzymatic lysis, which depends on the activity of enzymes like proteinase K, is usually chosen to avoid any interference during the DNA isolation process (Sambrook et al. 1989).

3. Removal of Contaminants

No doubt, the process of DNA precipitation using alcohol is highly efficient. Still, there is always a risk of coprecipitation of residual components of the lysate, such as RNA, proteins, and salts. These contaminants will make the DNA impure and also cause complications in subsequent steps. It is difficult but necessary to eliminate these impurities to get high-purity DNA. For this purpose, the following processes should be taken into account:

- *RNA removal*: It can be achieved using RNase, an enzyme, which will degrade predominantly RNA molecules and keep the DNA intact if the duration of treatment with RNase is optimized carefully and attentively (Wilfinger et al. 1997a).
- *Protein removal*: There are two approaches to eliminate protein contamination. Either of the two methods can be chosen depending on the desired result. One method employs a mixture of phenol and chloroform, which will discriminate proteins from isolated DNA based on the liquid-liquid extraction principle. This practice imposes the risk of organic solvent contamination. On the other hand, the second method uses a large quantity of salt like NaCl that bypasses the issue of residual contamination and provides highly pure DNA by efficiently removing contaminating proteins.

4. DNA Purification

Cell lysis releases all cellular components in the form of a mixture (lysate) containing different proteins, lipids, salts, and (of course) DNA. These biomolecular impurities must be eliminated first to extract an adequate quantity of pure DNA. For this purpose, several methods can be employed that are presented in the following:

- *Phenol-chloroform extraction*: As the name indicates, this technique utilizes a mixture of phenol and chloroform to purify or extract DNA from a mixture of biomolecules (lysate). Although this old-fashioned approach uses toxic chemicals and takes time, it works well and produces enough quantity of purified DNA (Wilfinger et al. 1997a).
- *Silica-based columns*: Silica-based spin columns, such as those made by Qiagen and Thermo-Fisher, are frequently utilized in currently available DNA purification techniques. When exposed to high salinity, the silica matrix captures DNA that can later be eluted using a low-salinity buffer solution (Boom et al. 1990). High-quality DNA can easily be obtained with the help of such methods, as they provide a quick and secure way to do so.
- *Magnetic beads-based methods*: In this method, DNA-binding ligands-coated magnetic beads are utilized to purify the DNA from the lysate. The isolation and recovery of bead-bound DNA are carried out using a magnetic field of appropriate

strength, which ultimately makes steps like elution and washing less labor-intensive and saves considerable time (Tobler et al. 1999).

5. DNA Precipitation

In one way or another, the success of the DNA extraction process depends on the appropriate precipitation of DNA, as it ensures that the downstream process is going well. For this purpose, a salt (ammonium acetate) aided by alcohol (i.e., isopropanol or ethanol) is a good choice, as it counterbalances the phosphate-induced negative charge and causes DNA to precipitate easily and efficiently. The mechanism of alcohol-facilitated precipitation and different factors that can influence the quality of DNA are discussed below:

(a) *Alcohol Precipitation of DNA*

The procedure of alcohol precipitation of DNA is among the most commonly implemented practices, which yields DNA in the form of threads or pellets that can easily be washed and resuspended for further applications. It includes:

- *Addition of salts*: Different combinations of acetate salt, such as sodium acetate or ammonium acetate, can precipitate DNA by preventing it from solubilization and neutralizing the negative charge imparted by abundantly present phosphate groups (Wilson 2001).
- *Addition of alcohol*: Isopropanol, usually added to about half the volume of the supernatant, aids the process of precipitation by making appropriate, visible, and long fibers of DNA that are subsequently pelleted with the help of centrifugation to carry out further steps effectively (Sambrook et al. 1989). A suitable volume and concentration of highly pure alcohol should be used for this process because it ensures the efficient precipitation of DNA strands.

(b) *Quality and Yield of DNA Postprecipitation*

Precipitation provides visible DNA fibers; the quality and yield depend on several different factors, including:

- *Alcohol type:* Two types of alcohol (i.e., ethanol and isopropanol) are the most widely used agents for DNA precipitation. When the aim is to precipitate shorter pieces of DNA, isopropyl alcohol takes the lead over ethyl alcohol (Wilfinger et al. 1997a).
- *Concentration of salt:* Adequate and precise concentration of salt should be maintained to obtain high-quality DNA, as high levels of salt provide more yield, but there is a risk of DNA contamination associated with high salt concentration.
- *Temperature:* After adding alcohol, a low temperature (i.e., −20 °C or −80 °C) must be maintained for about 12 h or so that will help DNA precipitate properly (Sambrook et al. 1989).

6. Washing the DNA Pellet

After the formation of DNA strands through precipitation, the solution is centrifuged at high speed to get a pellet that is subsequently washed using ethanol (70%)

to remove residual impurities and then air-dried to obtain high-quality DNA. However, be sure to eliminate all contaminants utilizing this procedure, as any residual impurity (i.e., organic molecules, salts, and alcohols) can decrease the quality of DNA and significantly affect DNA sequencing and amplification processes (Green and Sambrook 2012). Similarly, this step should be performed with great care as loss of the isolated DNA can occur during repeated washing and supernatant (containing impurities) removal process.

7. Resuspension of DNA
In this step, a buffer solution (usually Tris-EDTA) or nuclease-free distilled water is employed to dissolve the air-dried pellet of highly pure DNA. Although nuclease-free conditions will maintain the integrity of the extracted DNA, care must be taken in this case, as the addition of extra water or buffer and any mechanical damage can make a huge impact on the quality as well as quantity of DNA. It can also alter the result of the quantification and analysis process, i.e., electrophoresis or spectrophotometry.

8. DNA Quantification (Optional)
This step checks the quality and purity of the isolated DNA using any of the two methods (i.e., spectrophotometry and gel electrophoresis). In the spectrophotometric method, the sample is analyzed at 260 nm (to detect the presence and quantification of DNA) and 280 nm (to detect the presence and quantification of protein). The relative ratio of these two wavelengths (260/280) is a gauge of the purity of extracted DNA. The DNA having around the value of 1.8–2.0 is considered best to carry out studies (Wilfinger et al. 1997a). On the other hand, the gel electrophoretic method signifies (by the relative migration in the applied electric field) that the DNA fragments are broken or damaged during the DNA extraction and precipitation process.

1.4 Challenges and Solutions in DNA Extraction from Eukaryotic Cells

1. DNA Shearing
While handling and processing huge quantities or large-sized DNA molecules, mechanical damage to DNA is considered a common problem, as it may also cause the DNA to break into smaller pieces of irregular sizes and lengths that are useless for further downstream applications such as sequencing. The possible solutions to alleviate this problem are:

- *Gentle pipetting:* Unnecessary mixing and pipetting of extracted DNA must be prevented, after the lysis of the cell wall and cell membrane. Carrying out gentle pipetting can help against DNA shearing problems and cause the least damage to the isolated DNA.

- *Using low-shear buffers:* Buffers can also affect the integrity of DNA; therefore, using a buffer with low-shear capacity will help avoid degradation of DNA and maintain the quality of extracted DNA.

2. DNA Purity
The presence of various impurities, including RNA, protein, cell debris, and phenolic compounds, can compromise the purity of DNA. The impure DNA is not an ideal raw material for carrying out highly sophisticated processes like sequencing and PCR. This problem can be overcome by using commercially available automated kits (Sambrook et al. 1989), used for the process of DNA extraction and precipitation, or by using specially designed silica-based columns (Boom et al. 1990) for the efficient removal of any purity-compromising contaminant present in close association with the isolated DNA.

3. Cell Type Variability
It is a fact that the intracellular conditions and cell walls can vary from organism to organism and from cell to cell as they have different types of components that make their cell wall and the rate of deposition of material in the cell walls of various cells in the same organism can also vary. In this way, it is problematic and difficult to apply a generalized protocol for the extraction and precipitation of DNA to highly diverse cell types. Therefore, optimization of the generalized protocol for each cell type is necessary to obtain the desired results, as Aljanabi et al. (1999) devised procedures to isolate high-quality DNA from plant cells having resistant cell walls using additional steps like enzymatic degradation and efficient elimination of phenolic contaminants.

4. Recovery of Low-Input DNA
It is very difficult to isolate DNA from a source with a small quantity of DNA, as there are definite chances of DNA loss during the isolation process. If the amount of DNA is already minute, then how much DNA will be obtained after performing all the steps of the DNA extraction and precipitation process? To overcome this problem, technological advancements in microfluid dynamics and related fields are expected to resolve the issue and provide a way that will enable researchers to extract a sufficient amount of DNA from minimal available amounts.

1.4.1 Recent Advances and Applications

1. Automation of DNA Extraction
The manual DNA extraction method is prone to contamination, loss of quality and integrity, and yield reduction because of improper handling, measuring errors, and other related factors. On the other hand, automated DNA extraction kits such as Dneasy (developed by Qiagen), Genejet (designed by Thermo Fisher), and Quick DNA (invented by Zymo Research) made the process quicker, more efficient, more reliable, and high-yielding by circumventing the inaccuracies of the manually

operated process. Some systems utilize silica-based columns, while others prefer magnetic beads (Tobler et al. 1999) as the material of choice for DNA purification. Several companies are trying to improve the efficiency of automated systems. For example, kit that can manage several samples instantaneously, which is suitable for processing large numbers of DNA samples in a relatively shorter time. Thus, it is a more efficient and time-saving system that can also be used for commercial purposes. In this way, automation of the process has revolutionized the fields of molecular biology, forensic analysis, medical diagnostics, and clinical research.

2. Direct PCR and Next-Generation Sequencing (NGS)
With the advancement in technology, it is now possible to isolate DNA from any sample (i.e., whole organisms and complex samples) using a direct approach that will not require time-consuming steps such as grinding, centrifugation, washing, and purification. The fields of diagnostics and forensic science, which depend on the isolation of high-quality DNA, can now benefit from these highly effective methods (Schmidt et al. 2014). Similarly, methods like next-generation sequencing (NGS) require highly pure DNA, which is essential for its efficient operation and additionally for performing sequencing related to DNA present in a single cell and the large genomes of multicellular organisms.

The process of DNA isolation and precipitation from eukaryotic cells is crucial for the operational development of various research areas, including clinical physiology, diagnostics, forensics, and biotechnology. With time, subtle changes in the protocol have shaped the currently available procedure. There are high hopes that the process will get better in the coming years. Despite having drawbacks, each method offered distinct novelty and paved the way for technological advancements such as automation and next-generation sequencing, and helped a lot in shaping a quicker, more efficient, reproducible, and high-yielding method. However, it presents some challenges and hurdles that require careful handling to obtain an adequate quantity of high-quality DNA from any sample, regardless of its type (from plant cells with resilient cell walls to animal cells with no cell wall at all). On the whole, the imperative nature of the process is unquestionable as high-quality pure DNA is used as raw material for genome sequencing, genetic cloning, disease profiling, drug designing, and other related applications.

1.4.2 Plasmid DNA Isolation

Plasmids are small, independent, circular DNA molecules. They are extrachromosomal heredity determinants usually found in bacteria, but also exist in yeast and other organisms. They carry genes conferring antibiotic resistance and encoding virulence factors. Plasmids are being used as valuable tools in the vast field of recombinant DNA technology. Hence, the isolation of plasmid DNA with purity is crucial for downstream applications, including molecular cloning, gene expression analysis, etc.

1.4.2.1 Early Methods for Plasmid DNA Isolation

Ames et al. first devised the protocol for plasmid DNA isolation from bacterial cell culture in 1965. They used a simple alkaline lysis approach to extract plasmid DNA from bacteria. Birnboim and Doly later modified and refined that protocol in 1979, which proved to be the cornerstone for plasmid DNA extraction from bacterial cell culture. They offered a simple and efficient approach for plasmid DNA preparation using an alkaline solution for bacterial cell lysis, neutralization, and plasmid DNA precipitation.

Cell lysis: In this step, harvested bacterial cells are treated with lysis buffer. The lysis buffer, containing NaOH and SDS, executes the denaturation of chromosomal DNA and proteins as the cell lysate is formed upon breaking open the cell.

Neutralization: The denatured chromosomal DNA and proteins precipitate out by neutralizing the pH of the cell lysate with potassium acetate, resulting in the plasmid DNA in solution.

Purification: Phenol chloroform extraction and alcohol precipitation are employed to recover and purify the plasmid DNA.

Advantages: a simple, rapid, and cost-effective method for routine practice in molecular cloning.

Disadvantages: less purity due to genomic DNA and proteins, and inefficiency when working with large bacterial cultures.

1.4.2.2 Modern Methods for Plasmid DNA Isolation

Traditional alkaline lysis methods have been modified and refined over the years to enhance plasmid purity, quality, and yield. Since then, several methods have been developed for plasmid DNA isolation and purification, including density gradient centrifugation and commercially available kits based on silica-based column methods, which vary in terms of purity, scalability, and efficiency.

Scales for plasmid DNA isolation There are different scales for isolating plasmid DNA based on the amount required. Different commercially developed miniprep kits offer efficient, reliable, and rapid protocols for extracting plasmid DNA from small bacterial cultures. However, midiprep and maxiprep protocols yield a huge quantity of DNA. The fundamental principles of all protocols are similar except for the discrepancy in reagent quantity to deal with the high volume of bacterial culture.

1.4.2.3 Alkaline SDS-Based Plasmid DNA Purification

This method is similar to alkaline lysis for plasmid DNA isolation. The sodium dodecyl sulfate (SDS) denatures the protein and solubilizes the cell membrane. This modified method enhances the purity of extracted plasmid DNA by efficiently removing protein and lipid contamination (Sambrook and Russell 2001). Isolated plasmid DNA is purified using ethanol precipitation and column chromatography.

Advantages: Fast, large quantity of plasmid DNA.
Disadvantages: Residues of SDS CA impede downstream applications (Sambrook and Russell 2001), such as sequencing and enzyme digestion.

1.4.2.4 Cesium Chloride (CsCl) Density Gradient Centrifugation

Cesium chloride (CsCl) density gradient centrifugation is a classical technique with gold standards for obtaining high-quality, pure plasmid DNA. It is based on the buoyant density of chromosomal DNA and plasmid DNA. After the cell alkaline lysis, the plasmid DNA is separated against its buoyant density established with CsCl through ultracentrifugation.

Advantages: Plasmid DNA of high level of purity.
Disadvantages: Require highly specialized equipment, labor-intensive, and more processing time.

1.4.2.5 Silica Membrane-Based Column Purification

This technique involves binding plasmid DNA to a silica membrane under high-salt conditions. Nucleic acid (DNA and RNA) and protein contamination are removed, and pure plasmid DNA elutes under a low-salt condition.

Advantages: Automated for large-scale production, efficiently employed in various commercially available kits (Boom et al. 1990).
Disadvantages: plasmid DNA yield varies due to bacterial strain and plasmid size

1.4.2.6 Magnetic Bead-Based Plasmid DNA Purification

This method is an alternative to silica matrix-based column purification. It involves the use of magnetic nanoparticles coated with DNA-binding chemical agents. The plasmid DNA from the cell lysate binds to magnetic beads and separates from contaminants using a magnetic field.

Advantages: Offer automation and high-throughput applications (Anderson et al. 1999).
Disadvantages: Costly, needs equipment.

1.4.2.7 Anion Exchange Chromatography

This method requires a column with anion-exchange-charged resins to capture and separate the plasmid DNA from the cell lysate. The resins bind the negatively charged plasmid DNA, washing off impurities.

Advantages: Purity, scalability, and plasmid isolation at large scale for in vivo applications such as vaccine development and gene therapy, etc.
Disadvantages: More complex, costly, and requires highly specialized equipment.

1.5 Nucleic Acid Analysis

1.5.1 Techniques of Nucleic Acid Analysis

Nucleic acid analysis techniques have revolutionized various research areas, including forensics, genetics, oncology, molecular biology, developmental science, and biotechnology. The development of different diagnostic approaches helps a lot in studying and devising treatments for several genetic disorders, detrimental genetic mutations, and other related ailments. The technological advancements are making us hopeful about the betterment of these analysis techniques, opening the gateway to emerging technology that might pave the way for further development in the areas of pharmacogenetics, gene therapy, and patient-specific treatments.

1. Polymerase Chain Reaction (PCR)
In molecular biology, the polymerase chain reaction (PCR) is among the most hailed additions that revolutionized the field in such a way that paved the way for other astonishing discoveries. In 1983, Kary Mullis invented this technique that made it possible to copy a small quantity of nucleic acid millions of times in less than an hour. In this way, the copy number of a minute amount of DNA or RNA in the sample can be enhanced according to the needs of the subsequent analysis. This technique is based on repeated cycles of three basic steps: denaturation (unwinding of the double helix), annealing (primer-template interaction), and elongation (the extension of the polynucleotide chain). Using a small quantity of the initial sample enables the researcher to increase the number of copies at an exponential rate, as required for specific applications such as forensic analysis, genetic cloning, diagnosis, and gene expression analysis (Mullis et al. 1986).

- *Advancements:* With the advancement in technology and changing demands, several innovations have been made to the conventional technology of polymerase chain reaction (PCR). It has led to the development of different variants of PCR such as quantitative PCR (qPCR) which gives an account of gene expression in real-time (Heid et al. 1996), reverse transcriptase PCR (RT-PCR) eases the analysis of viral RNA by first converting it to complementary DNA (cDNA)

and then amplifying it (Tusher et al. 2001), multiplex PCR that saves time by analyzing several samples at a time, and these are the PCR variants to name a few.
- *Limitations:* While dealing with a small quantity of DNA or RNA, polymerase chain reaction (PCR) faces difficulties such as sample contamination, misleading diagnosis results such as false negative or false positive, and nonspecific amplification (Feng et al. 2019).

2. DNA Sequencing

Nucleic acid sequencing, especially DNA sequencing, is a widely used approach in molecular biology, forensic studies, and biotechnology. It involves the precise and accurate determination of the correct sequence of monomeric units (nucleotides) in a polynucleotide chain of DNA. F. Sanger developed this technique in the 1970s, and he employed a dideoxynucleotide-based chain-termination mechanism, which caused the termination of a particular polynucleotide chain whenever a dideoxynucleotide was added to it (Sanger et al. 1977). In this way, the sequence of the nucleobases of DNA is determined by combining the sequences of all the terminated chains.

- *Next-generation sequencing (NGS).* In the 200s, an important milestone in the field of sequencing was marked by the invention of the next-generation sequencing (NGS) approach, which enabled researchers to sequence the entire genome of an organism within a few hours. For rapid and precise sequencing, NGS employs the innovative services of different technologies, such as Oxford nanopore as well as Illumina sequencing. The usage of helper techniques made NGS more reliable, specific, rapid, cost-effective, reproducible, and suitable for performing large-scale analyses (Mardis 2008).
- *Applications:* Next-generation sequencing (NGS) has several applications in molecular biology, forensics, and diagnostics. For example, it is a widely used approach for metagenome and transcriptome analysis, sequencing of the entire genome of an organism (Metzker 2010), diagnosis of different genetically inherited diseases such as thalassemia and cystic fibrosis, and the detection of genetic variations (Ng et al. 2010).
- *Limitations:* Despite its promising advantages, NGS faces several challenges, as it generates a vast amount of data that requires fast and high-throughput computational systems for efficient analysis and sequencing of genomic data. It faces another major drawback, which is the difficulty of combining a large number of terminated chains to get the ordered sequence of the whole genome. This joining step is error-prone, as it carries a higher risk of nucleotide addition or deletion, which can lead to a misleading sequence of the DNA sample under study (Baker 2012).

3. Gel Electrophoresis

Gel electrophoresis is an effective approach for the appropriate separation of bands of DNA or RNA, along the horizontal axis of the solidified gel, based on their relative size and charge. The applied electric field of suitable strength forces the charged fragments to move through the gel matrix that is composed of either agarose or polyacrylamide. The resolution of the nucleic acid (DNA or RNA) bands occurs according to the pore size of the gel, which segregates the fragments in such a way that smaller bands move and cover more distance. In contrast, larger bands travel less due to the hindrance caused by the porosity of the gel.

- *Applications:* Gel electrophoresis is a valuable technique for analyzing the integrity, quality, concentration, and purity of nucleic acid samples. Similarly, it can also be utilized effectively to access the size of the PCR amplicon. After performing gel electrophoresis, the detection of a particular nucleotide sequence can be carried out through probe hybridization (Sambrook and Russell 2001).
- *Limitations:* Like all other techniques, gel electrophoresis struggles to analyze DNA samples that contain either very small or very large fragments. While working with such samples, gel electrophoresis is the least recommended method due to its low sensitivity (Prowse et al. 2019).

4. Blotting Techniques (Southern and Northern Blotting)

- *Southern blotting:* In 1975, Edwin Southern devised a technique for the efficient detection of a particular nucleotide sequence of DNA by its subsequent hybridization with a radioactive or fluorescent-labeled probe with a target complementary nucleotide sequence. After the electrophoretic separation of DNA fragments, the bands are moved to a desired membrane of suitable material (usually nitrocellulose). Afterward, the DNA bands containing membrane are immersed in probe solution for hybridization that is followed by the band visualization under x-rays (in the case of the radioactively labeled probe or using fluorescent microscopy (in the case of fluorescent-labeled probe) (Southern 1975).
- *Northern blotting:* Northern blotting operates on the same principle as Southern blotting. The major difference between these two processes is that Southern blotting is employed for the successful detection of DNA only, whereas Northern blotting is effective for the detection and analysis of RNA samples. With the help of this technique, the expression levels of different genes can be easily and efficiently analyzed by examining their transcriptional abilities (Alwine et al. 1977).
- *Advancements:* Although modern approaches like polymerase chain reaction (PCR) and next-generation sequencing (NGS) have taken over, blotting techniques still find their applications in various research areas such as gene expression analysis, detection of specific variations in nucleotide arrangement, and analysis of genetic mutations (i.e., insertion, deletion, and inversion) (Basu et al. 2009).
- *Limitations:* Despite several advantages and applications of blotting techniques, they face difficulty when analyzing large DNA or RNA samples. The sensitivity of these hybridization-dependent techniques is also lower, and they take more time to carry out sequence detection when compared with novel complementary approaches (Hegde et al. 2008).

5. Microarray Analysis

The microarray technique is a revolution in the fields of molecular biology and forensics, as it enables researchers to analyze multiple samples at once. In this method of nucleic acid analysis, the complementary probe is fixed at the bottom of a microarray plate having a varying number of wells (i.e., 96 or 384), and then an RNA or DNA sample, labeled with a fluorochrome or dye, is added to each well. Hybridization occurs subsequently, producing fluorescence that is detected by the detection system (i.e., fluorescent microscopy).

- *Applications:* This approach is extensively employed for analyzing the regulation of gene expression at various levels to assess the physiological conditions of the organism under study. Similarly, the incidence rate and severity of different genetically inherited disorders can also be investigated using this novel diagnostic approach (Schena et al. 1995).
- *Limitations:* No doubt, microarray analysis is an advanced and effective technique, but it has an analysis spectrum of a shorter range as compared to other complementary approaches such as blotting and sequencing. Microarrays also lag in different areas, such as the probes for all the potentially available genes may not be attached to the microarray plate, and the analyzer may miss the detection of certain less-expressed genes.

6. Fluorescence In Situ Hybridization (FISH)

Fluorescent in situ hybridization (FISH) is a technique that employs, in its operation, the fixation of a particular slice of the target tissue or unruptured cells to a stage, followed by the addition of a target sequence-specific fluorescent probe. Hybridization between the two (i.e., probe and the target sequence) occurs, which emits fluorescence of a certain wavelength that is either captured by a detection system or visualized through fluorescent microscopy. In this way, this technique is less labor-intensive when compared to other complementary hybridization-based approaches.

- *Applications:* In cancer-related studies, FISH is a common laboratory practice as it is an efficient technique for the detection and analysis of genetic mutations (i.e., insertion, deletion, inversion, and duplication) and chromosomal aberrations (Lichter et al. 1992).
- *Limitations:* Although FISH is a highly recommended methodology in cytogenetic analyses, the time-consuming nature of this technique cannot be ignored, as it depends on designing a suitable complementary probe that is specific for the target sequence. The resolving power of FISH is also lower than that of other hybridization techniques, such as microarray and blotting (Verma and Babu 2007).

7. CRISPR-Cas9 Gene Editing

Clustered regularly interspaced palindromic repeats (CRISPR)-associated protein 9 (Cas9) is a highly useful addition in the field of biomedical science. It enables researchers to target a specific polynucleotide sequence in the entire genome of an organism using guide RNA (gRNA), a crucial component of the CRISPR-Cas9 system. Then, the endonuclease (Cas9- the second component) enzyme-mediated process creates breaks in the DNA duplex by eliminating the target sequence. In this way, the target gene can be altered (either turned on or off) according to requirement (Razzaq et al. 2022).

- *Applications:* CRISPR-Cas9 is an extensively employed approach for the introduction of appropriate genetic modifications in various model organisms (Ijaz et al. 2023). CRISPR-Cas9 is an efficient gene-editing approach; therefore, it also finds applications in the development of genetically modified organisms (GMOs) and in the rectification of human cancerous cells to reverse the process of tumor formation (Jinek et al. 2012).
- *Limitations:* CRISPR-Cas9 can introduce mutations in genes other than the targeted ones, resulting in unintentional alterations and producing off-target effects, which is marked as a major drawback of this technique. Technological advancements are trying to resolve this issue and make the process more specific and target-oriented (Fu et al. 2013).

8. Single-Cell RNA Sequencing (scRNA-Seq)
Single-cell RNA sequencing (scRNA-Seq) is a crucial approach that enables the examination of the genetic characteristics of individual cells, rather than sequencing and analyzing the genome of a whole tissue or organism to obtain the required information. This cell-level approach helps in developing an overall scenario of the genetic complexity of that particular cell or targeted area.

- *Applications:* scRNA-Seq is an extensively utilized technique in various research areas, including molecular and developmental biology, genetic engineering, cancer-related studies, and gene expression analysis. Using this approach, the researchers were able to analyze individual cells among complex populations, and the information generated by the scRNA-Seq process helped them a lot in developing the profiles of immune functions of different organisms (Trapnell et al. 2014).
- *Limitations:* Despite the target-oriented nature of scRNA-Seq, it also faces some challenges and limitations regarding the analysis and interpretation of large volumes of highly valuable data. Additionally, the detection of severely low expression of target molecules (mRNA transcripts) is another critical issue. The expensive nature is considered a major hurdle in the large-scale adoption and implementation of this technique.

1.5.2 Applications of Nucleic Acid Analysis

1. Genetic Disease Diagnosis
The diagnosis of genetic disorders is a primary function of many of the commonly used nucleic acid analysis techniques, especially microarray, NGS, FISH, and PCR, because of their efficient analysis of genetic expression. By generating valuable information based on the mutation analysis and abundance of genetic markers, these techniques help the researchers in disease identification (i.e., cystic fibrosis), which subsequently leads to the development of prevention as well as treatment methods for that particular ailment (Ginsburg et al. 2009).

2. Cancer Genomics

The detection of genetic markers and variations linked to various genetic disorders, including cancer, is the ultimate goal of various routinely employed laboratory approaches, such as FISH, microarray, PCR, and NGS. With the help of these analyses and detection techniques, researchers are able to target specific tissues and cells to study and identify the expression patterns of various cancer-causing genes (oncogenes). Ultimately, this useful information facilitates the development of patient-specific (personalized) cancer treatment methods, as individuals within the same population vary in terms of their genetic composition (Garofalo et al. 2015).

3. Forensic DNA Analysis

In terms of traits and genetic composition, every living being differs from the individuals of other populations or even the same population, and the occurrence of unique genetic profiles is the main reason behind this individuality. It is a fact that 99% of the DNA of all humans (the whole population of the world) is similar, whereas they differ in only 1%, which is composed of short tandem repeats (STRs) and variable number tandem repeats (VNTRs). Therefore, for resolving paternity issues and identification of criminals, forensic investigations rely on PCR-based STR and VNTR analyses (Butler 2005).

4. Microbial Genomics

One of the most important applications of nucleic acid analysis techniques is the detection and identification of abundantly available microbial populations. Various methodologies, such as sequence analysis of 16S rRNA, microarray analysis, and probe-based hybridization, make it easy to study and analyze different microbial species that are related to the health and benefits of humans as well as their surrounding environment. Owing to the availability of highly sensitive and target-specific approaches, it is now possible to identify the pathogenic microbes among high loads of microbial populations (Turnbaugh et al. 2007).

1.6 Protein Isolation and Analysis

1.6.1 Protein Isolation

Protein isolation and analysis are integral components of various laboratories, including medical diagnostics, molecular biology, and forensics. The combination of advanced techniques such as nuclear magnetic resonance (NMR) spectroscopy, mass spectrometry (MS), enzyme-linked immunosorbent assay (ELISA), and X-ray crystallography with conventional protein isolation and analysis approaches has revolutionized several protein-related research areas, such as the identification of structural and functional domains, proteome analysis, and the study and understanding of protein-protein interactions. Recent innovations and technological advancements in the fields of protein extraction and analysis are successfully addressing challenges such as sample denaturation, degradation, low output, and the resolution of complex samples, enabling the fine-tuning of these processes to gain better insights into the protein of interest.

1.6.2 Protein Extraction from Biological Samples

Depending on availability, different components can be used as potential sources, such as isolated cells, cell clusters (tissues), and biological fluids, for the extraction of protein samples. The quality, quantity, and purity of the extracted sample highly depend on the method employed for protein extraction. Therefore, it must be chosen with great care for efficient isolation and preservation of the protein to carry out subsequent analysis. Commonly used extraction protocols include:

(a) Cell Lysis

The first step for the efficient extraction of proteins from tissue or cells involves the disruption of the cell membrane and cell wall, in some cases, to make the cytoplasmic content, which also contains cellular proteins. The lysis of cell membrane can be carried out by one of the following methods:

- *Physical methods:* Frequently repeated cycles of freezing and thawing, grinding (using a mortar and pestle), sonication, and homogenization are some of the commonly used physical methods for the efficient breakdown of cell walls as well as cell membranes. In sonication, high-frequency and low-wavelength sound waves are utilized to achieve the efficient lysis of the cells (Rao et al. 1995).
- *Chemical methods:* During the chemical method of cell membrane disintegration, various detergents such as Nonidet P-40 (NP-40), sodium dodecyl sulphate (SDS), and Triton X-100 are usually preferred due to efficiency and effectiveness (Peters et al. 1991).
- *Enzymatic methods:* As the name indicates, this lysis approach uses different enzymes (individually or as a mixture) such as collagenase (in the case of animal cells), a mixture of cellulase, xylanase, and pectinase (in the case of plant cells), and lysozyme (in case of bacterial cells) to release the intracellular components (Gould et al. 1991). Enzymatic methods are preferred over physical and chemical processes of cell lysis because the off-target effects are the least in the case of using enzymes for the disruption of cell membranes.

(b) Protein Solubilization

The next important step, after the successful extraction of proteins from living cells or tissues, is the preservation of those extracted protein molecules from coagulation as well as degradation by solubilizing them in a buffer solution of suitable composition. Sodium chloride (NaCl), beta-mercaptoethanol, dichlorodiphenyltrichloroethane (DDT), and lopinavir (a potent protease inhibitor) are the most common components of solubilization buffer (Sambrook and Russell 2001).

(c) Protein Precipitation

The solubilization of extracted protein in buffer solution is followed by its precipitation, which involves the removal of contamination and increasing the density of the sample. It can be achieved by one of the following methods:

- *Ammonium sulfate precipitation:* In this method, a high concentration of inorganic salt of sulfate (ammonium sulfate) is mixed with the buffer solution containing extracted proteins. Ammonium sulfate reduces the solubility of proteins in the solution, ultimately causing them to precipitate (Scopes 1987).
- *Trichloroacetic acid (TCA)/acetone precipitation:* Alternatively, a mixture of trichloroacetic acid (TCA) and acetone is employed for the efficient precipitation of proteins, but this method is labor-intensive and time-consuming as compared to the ammonium sulfate-based method because it requires a subsequent washing step for the removal of contamination (Wilkins et al. 1996).

1.6.3 Protein Purification

With the efficient precipitation of proteins from crude extract, there arises a need for further purification of the sample for the exclusion of other contaminating ingredients such as cell lysates. For this purpose, the purification method is selected according to the binding affinity, nature, size, hydrophobicity, and charge of the proteins to be purified. Some of the most common purification techniques include:

(a) Chromatography

Chromatography-based protein purification technique is one of the simplest and widely employed approaches because it separates proteins from the residual components using the interactional features of proteins. Several variants of chromatography can be used for purifying proteins, and those variants are:

- *Ion exchange chromatography:* In this approach, the proteins are purified based on their net charge by passing them through a charged (either negative or positive) resin-filled column to which oppositely charged proteins bind reversibly, which are then eluted either by altering the pH of the buffer or by increasing the concentration of salt such as NaCl (Carter 1984).
- *Size exclusion chromatography:* This method purifies the proteins based on their relative size in such a way that proteins with larger sizes get eluted first, followed by medium-sized proteins, while the elution of smaller proteins, including oligopeptides, occurs at a late stage of the process (Arakawa et al. 2007).
- *Affinity chromatography:* Unlike other chromatography-based techniques, which separate conventional and naturally occurring proteins, this unique variation of chromatography purifies recombinant proteins on the basis of affinity-based interactions of protein with the ligand molecules, such as tags (His-tag or GST-tag) or antibodies (Nicolson et al. 1983).

(b) Electrophoresis

Electrophoresis resolves or purifies the fragments of protein according to their charge, shape, and size. The following electrophoretic approaches are extensively employed for protein purification:

- *SDS-PAGE:* Sodium dodecyl sulfate-polyacrylamide gel electrophoresis (SDS-PAGE) is a method of choice for protein purification, resolving its negatively charged fragments formed by SDS denaturation based on molecular weight and size (Laemmli 1970).
- *Isoelectric focusing (IEF):* Two-dimensional polyacrylamide gel electrophoresis (2D-PAGE) uses the service of isoelectric focusing (IEF), which separates and purifies proteins from cellular debris by accumulating the neutral proteins at a specific pH (isoelectric point or pI). In this way, proteins get purified easily and can efficiently be utilized to carry out subsequent analyses (Righetti et al. 1993).

(c) Ultracentrifugation

It is a density-based purification technique that forms a density gradient, separating proteins and other cellular components into layers. The densest macromolecules (i.e., proteins) form a pellet, while remaining contaminants form subsequent layers in a decreasing order of density from bottom to top of the supernatant. In this way, centrifugal force-derived separation of large biomolecular compounds can be carried out easily and effectively (Peters et al. 1991).

1.6.4 Protein Quantification

Protein quantification is crucial for analyzing the efficiency of the isolation process, especially to assess the purity of the isolated protein. For this purpose, some of the important methods are:

(a) Bradford Assay

This method is based on the staining of isolated protein with a protein-specific dye (Coomassie Brilliant Blue), which specifically binds with the proteins and emits blue color fluorescence whose intensity directly correlates with the quantity of protein present in the solution. Bradford assay is an extensively employed approach for protein quantification because of its prominent features like high speed, accuracy, and simplicity (Bradford 1976).

(b) BCA Assay

Unlike the Bradford assay, the bicinchoninic acid (BCA) method needs alkaline conditions for protein quantification. In this method, the BCA reagent is used to react with the cupric ions formed in the presence of proteins. As a consequence of this reaction, the color of the entire sample solution changes to purple. The concentration of protein is estimated based on the intensity of the purple color (Smith et al. 1985).

(c) UV Absorbance
This method relies on the absorption of ultraviolet (UV) light of a specific wavelength (i.e., 280 nm) by amino acid residues (such as tyrosine, tryptophan, and phenylalanine) of protein. The amount of UV light absorbed quantifies the relative concentration of protein in the sample solution. In this way, this technique solely depends on the absorption coefficient of the target protein (Wilfinger et al. 1997b).

1.6.5 Protein Analysis

Different analyses are carried out to determine the structural conformation, functionality, and posttranscriptional modifications (PTMs) of the efficiently isolated and purified proteins. Mainly employed protein characterization approaches include:

(a) Mass Spectrometry (MS)
Mass spectrometry (MS) is a common and efficient laboratory technique for detecting and identifying proteins and peptides present in a sample solution. It detects the mass-to-charge ratios of different components of the sample and presents them as peaks. Similarly, it can also be used to estimate the molecular weight of the target protein. Other major applications of mass spectrometry include the analysis of post-translational modifications (PTMs), proteome analysis, and the identification of target proteins (Anderson and Polson 2009). The following two variants of mass spectrometry are extensively utilized for carrying out protein analysis:

- *MALDI-TOF (matrix-assisted laser desorption/ionization-time of flight):* In this technique, a laser beam ionizes the proteins and peptides, and the time of their respective flight through the tube is calculated when they strike the detector. In this way, this technique mainly relies on the calculation of time of flight for the accurate determination of the mass-to-charge ratio of each component (i.e., protein or peptide) of the sample (Karas et al. 1988).
- *LC-MS/MS (liquid chromatography-tandem mass spectrometry):* This method combines two powerful analytical techniques. One of these is liquid chromatography (LC), which enables the separation of sample constituents based on their chemical properties. The second component is mass spectrometry (MS), which takes those already separated constituents and accurately determines the mass-to-charge ratios. In this way, LC-MS is a highly recommended variant of MS for the detection and identification of proteins present in complex mixtures (Domon and Aebersold 2006).

(b) Western Blotting
Similar to nucleic acid analysis, proteins can also be analyzed based on their relative rates of hybridization with sequence-specific antibodies through the Western blotting method. For this purpose, SDS-PAGE resolved fragments of proteins are transferred to a membrane of suitable composition (i.e., nitrocellulose) followed by its immersion into an antibody-containing solution. In the end, the results of the

hybridization reaction are visualized with the help of a detection system. Therefore, Western blotting is a powerful method for the detection of particular proteins as well as for their efficient quantification (Towbin et al. 1979).

(c) X-Ray Crystallography
X-ray crystallography is a useful approach for the analysis of the three-dimensional (3D) structure of target molecules and biomolecules (i.e., nucleic acids and proteins). It relies on the diffraction pattern analysis of X-rays by the high-quality crystals of the target protein present in the sample solution (Rupp 2010). Using different bioinformatics tools, structural information can then be utilized to determine the functional properties of the protein of interest.

(d) Nuclear Magnetic Resonance (NMR) Spectroscopy
Nuclear magnetic resonance (NMR) spectroscopy is another powerful analytical technique that is widely used to understand the three-dimensional (3D) structural conformation of different molecules and biomolecules (nucleic acids and proteins) based on their magnetic properties. It helps study the dynamic nature of the target protein as well as its interactions with other molecules in the biological setting. The requirement for a large quantity of highly pure protein samples poses a major challenge to NMR, which hinders its widespread use (Wüthrich 1986).

(e) Enzyme-Linked Immunosorbent Assay (ELISA)
Enzyme-linked immunosorbent assay (ELISA) is extensively employed to detect the presence of a particular protein, followed by the efficient quantification of protein-antibody interactions through the measurement of chemiluminescence or fluorescence, which is emitted during the reaction of antibody-linked enzyme and its substrate (Engvall and Perlmann 1972).

1.6.6 Challenges in Protein Isolation and Analysis

1. *Protein degeneration and degradation:* During the efficient process of extraction and isolation of proteins, there are chances of protein denaturation (by high temperature or altered pH), protease-mediated degradation (proteolysis), and temperature or pH-triggered aggregation. Therefore, the composition of different constituents (i.e., NaCl, DDT, lopinavir) of the solubilization buffer and other conditions (temperature and pH) should be maintained in such a way that they do not interfere with the efficiency of the isolation process (Sambrook and Russell 2001).
2. *Low-yield proteins:* The process of protein isolation is slightly inefficient when dealing with the extraction of low-quantity proteins. Similarly, the retrieval of membrane-associated proteins is also challenging. The utilization of highly sophisticated techniques, such as affinity chromatography (AC) and the selection of tissues with a high abundance of target protein, is considered a reliable option to resolve the aforementioned issues (Carter 1984).

3. *Complexity of samples:* The purity of the protein sample is critically important for carrying out subsequent analyses because an impure and contaminated sample can decrease the efficiency and reliability of protein extraction and analysis, which require additional steps of purification and contamination removal (Wilkins et al. 1996).

1.7 Polymerase Chain Reaction: Principle and Application

Owing to the higher efficiency, specificity, reliability, sensitivity, and reproducibility, polymerase chain reaction (PCR) is considered a gold standard technique, which has revolutionized several disciplines, including molecular biology, genetics, medical and clinical diagnostics, biotechnology, forensics, biochemistry, and other biological research areas. PCR can sense and amplify the minimum quantity of DNA present in any mixed or even degraded samples. Different variants of PCR, such as reverse transcription PCR (RT-PCR), quantitative PCR (qPCR), digital PCR (dPCR), and nested PCR, emerge by slightly modifying the design and functionality of standard PCR. These variations are routinely utilized in various laboratories for disease diagnosis, criminal identification, pathogen detection, and other such applications. PCR is also facing a few shortcomings that must be overcome, some of which have been addressed by making technological innovations in the existing technology to maintain the reliability and effectiveness of the PCR-based processes.

1.7.1 Principle of Polymerase Chain Reaction

Polymerase chain reaction (PCR) works on the same principle of DNA replication, which is employed by the cells of a living organism. The major difference between the processes (i.e., natural and PCR) is that the enzyme (DNA polymerase) utilized by PCR does not degrade and performs its activity even at high temperatures. In contrast, the polymerase found in humans and other terrestrial animals degrades easily at a slightly higher temperature than the normal body temperature. The thermostable enzyme amplifies the nucleotide sequence of interest by naturally synthesizing its complementary strand.

The following are the main ingredients utilized in PCR:

1. *Template DNA:* A sufficient quantity of DNA sample having the gene of interest, which is copied and elongated by the polymerase enzyme.
2. *Primers:* These are the short chains of nucleotides, usually 18–22 bp but can vary according to the length of the target sequence, which serve as points for the DNA polymerase to initiate the synthesis of complementary strands. Shorter

primers are preferred over longer primers because they can bind with the target sequence more specifically.

3. *Thermophilic DNA polymerase:* It is a thermostable enzyme, which is usually derived or isolated from hydrothermal vent-dweller microbe (*Thermus aquaticus*), therefore, the enzyme is named as Taq polymerase. It is suitable for performing DNA replication at high temperature (i.e., 94 °C).
4. *Nucleotides* are the basic monomeric units that serve as the building blocks during the process of DNA replication, which is catalyzed by the thermophilic Taq polymerase. For DNA synthesis, four commonly used deoxyribonucleotide triphosphates are deoxyadenosine triphosphate, deoxyguanosine triphosphate, deoxycytidine triphosphate, and deoxythymidine triphosphate.
5. *Buffer solution:* It is a mixture of different salts (e.g., $MgCl_2$, which serves as a cofactor for polymerase), a pH-stabilizing component (e.g., Tris-HCl), and a reaction enhancer (e.g., gelatin, which prevents both the template and newly synthesized DNA from degradation). It usually maintains the working environment for getting optimal activity of the enzyme.

1.7.2 PCR Process: Three Main Steps

For the amplification of the gene of interest, PCR works in a cyclic route having three steps in common. Typically, 25–40 cycles are required to get an ample quantity of DNA from a small amount of template DNA samples. The main stages of the PCR process include:

1. *Denaturation* (94–98 °C): This step is required to denature the DNA duplex, with the help of high temperature (i.e., 94–96 °C), and make it single-stranded so that PCR primers can bind with each separated strand easily and efficiently.
2. *Annealing* (50–65 °C): In this step, short synthetically designed sequences (primers) scan the separated strands of denatured DNA to find their complementary partners. Once found, primers bind with them via hydrogen bonding at relatively low temperature (typically 50–65 °C, depending on the length of primers) and provide the points for the Taq DNA polymerase to start the process of replication.
3. *Extension* (68–75 °C): The final step marks the elongation of the primers, from which the enzyme initiates its operation, in such a way that the polymerase enzyme synthesizes a complementary copy of each of the separated strands.

Subsequently, the repeated cycles of heating and cooling take over the charge and start synthesizing copies of target DNA molecules at an exponential rate, which means the copy number doubles with each passing cycle (Mullis et al. 1986).

1.7.3 Variations of the Polymerase Chain Reaction

To meet the operational demands and specific needs of various areas of molecular biology and forensic science, several innovations have been introduced to the conventional PCR setup, making it applicable in these areas of research and investigation.

1. Reverse Transcription PCR (RT-PCR)
Reverse-transcription PCR (RT-PCR) is an innovative variant of conventional PCR, which uses RNA as a template and the enzyme reverse-transcriptase is an additional component, when compared to standard PCR, that converts RNA into complementary DNA (cDNA). This product of reverse transcription is then utilized as a template for the conventional DNA polymerase enzyme, and the subsequent process occurs according to the standard operating procedure of traditional PCR.

- *Applications:* This modification of PCR is extensively employed for the detection and estimation of viral RNA (i.e., HIV, HCV, or HBV) in the sample, gene expression analysis, validation of mRNA load in the sample, and studying the composition and variations in the transcriptome (Sambrook and Russell 2001).
- *Limitation:* The efficient working of RT-PCR requires an additional and preliminary step of reverse transcription, which makes the process somewhat lengthy compared to the standard method and error-prone if the setup is not optimized effectively and precisely.

2. Quantitative PCR (qPCR)/Real-Time PCR
Quantitative PCR (qPCR), also known as real-time quantitative PCR, is a useful variation of PCR that provides continuous evaluation as the template DNA is copied in each PCR cycle. This process involves the utilization of a label, either with a fluorescently-labeled probe or with a fluorescent dye, to one end of the DNA duplex. In contrast, the other end of the duplex contains a quencher, which functions to suppress the fluorescence in the absence of amplicons. In this way, the detection system estimates the quantity of amplified DNA based on the intensity of the fluorescent signals it receives. Hence, it provides real-time information about the amplification of the DNA of interest.

- *Applications:* It is a widely used technique in forensics, medical diagnostics, and genetics for the appropriate analysis of genetic variations, studying and estimation of viral agents, real-time detection of other pathogenic microbes (i.e., bacteria), and accurate monitoring of gene expression (Heid et al. 1996).
- *Advantages:* By providing real-time access to the quality as well as quantity of the amplified DNA, it bypasses the need for further analysis techniques such as mass spectrometry and gel electrophoresis, which are utilized for estimating the amount and quality of the copied DNA product, respectively.
- *Limitation:* It is a highly sensitive approach, and it requires sophisticated and high-quality chemical reagents, well-calibrated apparatus, and equipment for its efficient operation.

3. Nested PCR

Nested PCR is a highly recommended variant of conventional PCR because of its enhanced sensitivity, reliability, and specificity. In its operation, it employs two sets of primers in such a way that one set of primers binds and amplifies the large portion of the DNA sample nonspecifically, in the sense that it directs the polymerase to amplify nontarget regions of DNA that are present around the target sequence of DNA. The second set of primers only binds and amplifies the target sequence of DNA (i.e., DNA region or gene of interest), which is present in the initially amplified PCR product (amplicon). In this way, it can detect the presence of a minute quantity of target DNA present in the sample, and nested PCR can easily discriminate the DNA of interest among different sequences of DNA found in the sample, specifically increasing the copy number of that particular region.

- *Applications:* It is a highly useful approach for the efficient detection of pathogenic microorganisms such as bacteria and viruses, which are present in minimal concentrations (Robinson et al. 2000). Therefore, nested PCR is routinely employed as a common technique in medical diagnostic laboratories.
- *Limitation:* In this variation of PCR, there is a high risk of sample contamination because it takes several cycles of continuous amplification, and if an error occurs in any of the cycles, it will continue to replicate and amplify with each passing cycle.

4. Digital PCR (dPCR)

Digital PCR (dPCR) is an upgraded and better form of quantitative PCR (qPCR), and it produces more accurate and reliable results and gives precise information about the concentration of amplified target sequence of DNA by measuring the number of molecules per droplet. The working principle of dPCR differs from that of qPCR in that it requires sample DNA (template) to be divided into small droplets, which are placed in different portions. Afterward, the PCR process occurs in each of those portions separately. The quantity of amplified DNA sequence is directly proportional to the number of portions, having droplets of template and other PCR ingredients, showing positive results.

- *Applications:* Digital PCR (dPCR) is a widely used approach for the efficient detection and measurement of nucleic acids (DNA and RNA), less common genetic variations, discrepancies in the copy number of PCR amplicons, and precise and accurate monitoring of the sample under study (Hindson et al. 2011).
- *Limitation:* Owing to the advanced nature of digital PCR (dPCR), it requires highly sophisticated and well-calibrated laboratory equipment for its proper functioning. The high cost of reagents and other necessary materials makes this technique expensive.

1.7.4 *Applications of Polymerase Chain Reaction*

The application spectrum of PCR is so vast that it encompasses several areas of medical science and research, such as molecular biology, genetics, diagnostics, biotechnology, biochemistry, and forensics. The major applications of PCR in these fields include:

1. Clinical Diagnostics

- *Applications:* The field of clinical diagnostics is hailed with the services of PCR technology because it uses the detection and identification abilities of PCR in the diagnosis of different genetically inherited disorders, viral (i.e., HIV) and bacterial (i.e., tuberculosis) infections, performing prenatal screening (to determine the health condition of the fetus), detection of pathogens (either having DNA or RNA genomes), and diagnosis of various diseases, particularly cancer.
- *Example:* PCR, more precisely reverse transcription PCR (RT-PCR), plays a pivotal role in the efficient detection and diagnosis of infections caused by novel viral and bacterial strains that have RNA as their genetic material. In recent times, the detection of coronavirus (SARS-CoV-2) has been extensively performed using RT-PCR (Corman et al. 2020).

2. Forensic DNA Analysis

- *Applications:* PCR finds its applications in various investigations of forensics ranging from resolving paternity issues (through DNA fingerprinting) to criminal identification (by collecting the DNA samples from crime scenes and then using human-specific markers of DNA like short tandem repeats (STRs) or variable number tandem repeats (VNTRs) to identify the culprit among different suspects) to the recovery of information from the DNA or RNA samples found at archeological sites as well as at fire-affected areas (Butler 2005).
- *Advantage:* Several variants of PCR, such as qPCR and dPCR, are highly sensitive and efficient as they can detect minute quantities of nucleic acid (DNA or RNA) present in a sample of any nature (normal or even degraded).

3. Genetic Research

- *Applications:* In genetics, PCR is a routinely employed laboratory practice for the appropriate analysis of genetic variations, studying host-pathogen interactions, transcriptome analysis, efficient cloning of nucleotide sequences of interest, understanding the mechanisms of disease occurrence, analysis of gene expression, and integration of cloned genes into a suitable vector, which is employed for developing genetically modified organisms (GMOs).
- *Example:* Several transgenic plants and animals have been developed using amplification strategies that are powered by standard polymerase chain reaction (PCR) and many of its variants, along with the highly effective nucleic acid analysis techniques such as sequencing and hybridization.

4. Environmental and Agricultural Applications

- *Applications:* PCR-based technology is also serving areas such as the environment and agriculture, which humans and other living beings encounter daily. Environmental studies utilize PCR methodology for the detection and estimation of microbial loads in different samples, including water, soil, and air, to ensure the quality of each, as they are fundamental ingredients for the sustainability and survival of life on Earth. Similarly, PCR is used in agriculture-related research to detect the pathogenic microbes affecting the yield of crops, validate the purity and effectiveness of newly-developed seeds, estimate the presence of multiple insects and related creatures (to ensure the biodiversity of a particular area), and assess the condition of transgenic crops.
- *Example:* Along with standard PCR, different variants of PCR are used for studying the composition of agricultural and environmental samples. They are also extensively employed to detect the presence of genetically modified organisms (GMOs) as well as pathogens in different agricultural samples and agriculture-related edible products (Mikami et al. 2002).

5. Food Safety and Quality Control

- *Applications:* Concerning human health, PCR plays a unique and commendable role in checking and controlling the quality of food produced by various industries worldwide. Food quality control units and food safety regulation authorities ensure that the food products (made by multiple companies and subsequently consumed by humans and other animals) are pure, safe, and pathogen-free. They provide food safety and authenticity using different PCR testing approaches to assess the physiological conditions of various food samples. PCR is a more reliable, specific, rapid, and sensitive approach as compared to conventionally used complementary techniques such as culture-sensitivity tests.
- *Example:* PCR is a widely used technique for the efficient detection and analysis of food borne pathogenic microbes such as *Salmonella* spp., *Listeria monocytogenes*, and *Escherichia coli.* Food safety laboratories are also detecting the presence of different allergens in food-based samples and other related products.

1.7.5 Limitations of Polymerase Chain Reaction

Although polymerase chain reaction (PCR) and its several variations have widespread applications in almost every field of research and medical science, they also encounter some drawbacks that include:

1. *Sensitivity to concentration:* Despite the high sensitivity of PCR variants, there is always a chance of sample contamination through several sources, such as improper handling, using contaminated reagents, using inefficient equipment, and the presence of lysogenic nucleic acid (prophage DNA or RNA).
2. *Primer design:* Appropriate primer design is an integral and fundamental part of the PCR process, as primers are the main components that ensure the quality and purity

of the PCR product. On the other hand, inefficiently designed primers face the problem of nonspecific and off-target amplification, or they may produce a lesser amount of low-quality product, leading to the wastage of both resources and time.

3. *Inhibition:* The polymerase chain reaction (PCR) is prone to inhibition that may lead to the failure of the overall process. In diagnostics and clinical testing, inhibition of PCR can occur due to the presence of different inhibitor components, such as phenolic compounds, proteins, urea present in urine, hemoglobin found in blood samples, and contaminating ingredients, including different detergents (Wilson 1997).
4. *Quantification challenges:* The variants of PCR, such as quantitative PCR (qPCR) and digital PCR (dPCR), which provide accurate quantitative and real-time information about the concentration of the amplified product, require highly sophisticated, precisely calibrated, and properly maintained controls to perform their assigned task of nucleic acid quantification efficiently.

1.8 Gel Electrophoresis: Principles and Methods

Gel electrophoresis is one of the most important and useful techniques for resolving nucleic acids and proteins, as evidenced by its widespread applications in biochemistry, forensics, genetics, oncology, molecular biology, and biotechnology. The advancements in the areas of genome and proteome analysis have become possible only because of the highly versatile features of gel electrophoresis, such as high resolution, cost-effectiveness, and easy operational demands. However, low sensitivity for less-abundant samples, poor processing of larger samples, and sample degradation issues always remain major drawbacks, which have been continuously addressed by slightly modifying the gel composition, replacing the existing technology with novel and more reliable methods, and integrating other techniques such as mass spectrometry (MS) to boost the efficiency, reliability, sensitivity, and reproducibility of the gel electrophoresis process.

1.8.1 Principles of Gel Electrophoresis

The basic working principle of gel electrophoresis is the migration of negatively charged fragments of nucleic acids (DNA and RNA) or proteins towards the positive electrode, through the pores of the gel, under the influence of a low-voltage electric field. Biomolecular fragments move across the horizontal axis of the gel according to their respective charge, shape, size, and nature of the gel.

(a) Electric Field

The application of an electric field of suitable strength (usually, a low voltage is preferred) is the most important driving force behind the process of electrophoresis.

According to Coulomb's law, the separation of molecules occurs due to attraction caused by the electrodes having opposite charges compared to the charge on the separating molecular fragments. In this way, bands with a positive charge are attracted to the negatively charged electrode (cathode) and vice versa. Charge, conformation, size, porosity of the gel, and the strength of the electric field are among the major factors that determine the migration rate of different macromolecular (i.e., DNA, RNA, or protein) bands (Sambrook and Russell 2001).

(b) Separation Based on Size and Charge

- *Size:* The size of the migrating molecule is one of the most important factors that can affect the efficiency of the gel electrophoresis process. Due to the sieving property of the gel matrix, small-sized fragments migrate faster and travel farther, whereas larger fragments move more slowly and cover a shorter distance across the gel.
- *Charge:* For the efficient separation of biomolecules, gel electrophoresis follows the principle of Coulomb's law, which states that opposite charges attract each other. Therefore, molecules with a negative net charge, such as nucleic acids (DNA and RNA), migrate towards the positive electrode (anode). In contrast, the charge on different proteins can vary because the net charge of each protein is determined by the combination of charges of its amino acid side chains, as well as the pH of the buffer solution in which the gel is placed during electrophoretic separation (Laemmli 1970). Hence, a protein having a positive charge migrates towards a negative electrode (cathode) and vice versa.
- *Shape:* Similarly, the rate of macromolecular separation also depends on the shape of the migrating molecule. Despite having similar sizes, fragments with regular and linear symmetry travel more quickly through the gel compared to those with irregular and globular symmetry.

(c) Gel Matrix

The porosity and sieving properties of the gel matrix hold crucial importance for the appropriate resolution of bands of DNA, RNA, or protein. Acting as a molecular sieve, a gel matrix allows the free movement of smaller molecules while it offers resistance to the movement of larger molecules. The choice of gel matrix to be used for electrophoretic separation depends on the type, suitability, and other separation-based requirements of the sample. Therefore, a gel matrix made up of agarose is preferred for the resolution of nucleic acids, while a polyacrylamide gel matrix is mostly employed for the efficient resolution of protein samples.

1.8.2 Types of Gels Used in Electrophoresis

(a) Agarose Gel Electrophoresis

Agarose gel is composed of seaweed-extracted polysaccharide, which has solidification ability and forms a porous structure when cooled after heating at high temperatures, such as 121 °C. Agarose-based gel electrophoresis is routinely used for resolving nucleic acid (DNA or RNA) bands. The interconnected network of agarose gel forms pores that only permit molecules with a certain size limit to pass through, resolving the fragments based on their relative sizes. Depending on the size of the migrating nucleic acid fragments, the extent of porosity and the size limit of pores can be adjusted by altering the final concentration of agarose in the buffer solution. 1–2% solution of agarose in the buffer is usually preferred to separate the bands having sizes up to a few kilobases (Sambrook and Russell 2001).

Applications

- Different DNA-based analyses, such as Southern blotting, fluorescent in situ hybridization (FISH), polymerase chain reaction (PCR), and restriction fragment length polymorphism (RFLP), use agarose gel electrophoresis as a method of choice.
- Similarly, several RNA analysis techniques, including Northern blotting and single-cell RNA sequencing (scRNA-Seq), also benefit from the services of agarose gel electrophoresis.
- Agarose gel electrophoresis is a widely used method for efficiently resolving DNA fragments and plasmids, which are commonly used in genetic engineering experiments.

(b) Polyacrylamide Gel Electrophoresis (PAGE)

Polyacrylamide gel electrophoresis (PAGE) is extensively employed for the appropriate and effective separation of protein fragments. However, DNA or RNA fragments with a small size, which are poorly resolved by agarose gel electrophoresis, can also be separated using this technique because the pore size of acrylamide-based gel is relatively smaller when compared to the agarose-based gel having a similar concentration. However, it resembles structural conformation with agarose gel because polyacrylamide gel also forms an interconnected network, just like agarose gel does. The main advantage of using PAGE is that it can efficiently resolve the different proteins with nearly similar molecular weights (Laemmli 1970).

(c) Types of PAGE

- *Native PAGE:* It is employed for resolving bands of protein samples according to their respective size, charge, and shape while their structural conformation and functionality properties remain intact and undisturbed.
- *SDS-PAGE:* This variant of polyacrylamide gel electrophoresis (PAGE) employs a detergent, which is sodium dodecyl sulfate (SDS), for the denaturation of the protein sample. The main function of SDS is to convert large and complex proteins into smaller, easily separable, and uniformly charged fragments. Afterward, the application of an electric field separates the fragments only based on their relative sizes (Laemmli 1970).
- *Acrylamide gel for nucleic acids:* Instead of the agarose-based electrophoresis method, small DNA fragments such as oligonucleotides and products of restriction fragment length polymorphism (RFLP) are preferably separated using a polyacrylamide-based electrophoresis method because of its enhanced resolution power (Sambrook and Russell 2001).

1.8.3 Method of Gel Electrophoresis

(a) Preparatory Steps for Gel Electrophoresis

The formation of a gel matrix is a prerequisite for initiating the process of gel electrophoresis, as electrophoretic separation cannot be carried out without one. The next step is loading prepared samples of nucleic acid or protein into the appropriate wells that have been created in the gel by placing a comb into the gel solution during its solidification process. Then, the solidified gel is placed in a running buffer of suitable composition.

- *Gel preparation*: The size of the sample fragments determines the concentration of the gel to be used. For the preparation of agarose gel, the gelling agent (i.e., agarose) is first dissolved in a suitable buffer, followed by heating at high temperature and then gradually cooling it down for solidification, making a gel matrix. On the other hand, polyacrylamide gel is prepared by interlinking the gelling agent (i.e., acrylamide) with other molecules such as bis-acrylamide and then allowed to polymerize at room temperature.
- *Sample loading:* In this step, samples are usually mixed with the loading buffer, which has a dye like bromophenol blue to visualize the progress of the electrophoresis process. The addition of dye serves various functions, including increasing the density of the sample so that it does not spill out of the loading well and facilitating the visualization of the sample. In the case of nucleic acid separation, a reference sample is also loaded in the form of DNA (for DNA samples) or RNA (for RNA samples) ladders, which serve as markers of known molecular weight, to determine the length of the separated bands.
- *Electrophoresis setup:* This step involves transferring the solidified gel into an electrophoresis tank containing a suitable running buffer solution. The primary purpose of using a running buffer is to provide an efficient conducting medium for the electric current to pass through, which is crucial for resolving fragments of the loaded samples. The stabilization of the pH (ionic strength) of the medium is another important function of the buffer solution. The power supply connected to the electrophoresis tank is adjusted to a suitable voltage.

(b) Running the Gel

- *Application of electric field:* After setting up the apparatus and placing the sample-containing gel in the electrophoretic tank, an electric field of low voltage is applied for efficient and more reliable separation of molecules because higher voltage may produce excessive heat and even liquify the gel matrix, which is one of the most avoided situations. Electric current causes the charged fragments to move through the gel matrix towards the opposite electrode, which means molecules having a positive charge migrate to the cathode and vice versa.
- *Separation of molecules:* Under the influence of an electric field, the molecules in the sample are separated in such a way that larger fragments migrate more slowly, while smaller fragments migrate farther and faster. The resolution of bands takes place according to their charge and size. In the case of agarose gel

electrophoresis, the separation of fragments occurs along the horizontal axis, while polyacrylamide gel electrophoresis separates the pieces along the vertical axis.

(c) Visualization of Results

After the completion of electrophoretic separation, the gel is mixed with dyes for efficient visualization of resolved fragments:

- *DNA visualization:* Ethidium bromide-based, which is a DNA intercalating agent and emits fluorescence under ultraviolet (UV) light (Sambrook and Russell 2001), staining method is currently not very common as it poses health hazards to researchers because of its mutagenic nature. Alternatively, gels are now stained with an adequate quantity of loading dye such as SYBR Green or bromophenol blue because of their lower toxicity and less hazardous nature.
- *Protein visualization:* Silver staining and Coomassie Brilliant Blue are two widely used staining methodologies for the efficient visualization of separated protein bands. Coomassie Brilliant Blue has a higher binding affinity for the protein molecules, making them easily visible under ultraviolet light after electrophoretic resolution (Wilkins et al. 1996).

1.8.4 Applications of Gel Electrophoresis

The application spectrum of gel electrophoresis covers a wide variety of research areas such as biochemistry, proteomics, forensics, molecular biology, genetics, and biotechnology. Major applications in these fields include:

(a) DNA Analysis

- *Genotyping:* With the help of gel electrophoresis, researchers can accurately determine the genetic composition of various organisms, including humans, model plants, and animals. Now, they can easily locate the positions of genetic mutations (i.e., single-nucleotide polymorphisms and restriction fragment length polymorphisms (RFLPs)) using organism-specific genetic markers.
- *PCR product analysis*: Agarose gel electrophoresis aids in the analysis of PCR-amplified DNA products by providing an efficient separation and detection platform for resolving DNA molecules of varying sizes (Sambrook and Russell 2001).
- *Southern blotting:* Similarly, the process of gel electrophoresis is also employed as a prerequisite approach for carrying out probe-based hybridization, which binds specifically with the complementary DNA sequence (Southern 1975).

(b) RNA Analysis

- *Northern blotting:* Gel electrophoresis also finds its applications in transcriptome analysis and gene expression studies through its efficient operation of resolving RNA bands followed by the binding of sequence-specific fluorescent probe, which enhances the reliability of the subsequent analysis (Alwine et al. 1977).

- *RNA quantification:* The versatile approach of gel electrophoresis is also commonly used for the estimation of the quality and purity of the separated nucleic acid, especially RNA fragments. The size of each resolved band can be analyzed by comparing it with the size of the corresponding band of the RNA ladder (reference sample).

(c) Protein Analysis

- *Protein purification:* Different variants of polyacrylamide gel electrophoresis (PAGE), especially the Sodium dodecyl sulphate (SDS)-aided version of PAGE, are highly recommended for carrying out product purification (i.e., isolated protein) analysis. Based on SDS-PAGE analysis, subsequent studies can effectively detect contamination and accurately determine whether the isolated protein sample is contaminated or not (Laemmli 1970).
- *Western blotting:* Similar to nucleic acid analysis, gel electrophoresis is also considered a gold standard technique for analyzing protein samples because it separates protein fragments, which undergo hybridization with sequence-specific antibodies that facilitate their better detection and visualization (Towbin et al. 1979).

(d) Proteomics

Several novel research areas, including proteomics, come into existence as a result of technological advancement and the development of different protein analysis techniques such as 2-dimensional polyacrylamide gel electrophoresis (2D-PAGE) and mass spectrometry (MS). 2D-PAGE is a combination of SDS-PAGE and isoelectric focusing, which resolves proteins according to their isoelectric points (pH at which they are electrically neutral). By employing 2D-PAGE, it is now an easy task to isolate and identify several proteins present in a complex mixture of samples simultaneously (O'Farrell 1975).

1.8.5 Advantages and Limitations of Gel Electrophoresis

Advantages

- *High resolution:* The higher resolution power of gel electrophoresis, especially the polyacrylamide-based method that is utilized for resolving protein samples, discriminates it among other detection, identification, and purification techniques.
- *Versatility:* The versatile nature of gel electrophoresis makes it applicable for the efficient separation and detection of important biomolecules such as nucleic acids (DNA and RNA) and proteins.
- *Ease of use:* Owing to its ease of handling, efficiency, and low operational cost, it is a widely available technique in molecular biology, genetics, medical diagnostics, and other related laboratories around the globe.

Limitations

- *Limited sensitivity:* While working with a minute quantity of nucleic acid or protein sample, gel electrophoresis lags and gives less accurate results, which questions the sensitivity of this approach. Therefore, such a sample needs pre-treatment modification (such as concentrating the sample using an appropriate material) to get a desirable outcome.
- *Size limitations:* The standard gel electrophoresis method faces difficulty in resolving samples having large molecular fragments, which encounter considerable resistance during migration. To overcome this problem, another variant of gel electrophoresis, pulse field gel electrophoresis, is employed for getting better results (Sambrook and Russell 2001).
- *Potential for sample degradation:* Improper handling poses a continuous risk of sample degradation, as even a minor mistake at any point can lead to the failure of the entire process. Therefore, gel electrophoresis must be performed with great care to prevent the loss of both time and resources.

1.9 Agarose Gel Electrophoresis

Agarose gel electrophoresis is a simple, flexible, reliable, highly effective, cost-effective, and versatile approach for the separation and analysis of nucleic acids, preferably, proteins. These attributes of agarose gel electrophoresis make it rank higher among common laboratory techniques in various fields of medical research, such as forensic sciences, molecular biology, biotechnology, and biochemistry. It has an expanded applicability due to its wide range of combinations with different components, such as buffer solutions, gel compositions, and various staining agents, which aid in better visualization of migrated fragments when the gel is placed under UV light. With time, the technique continues to evolve according to the changing demands of the research work by overcoming the possible limitations, such as ethidium bromide (mutagenic) being replaced with SYBR Green (less toxic). Agarose gel electrophoresis struggles with the resolution of larger molecules and complex samples and, as a result, urges scientists and technical workers to develop novel approaches (i.e., pulsed-field gel electrophoresis (PFGE) and capillary electrophoresis (CE)) that can address the current shortcomings and enhance the resolution of larger and more complex sample mixtures.

1.9.1 Principles of Agarose Gel Electrophoresis

The working principle of agarose gel electrophoresis depends on the separation of different bands of nucleic acid (i.e., DNA or RNA) or protein after the application of an electric field of suitable strength. Under the influence of voltage, the fragments of DNA, RNA, or protein migrate through an agarose matrix according to their size-to-charge ratio. In this way, the pieces with shorter sizes move farther away from the sample loading wells than fragments with larger sizes. Sieves of varying sizes form within the gel matrix, which allows the movement of only those molecules having a size smaller than the sieve size, causing hindrance for the larger molecules. As a result, larger pieces of migrating molecules form thick bands near the loading well while smaller pieces form distinct bands of variable length along the horizontal axis of the gel matrix (Sambrook and Russell 2001).

1.9.2 Types of Agarose Gel Electrophoresis

1. Horizontal Agarose Gel Electrophoresis
Among different types of agarose gel electrophoresis, the horizontal setup, especially for the separation of bands of DNA or RNA samples, is usually preferred over the vertical one. However, it works on the general principle of fragment migration through agarose sieves under the action of an electric field that is applied horizontally.

2. Vertical Agarose Gel Electrophoresis
The basic working of this setup is similar to that of the horizontal one, but it differs only in the orientation of the gel, the direction of the electric field, and usage for the separation of bands of proteins, not of DNA or RNA.

1.9.3 Applications of Agarose Gel Electrophoresis

The following are some important applications of agarose gel electrophoresis:

1. DNA Analysis
Agarose gel electrophoresis is employed to visualize the migrating nucleic acid, especially DNA bands along the horizontal axis of the gel matrix. It is critically important for several genetic and molecular studies, including gene expression analysis, PCR reliability confirmation, analysis of restriction enzyme-digested fragments, DNA sequencing, and many more. In forensic studies and medical research, DNA bands of the sample can be compared to the bands of a maker or ladder (having bands of known weight and size) to carry out precise and accurate analyses, such as for resolving paternity issues and identification of the criminal (Mullis and Faloona 1987).

2. RNA Analysis
Much like DNA, RNA can also be precisely and accurately analyzed using agarose gel electrophoresis to check whether the RNA present in the sample is useful for further studies, such as gene expression analysis. The effectiveness of RNA is determined based on the purity and integrity of the RNA sample. Unlike DNA, which can

be degraded under harsh conditions, RNA is prone to degradation even under moderate cellular conditions because of its single-stranded structure. Various downstream applications (i.e., RT-PCR, microarray analysis, Northern blotting, and RNA sequencing) demand the analysis of RNA through agarose gel electrophoresis as a prerequisite (Wilfinger et al. 1997b).

3. Protein Analysis
Agarose gel electrophoresis is preferably used for the analysis of nucleic acids (DNA and RNA), but it can be used for the accurate analysis of large-sized and high molecular weight protein samples present in their inherent form (Kirkland 2001). Although it can easily be visualized using a protein-specialist electrophoretic technique, SDS-PAGE (sodium dodecyl sulfate-polyacrylamide gel electrophoresis). In this way, the analyzed sample will then be utilized for various applications, including disease diagnosis, drug development, expression analysis, and protein engineering.

4. DNA Fingerprinting and Genotyping
DNA fingerprinting and genotyping are among the most versatile and demanding applications of the agarose gel electrophoresis technique, as it enables the researcher to visualize and compare separated DNA bands from different samples and identify a particular pattern based on the similarity in the lengths and sizes of different bands on the gel matrix (Butler 2005). It is a very convenient technique and is an integral part of genetic testing and forensic analysis laboratories because of its reliability, easy handling, and high accuracy.

1.9.4 Factors Affecting Separation in Agarose Gel Electrophoresis

1. Agarose Concentration
The concentration of agarose in the gel matrix plays a crucial role in determining the size of the molecule or fragment that can pass through. Here, the sieving property of agarose gel comes into play as the concentration of agarose in the gel is inversely proportional to the pore size and size of the migrating fragment. For example, a higher concentration of agarose (usually 1.5–3%) creates smaller pores, allowing the movement of small-sized pieces, and vice versa (Sambrook and Russell 2001).

2. Buffer Type and pH
A buffer is a solution that tries to maintain the overall pH of the system in which it is present by stabilizing the ionic strength. In this way, the type of buffer will have a huge impact on the separation of nucleic acid or protein fragments, as it controls the pH as well as the ionic strength of the solution in which the agarose gel is placed. For the resolution of nucleic acids (DNA and RNA), TBE (Tris-borate-EDTA) and TAE (Tris-acetate-EDTA) buffers are mostly preferred. In the case of

protein samples, SDS is an additional component in either of the mentioned buffers (Mullis and Faloona 1987).

3. Voltage and Running Time
For the proper resolution of nucleic acid or protein fragments, appropriate voltage and optimal running time matter as they are the determinants of migration rate. For example, high voltage can be applied to make the electrophoresis process faster, but it will melt the solidified gel due to continuous heat dissipation. Another issue with higher voltage is that it prevents the bands or fragments from separating properly. Hence, the resolution of the sample is compromised at a higher voltage. Therefore, lower to moderate voltage (50–150 V) is preferred and optimally employed to resolve the bands of the sample for further analysis (Sambrook and Russell 2001).

4. Staining and Visualization
The nature of the staining agent determines the visualization power of the bands that are resolved on agarose gel. Different types of dyes and intercalating agents are used for the staining of nucleic acids and proteins to make them visible under ultraviolet (UV) light. For staining DNA or RNA, ethidium bromide (a potential mutagen) was previously used, but now it has been replaced with SYBR Green dye, which is highly sensitive and less toxic. On the other hand, proteins (which are less common than nucleic acids) are stained with silver ions (Wilfinger et al. 1997b), and the gel is subsequently placed in a UV transilluminator for the visualization of separated protein fragments in the form of an array of bands along the horizontal or vertical axis.

1.9.5 Limitations of Agarose Gel Electrophoresis

1. Resolution for Large Molecules
A major drawback of agarose gel electrophoresis is that it cannot efficiently resolve nucleic acid or protein molecules, which are larger than 20 kb in size. Smear formation occurs due to a low migration rate of very large DNA, RNA, or protein fragments. While dealing with nucleic acids or proteins of a larger size, pulsed-field gel electrophoresis (PFGE) is preferred over conventional agarose gel electrophoresis. PFGE uses short and momentary pulses of high voltage that help larger fragments to resolve properly and efficiently (Schwartz and Cantor 1984).

2. Poor Resolution for Complex Mixtures
While working with complex mixtures containing components of different natures, agarose gel electrophoresis causes poor resolution of the fragments and, in some cases, causes smear formation or does not resolve the components at all. To overcome this limitation, researchers use alternate techniques such as 2D gel electrophoresis (2D-GE), capillary electrophoresis (CE), or polyacrylamide gel electrophoresis (PAGE) to obtain better resolution and appropriate separation of the components of complex samples.

1.10 Polyacrylamide Gel Electrophoresis (PAGE)

Polyacrylamide gel electrophoresis (PAGE) is an indispensable tool for carrying out accurate and reliable assessment of intrinsic features (i.e., size and molecular weight) of different biomolecules such as proteins and nucleic acids (DNA and RNA), which perform biologically important functions in the bodies of living organisms. Due to its laudable features (i.e., enhanced resolution power, reproducibility, efficiency, and accuracy), it has a broad range of applications in various research areas, including proteomics, metabolomics, molecular biology, diagnostics, biochemistry, and biotechnology. Current challenges and limitations, regarding the effective resolution of larger biomolecular samples and limited sensitivity for proteins with low expression levels, cause hindrances in its large-scale adoption. These issues can be tackled by integrating an efficient staining approach, using an advanced and modified version of PAGE, which is 2D-PAGE, and other such innovations with the existing technology, making it accessible and easy to use.

1.10.1 Principles of Polyacrylamide Gel Electrophoresis

Polyacrylamide gel electrophoresis (PAGE) works on the general principle of electrophoresis, which resolves the bands of biomolecules (i.e., proteins, DNA, and RNA), on the application of a low-voltage electric field, with the help of a matrix that is made of a suitable material (in this case, polyacrylamide). Depending on size and charge, macromolecular fragments move through the pores of the gel matrix, which are created by an applied electric field. Owing to the small size of some fragments, they encounter less resistance from the matrix as compared to large pieces. That is why smaller fragments appear at the bottom (far away from the loading well) when visualized under ultraviolet (UV) light.

(a) Polymerization of Acrylamide

Unlike agarose gel, which is prepared by dissolving it in a suitable buffer solution, the preparation of polyacrylamide gel requires the polymerization of acrylamide units with bis-acrylamide (N, N′-methylenebisacrylamide which is a cross-linking agent. The range of resolution and size-based separation primarily depends on the quantities of the polymerizing molecules (i.e., acrylamide and bis-acrylamide), as their relative concentrations are crucial in determining the porosity of the gel through which the separating molecules must pass. Due to the lower porosity of thick gels (having a high concentration of acrylamide and bis-acrylamide), they are suitable for the efficient resolution of small-sized molecules (i.e., nucleic acids). On the other hand, molecules of large size can be effectively separated using a gel with low concentrations of acrylamide and bis-acrylamide (Laemmli 1970).

(b) Electric Field and Separation Mechanism

The application of a low-voltage electric field separates the charged molecules according to the general principle of electrostatics, which states "like charges repel each other while opposite charges attract each other". Following the rules of electrostatics, positively charged proteins and nucleic acids (DNA and RNA) get attracted towards the negatively charged electrode (cathode) while biomolecules having a negative charge on them migrate towards the positively charged electrode (anode). The resolution ability of each separating molecule is unique because the charge and size of that molecule determine it. The porosity of gel obstructs the free and easy movement of large-sized fragments, forcing them to resolve according to their size (Sambrook and Russell 2001).

(c) Separation Based on Size and Charge

- *Proteins:* The separation of migrating biomolecular samples in native polyacrylamide gel electrophoresis (PAGE) depends on two of their most important intrinsic factors (i.e., size and charge) because, in this technique (native PAGE), the structural and functional features of the molecules remain undisturbed. In sodium dodecyl sulfate-polyacrylamide gel electrophoresis (SDS-PAGE), proteins are denatured by SDS, losing their native charge, which is imparted by their symmetrically arranged amino acid residues. Therefore, the resolution of protein fragments occurs only based on their size.
- *Nucleic acids:* It is a well-known fact that the resolution power of polyacrylamide gel electrophoresis (PAGE) is higher than that of agarose gel electrophoresis (AGE). Owing to this, small nucleic acid components (i.e., oligonucleotides and small plasmids) can efficiently be separated and analyzed with the help of polyacrylamide gel electrophoresis (Sambrook and Russell 2001).

1.10.2 Types of Polyacrylamide Gel Electrophoresis

(a) Native PAGE

As the name implies, innate features such as the structure and function of proteins are retained during gel-aided size and charge-based separation. Denaturation of protein in the sample is not an operational requirement of native PAGE. Along with charge and size, the structural conformation (either fibrous or globular) of the biomolecules (especially proteins) plays a key role in their efficient electrophoretic resolution.

Applications

- It is extensively used for studying and understanding intermolecular interactions (e.g., protein-protein and protein-ligand).
- Most of the enzymes are proteins. Therefore, native PAGE can be employed for the estimation and analysis of the biological activity of different enzymes.
- Using this technique, complex mixtures of proteins can easily be resolved and analyzed for the efficient isolation of native proteins (Wang et al. 2016).

(b) SDS-PAGE

In this technique, a known detergent, sodium dodecyl sulfate (SDS), is employed for unfolding the complex structures (i.e., tertiary and quaternary) into linear polypeptides and oligopeptides. The second major function of SDS is the distribution of the overall negative charge to the denatured protein sample. SDS-PAGE resolves the macromolecules according to their sizes only because SDS causes hindrance in the expression of their net charges. As there is no influence of charge, the determination of the exact molecular weight of each separated fragment becomes an easy task.

Applications

- With the help of this powerful approach, the molecular weight of resolved proteins can easily be estimated by comparing it with a reference sample (i.e., protein ladder).
- It keeps a check on the efficiency of purification processes by analyzing the purity of the isolated proteins.
- Subsequent analysis of proteins, such as Western blotting, which involves the transferring of separated protein bands onto a suitable membrane, can only become possible if SDS-PAGE resolves the protein fragments efficiently (Laemmli 1970).

(c) Two-Dimensional PAGE (2D-PAGE)

Two-dimensional polyacrylamide gel electrophoresis (2D-PAGE) is an upgraded version of PAGE, combining two different but complementary approaches, such as SDS-PAGE (which causes protein denaturation) and isoelectric focusing (IEF), which causes the proteins to accumulate by making them electrically neutral at a specific pH (isoelectric point). This technique is named 2D because it achieves two-dimensional sample resolution, where isoelectric point-based separation is a one-dimensional separation, followed by SDS-aided size-based electrophoretic separation (the second dimension). This innovative variant of polyacrylamide gel electrophoresis is widely used in proteome-related research because it can simultaneously handle a larger number, even thousands, of proteins present in complex mixtures.

Applications

- Through the efficient resolution of a complex amalgam of proteins, it provides a detailed and in-depth analysis of each component that is present in the sample mixture.
- 2D-PAGE is employed as a reliable method for the diagnosis of different genetically inherited (e.g., cystic fibrosis) as well as acquired diseases (e.g., cancer) by carefully analyzing the involved proteins.
- For the sustainability, survival, and proper functioning of proteins in cytoplasmic settings, they acquire various posttranscriptional modifications (PTMs), including acetylation, methylation, and glycosylation. 2D-PAGE aids in the identification and characterization of multiple PTMs that are crucial for proteins to carry out their biological functions (Görg et al. 2004).

1.10.3 Methodology of Polyacrylamide Gel Electrophoresis

(a) Gel Preparation

- *Acrylamide concentration:* Most commonly used polyacrylamide gels contain about 5 to 15% of acrylamide because the formation of small to medium-sized pores has been observed at this range of concentrations, which are suitable for resolving a wide variety of samples ranging from proteins to small nucleic acids. The pore size can also be adjusted by increasing or decreasing the relative concentration of acrylamide, as a thick gel containing a higher quantity of gelling agent is employed for the separation of large proteins. In contrast, small-sized proteins are resolved using gels having a low amount of acrylamide (Sambrook and Russell 2001).
- *Casting the gel:* This step involves the Preparation and solidification of polyacrylamide gel, which is achieved by adding appropriate concentrations of acrylamide, bis-acrylamide, tetramethylethylenediamine (TEMED), and ammonium persulfate (APS) to a suitable buffer. APS and TEMED complement each other, starting the process of gel polymerization. Afterward, the gel is poured into a casting tray to solidify, and sample loading wells are generated by placing a comb with equally spaced teeth.
- *Buffer system:* The selection of the running buffer is sample-dependent because compatibility of the sample and buffer is necessary for the appropriate resolution of the sample molecules (Laemmli 1970). For this purpose, Tris-glycine fortified buffer is usually employed. However, other buffer solutions (e.g., Tris-HCl and Tris-acetate) can be used as an alternative to Tris-glycine.

(b) Sample Loading and Electrophoresis

- *Sample loading:* Before loading the samples into the wells, they are thoroughly mixed with a desaturating agent (SDS), a fluorescently labeled dye (for visualization aids), and a reducing reagent (i.e., beta-mercaptoethanol). The addition of all these components in different Eppendorf tubes, containing the sample and loading buffer, aids in enhancing the density of the sample so that it cannot spill out of the Eppendorf loading wells. After thorough mixing with the help of a micropipette, the samples are precisely pipetted into the loading wells formed in the solidified polyacrylamide gel.
- *Running the gel:* In the next step, the gel with loaded samples is transferred to the electrophoresis containing an adequate volume of the sample-compatible running buffer of suitable composition (i.e., Tris-glycine or Tris-HCl). Afterward, the electrodes (i.e., anode and cathode) are connected to the battery, which sets a low voltage electric field across the gel and forces the macromolecular components (proteins) present in the wells to migrate in such a way that proteins with less molecular weight move quickly than the proteins with high molecular weight.

(c) Staining and Visualization

- *Coomassie brilliant blue:* After the successful electrophoretic separation of molecules, the gel is immersed in a tank containing staining dye, in this case,

Coomassie Brilliant Blue (CBB), which specifically binds and forms complexes with the complementary amino acid residues such as arginine, histidine, and lysine. It gives the separated bands a characteristic blue color of varying intensity, which is proportional to the concentration of the particular resolved protein fragment (Wilkins et al. 1996). This staining method is cost-effective, efficient, and quite simple, leading to its widespread applications.

- *Silver staining:* The sensitivity, and hence the reliability, of the silver staining technique is higher than that of the Coomassie Brilliant Blue-based method, making it a better option for the efficient detection of less abundantly available proteins in the sample mixture. Unlike the Coomassie Brilliant Blue method that uses dye, this method employs the services of silver ions that form a complex with the target (complementary) regions of the resolved protein sample, followed by subsequent visualization of complexes using formaldehyde-based (reducing agent) reduction processes (Heukeshoven and Dernick 1985).
- *Fluorescent dyes:* Several other staining dyes are also available in the market, which provide better visualization of separated protein bands and can bind more specifically to proteins. Therefore, the use of SYPRO Ruby and Deep Purple is the most sensitive choice for the visualization of resolved fragments of proteins when compared with other staining methods such as CBB and silver methods (Wilkins et al. 1996).

1.10.4 Applications of Polyacrylamide Gel Electrophoresis

(a) Protein Analysis

- *Molecular weight determination:* One of the major applications of polyacrylamide gel electrophoresis (PAGE), especially SDS-PAGE, is the estimation of the molecular weight of the separated protein bands by comparing their thickness (concentration) with that of the bands of ladder protein that is used as a reference sample (Laemmli 1970).
- *Protein purity:* Another important application of SDS-PAGE is the determination of the purity level of different protein samples following their efficient resolution on the gel. In this way, polyacrylamide gel electrophoresis keenly detects the contamination (e.g., chaperones and degraded sample components) and ensures the purity of the isolated proteins (Görg et al. 2004).
- *Western blotting:* For more reliable and accurate quantification of proteins, the separated bands of the samples are transferred to a membrane for their hybridization with sequence-specific antibodies, followed by analysis using an efficient approach such as densitometry (Towbin et al. 1979).

(b) Nucleic Acid Analysis

- *DNA fingerprint size analysis:* Along with analyzing protein samples, polyacrylamide gel electrophoresis (PAGE) is also considered a reliable approach for the appropriate resolution and analysis of nucleic acid (especially DNA) fragments that are hundreds to thousands of base pairs long. Owing to the lower resolution ability of agarose gel electrophoresis, it cannot efficiently analyze small-sized nucleic acid samples. Therefore, PAGE is employed as an alternative method for analyzing such samples (Sambrook and Russell 2001).

- *RNA analysis:* Similarly, agarose gel electrophoresis fails to provide accurate quantitative analysis of RNA samples with a shorter size range of molecules. Effective separation and quality-based analysis of different RNAs (such as lncRNAs, tRNAs, rRNAs, miRNAs, and snRNAs) can be carried out with the help of this powerful and versatile approach (Sambrook and Russell 2001).

(c) Proteomics

- *2D-PAGE proteomics:* Two-dimensional polyacrylamide gel electrophoresis (2D-PAGE) is a fundamental technique in the field of proteomics, as it provides a method for obtaining high-resolution images of a large number of proteins present in a complex mixture. The most interesting feature of this approach is its ability to handle thousands of proteins simultaneously, separating them according to their respective sizes and isoelectric points (pI). Therefore, it is a gold standard technique for understanding various fundamental processes like analysis of posttranscriptional modifications (PTMs), cellular signal transduction, and detection of important biomarkers (e.g., STRs and VNTRs) that are related to forensic investigations or disease diagnosis (Görg et al. 2004).

1.10.5 Advantages and Limitations of Polyacrylamide Gel Electrophoresis

Advantages

- *High resolution:* When it comes to dealing with biomolecular samples (such as protein, RNA, or DNA) having restricted or shorter length fragments, polyacrylamide gel electrophoresis (PAGE) takes the lead over agarose-based electrophoresis method because of its ability to resolve samples in the form of efficiently separated and clearly visible bands.
- *Flexibility:* There are different variants of polyacrylamide gel electrophoresis (PAGE), such as native PAGE, two-dimensional polyacrylamide gel electrophoresis (2D-PAGE), and sodium dodecyl sulphate-polyacrylamide gel electrophoresis (SDS-PAGE), that make this technique highly versatile. Owing to this, its application spectrum encompasses several areas of research, including proteomics, molecular biology, and biotechnology.
- *Quantification:* By precisely analyzing the protein fragments, which are resolved by PAGE, one can easily estimate the quantity of each. It provides a quantitative overview of the protein sample under study, which can be further utilized to advance subsequent analyses.

Limitations

- *Resolution limitations:* Polyacrylamide gel electrophoresis (PAGE) is a well-established method for the analysis of molecular fragments having a limited range of sizes, usually smaller ones. PAGE finds it difficult to separate larger molecules of proteins and nucleic acids (DNA and RNA). Pulse field gel electrophoresis (PFGE) addresses this limitation by resolving large biomolecular samples with the help of short and periodic pulses of high voltage.
- *Complexity of 2D-PAGE:* Two-dimensional polyacrylamide gel electrophoresis (2D-PAGE) is a highly reliable and innovative technique. It is not considered a good option to be employed at a large scale because of its labor-intensive and time-consuming nature. The separation of molecules depends on their isoelectric point (pI), which is specified by a certain pH value. Therefore, the pH of the overall setup, along with running time, needs proper optimization (Görg et al. 2004).
- *Sensitivity:* PAGE faces challenges during the quantification of small amounts of macromolecular samples. This limitation can be addressed by employing the services of more sensitive approaches, such as efficient staining methods and mass spectrometry, which are crucial for efficient visualization and quantification, respectively (Wilkins et al. 1996).

1.11 Electroelution

Electroelution is a process of great importance as it is crucial in making groundbreaking discoveries in other areas of research, such as genome expression analysis, proteome structural and functional analysis, drug development by interpreting valuable insights from antigen-antibody interactions, and product designing using purified molecules. It provides several advantages, such as moderate effects of buffer on biomolecules, cost-effectiveness, and retrieval of highly pure molecular fragments. The shortcomings and challenges, like contamination of products with gel residues, lower efficiency, and other related issues, must be addressed through technological innovations like automation of the process and integration of several high-throughput approaches along with the basic process. Despite all these issues, electroelution is frequently employed as a common technique in various forensic, molecular biology, and diagnostic laboratories.

1. Principle of Electroelution

The process of electroelution works on a similar principle to that which dictates the separation of biomolecular pieces through electrophoresis. The application of low-voltage electric current causes the pieces or fragments of the molecule under study to move through the gel matrix. Similarly, in the presence of an elution buffer, the extraction of resolved bands from a piece of gel by applying the electric field of suitable strength marks the central step of the electroelution process. The purified fragments can then be exploited for further analysis, such as sequencing, expression studies (northern, southern, or western), blotting, cloning, and other forensics-related applications.

Several determinants are essential for the successful working of the electroelution process. These factors include:

- *Electric field strength:* Among the major influencers of the process, electric field strength holds an integral importance as it drives the DNA, RNA, or protein fragments to move from the gel to the elution buffer (Blakeslee and Cummings 1971).
- *Buffer composition:* The nature of the buffer composition, which is ultimately determined by the pH and the ionic strength of the buffer, holds significance in making the molecules move effectively from the gel matrix into the elution buffer.
- *Gel type:* The concentration of the gel directly influences its pore size. In this way, the composition of the gel determines the size of the fragment that it allows to pass through. Hence, the elution process depends on the type of gel from which the molecules are to be eluted (Peters 1996).

2. Methodology of Electroelution

Electroelution is a multistep process, and each step is crucial for the efficient retrieval of the molecules from the gel. Different steps of the process include:

1. *Electrophoretic separation:* The first and foremost step of the process involves the migration of fragments of the sample, which is either nucleic acid (DNA and RNA) or protein, which is caused by an external force (applied electric field). The separation bands get separated according to their relevant charge, shape, size, pore size, and nature of the gel (forming the matrix).
2. *Gel slice preparation:* In the second step, the particular part of the gel, having efficiently resolved and separated bands, is excised cautiously to remove the undesirable area of the gel (Tu and Lee 1981).
3. *Elution chamber setup:* The next step involves the proper arrangement of the framework, which is crucial to the operational success of the process. For this purpose, a buffer of desired composition is poured into the elution chamber, which is followed by the appropriate positioning of the excised part of the gel containing the required fragments. The electric field is applied using electrodes having opposite charges (i.e., positive and negative).
4. *Elution process:* This is the main step of the process because, in this step, the charged fragments of the molecule of interest migrate from a gel into the surrounding buffer under the influence of applied electric current. The size of the part of the gel under study and the nature of the eluting molecule make it the most time-consuming phase of the process, and it may take a few minutes to several hours (Sambrook and Russell 2001).
5. *Collection of eluted molecules:* In the final step, the buffer containing the eluted fragments of the molecule of interest (eluate) is gathered from the elution chamber to carry out a subsequent assessment of the quality and purity of the collected material.

3. Applications of Electroelution

The application spectrum of electroelution covers several research areas such as genetic engineering, genome-related studies, gene expression analysis, protein structural and functional analyses, and antigen-antibody interactions. A few prominent applications of this process are presented in the following:

- *Protein Purification*

The resolved fragments of protein on either agarose or polyacrylamide gel can be recovered efficiently using the process of electroelution. Different protein extraction and isolation approaches, such as pulse field gel electrophoresis (PFGE) and 2-D gel electrophoresis (2-DGE), use this process as a preferred method of choice. The isolated protein fragments can then be utilized to advance subsequent studies such as Western blotting, mass spectrometry (MS), and proteome analysis (O'Farrell 1975).

- *Nucleic Acid Recovery*

Recovery of nucleic acid (DNA or RNA) from the excised gel, containing migrated pieces, is another important application of electroelution. The purified nucleic acid is further used for studying transcriptome, gene expression, genetic cloning, genomics, and sequencing. These analyses provide useful information that helps design targeted drugs and have vital applications in other areas of biomedical science (Aljanabi and Martinez 1997).

- *Peptide Recovery*

Small-sized proteins or even peptides, which are resolved by native gel electrophoresis, polyacrylamide gel electrophoresis (PAGE), or 2-D gel electrophoresis (2-DGE), can be eluted effectively using the protocol of electroelution. The peptides recovered using this technique can be utilized in diverse applications, including protein sequencing, blotting, drug designing, and performing other related analyses (Shevchenko and Wilm 1996).

- *Immunology*

Using the methodology of electroelution, several types of antigens and their relevant as well as novel antibodies can be purified from different kinds of gels. These are then employed to study antigen-antibody interactions and help in performing several assays crucial for the development of novel diagnostic approaches, such as the targeted use of vaccines and antibody-mediated therapies for patients with chronic diseases (Schnell and Miki 2006).

4. Advantages of Electroelution

Prominent areas of research, which require purified biomolecules, prefer the use of electroelution over other related purification methodologies because of the following reasons:

- *High purity recovery:* The quality and purity of the eluted fragments of biomolecules get enhanced when the usage of electroelution is paired with high-resolution gel electrophoresis, as the combination of both allows the efficient recovery of good-quality and high-purity nucleic acids (DNA and RNA) or proteins. These pure molecular pieces are highly valuable in forensic science, molecular biology, and biotechnology (Hames 1990).

- *Gentle on biomolecules:* The composition of the elution buffer is such that it does not produce any detrimental or deleterious effects on the eluted molecular fragments. Hence, degradation-prone or highly sensitive molecules can be recovered and purified using this technique, as it is gentle on biomolecules when compared to other complementary approaches (Williams and O'Leary 2000).
- *Simplicity and cost-effectiveness:* Electroelution provides such a methodology that is simple, efficient, easy to use, and manageable. The laboratory essentials needed to perform electroelution are cheaper than the components of other purification techniques, making it a suitable and desirable molecular biology as well as medical laboratory technique.

5. Challenges and Limitations

Although electroelution is hailed for having several advantages over other purification approaches, its drawbacks cannot be neglected. A researcher has to face several different challenges during the purification of the desired biomolecule, and those challenges include:

- *Recovery efficiency:* The effective recovery of the molecular fragments occurs only if all the required conditions are met properly. When any of the conditions are not maintained, such as irregular application of the electric field or inefficient retrieval of the fragments from the gel, as a consequence, the efficiency of the process may be compromised (Ghosh and Krishna 2004).
- *Gel residue:* The nature of the gel is very important for the proper functioning of electroelution. Gel-like polyacrylamide can cause contamination of the recovered product. Hence, the purification as well as the downstream applications of the recovered fragments will be affected adversely.
- *Optimization of buffers:* The electroelution process is time-consuming and labor-intensive because steps need to be aligned and optimized, especially the composition of the extraction or elution buffer, to achieve these results (Molinaro and Telford 2003).

6. Recent Developments in Electroelution

Like all other areas of life, the global wave of technological advancement requires electroelution to evolve and get better in terms of enhanced recovery, better efficiency, and a gradual shift from manual operation to the complete automation of the process. Concerning electroelution, current innovations are:

- *Microfluidic systems:* The integration of microfluidic systems within the conventional electroelution equipment made the process faster, more efficient, and more reliable than before. By doing so, the output of the process also gets enhanced, and the process takes less time than that of a conventional method (Cho and Zeman 2010).

- *Integrated systems:* This operational innovation enabled the use of electroelution in combination with other valuable techniques, such as sequencing and mass spectrometry (MS), which will definitely improve efficiency and reduce the overall time required for the process. The automated integrated systems provide better analysis with a minimum integrity loss of the purified molecules (Edwards and Patel 2008).

1.12 Formaldehyde Gel for RNA Electrophoretic Separation

Formaldehyde gel electrophoresis is a widely used technique in various research areas such as forensic science, molecular biology, biochemistry, and medical diagnostics. It is an essential approach with enhanced efficiency, reliability, sensitivity, and reproducibility, and these features make this technique a must to opt for a method for researchers working on projects that involve analysis of gene expression, verification of RNA integrity, getting insights into alternative splicing, and other related processes. Like all different techniques, it also faces some difficulties in terms of sample degradation because of prolonged incubation and time-consuming procedures that require the researcher to be patient. Other drawbacks, including the use of ethidium bromide and handling of larger RNAs, have been resolved by the introduction of novel staining agents such as SYBR Gold and automation of the overall process, respectively. In short, the continuous evolution of the process is making it more reliable, reproducible, effective, and suitable for resolving the fragments of nucleic acid, especially RNA samples.

1. Principles of RNA Electrophoresis in Formaldehyde Gels

In nature, RNA forms secondary as well as tertiary conformations because of the Watson-Crick base pairing between its residues, whereas efficient resolution requires RNA to be in a linear form that is essential for proper estimation of the sizes of the separated fragments. For this purpose, formaldehyde-based gel electrophoresis is carried out as it denatures RNA and prevents the reformation of secondary or tertiary structures, which ultimately helps in accurate resolution according to their respective charge and size.

The principle of formaldehyde gel electrophoresis is similar to that of gel electrophoresis, with an additional step of RNA sample treatment with formaldehyde to carry out its denaturation that is maintained by the buffer solution, even at elevated temperature, in which the gel is prepared. The denatured sample is then loaded into the wells of agarose gel, and the application of a low-voltage electric field resolves the bands of the sample. The migration of fragments occurs in inverse relation to the size of the fragments. Hence, larger fragments travel more slowly and cover less distance across the gel, whereas smaller fragment moves faster and travel longer along the horizontal axis of the gel.

Key Features

- *Formaldehyde denaturation:* The denaturation of RNA secondary or tertiary conformation, using formaldehyde, to yield linear nucleotide chains of RNA is the most distinguishing feature of the formaldehyde gel electrophoresis that will assist the efficient separation of the individual bands of varying length across the gel matrix (Alwine et al. 1977).
- *Electrophoresis under denaturing conditions:* Another reason for the uniqueness of formaldehyde gel electrophoresis is the loading of denatured dye-labeled samples instead of conventional dye-labeled samples, which helps avoid the reformation of complex structures and facilitates better and more efficient separation of resolved fragments (Sanger et al. 1977).

2. Methodology of Formaldehyde Gel Electrophoresis for RNA Separation

The main procedural steps involved in the efficient separation of the RNA bands using formaldehyde gel electrophoresis are:

1. *RNA isolation:* The extraction of RNA from targeted cells or tissues, using an extraction agent such as TRIzol, is a prerequisite for carrying out the process effectively.
2. *Formaldehyde treatment:* In the subsequent step, the secondary or tertiary structure of the isolated RNA sample is linearized using formaldehyde for its efficient resolution through a gel electrophoresis procedure.
3. *Gel preparation:* Formaldehyde gel electrophoresis employs a special type of gel, which contains agarose as a major constituent that is dissolved in a formaldehyde-fortified buffer. The composition of the gel is adjusted to maintain the RNA in a linear form at all times.
4. *Loading and running the gel:* In the preceding step, the previously denatured sample is loaded into the wells of the agarose gel that is immersed in a specific MOPS (3-(N-morpholino)propanesulfonic acid) buffer. Afterward, the application of an electric field across the gel causes the bands to migrate. All this occurs at a high temperature of about 60–70 °C to preserve the linearity of the sample and keep the RNA sample free of secondary or tertiary conformation (Maniatis et al. 1982).
5. *Visualization:* In the final step, the gel, containing the separated fragments, is stained using a nucleic acid staining dye (SYBR Green or methylene blue) or an intercalating agent (ethidium bromide). The resolved bands can then be visualized by placing the gel under ultraviolet (UV) light, which allows the researcher to estimate the appropriate size and length of each band according to its migration rate and thickness within the sample (Sambrook and Russell 2001).

3. Applications of Formaldehyde Gel Electrophoresis in RNA Research

For performing different analyses of RNA, formaldehyde gel electrophoresis is a common technique of choice. Important applications of this method include:

1. *RNA size estimation:* The general principle of gel electrophoresis is the separation of fragments based on their charge and size. In this way, formaldehyde gel electrophoresis is performed to get resolved bands for the appropriate estimation

of their sizes that have further applications in cellular RNA-related (i.e., rRNA, mRNA, and lncRNA) analyses, such as getting insights into cellular mechanisms (i.e., transcription and translation) and functionality determination (Alwine et al. 1977).

2. *mRNA analysis:* The application spectrum of this approach can be enhanced by coupling it with other high-throughput techniques, such as Northern blotting, which is based on the specific binding of the hybridization probe to its complementary sequence present in the target molecule or sample. It is useful in the quantification of the electrophoretically separated bands, and the obtained data can then be utilized for understanding complex biological phenomena such as posttranscriptional and posttranslational regulation of gene expression.
3. *RNA separation:* Formaldehyde gel electrophoresis also finds its application in the processing of ribosomal RNA (rRNA), which is an integral component of the ribosome. In this way, this technique enables the researchers to separate larger and smaller subunits of the ribosome to dig deeper into the composition and functionality of these ribosomal units and carry out subsequent studies (Inoue and Marshall 1984).
4. *Quantification of RNA:* Quantification of nucleic acid is the general application of the gel electrophoresis process. Due to this, formaldehyde gel electrophoresis can be used for the accurate estimation and quantification of RNA fragments by comparing their thickness with that of the reference sample (ladder). The resolved bands having suitable concentration can then be eluted using the electroelution process and further applied in various studies, such as reverse-transcriptase PCR (RT-PCR) to produce copies of the complementary DNA (cDNA) (Wakeley and Finch 1992).
5. *Detection of alternative splicing:* From the pattern developed by the resolved bands, formaldehyde gel electrophoresis also provides a better opportunity to visualize and identify different posttranscriptional modifications, such as integration of cap and tail, as well as alternative splicing of introns present in the mRNA molecules of eukaryotes (Fong and Chua 1999).

4. Advantages of Formaldehyde Gel Electrophoresis

- *Prevents secondary structure formation:* Formaldehyde linearizes RNA molecules and also prevents them from forming secondary structure again. This unique feature of this method is responsible for the separation of RNA fragments based on their sizes and charges.
- *Highly reproducible:* Owing to the specificity, reliability, and effectiveness of formaldehyde, this method is highly recommended as a prerequisite for carrying out any RNA-related studies. The results generated using this technique can be reproduced consistently without any discrepancy. That is why formaldehyde gel electrophoresis is also known to be a highly reproducible method (Berkhout and van der Weijden 1996).
- *Versatility:* The analysis of any RNA, regardless of its size and nature, implies the versatile nature of this process. Hence, messenger RNA (mRNA), ribosomal RNA (rRNA), small nuclear RNA (snRNA), and long noncoding RNA (lncRNA)

can be processed and analyzed using the generalized protocol of formaldehyde gel electrophoresis.
- *High resolution:* The resolving power of agarose is so high that it can separate fragments, which are even single or a few base pairs different in length from each other. Hence, this technique is highly reliable when the aim is to get a high resolution of the RNA sample under study (Sambrook and Russell 2001).

5. Challenges and Limitations

Along with several useful features, this technique also has a few drawbacks, just as there are two sides to a coin. Those limitations are:

- *Loss of RNA integrity:* Denaturation of the sample, without which this method does not hold any significance, is the starting point of this process. The importance of this step is undeniable, but improper handling and prolonged incubation time at higher than required temperature can lead to the complete degradation of the RNA sample rather than partial denaturation, which would result in the loss of integrity of the sample and ultimately would lead to the failure of the whole process (Vilfan and Moores 2007).
- *Difficulty in handling large RNAs:* Similar to simple and conventional agarose gel electrophoresis, formaldehyde gel electrophoresis also struggles to resolve larger RNA molecules. Hence, this technique is constrained to the separation of molecules with a limited range of sizes.
- *Time-consuming process:* In terms of gel preparation and setup of other essential components, this process is more labor-intensive and time-consuming when compared with complementary techniques such as reverse-transcriptase quantitative polymerase chain reaction (RT-qPCR), expression analysis, and other related studies (Sambrook and Russell 2001).
- *Ethidium bromide usage:* Ethidium bromide staining is a common laboratory practice for the efficient visualization of separated nucleic acids, in this case, RNA, bands on formaldehyde-fortified agarose gel matrix. It is a well-known fact that ethidium bromide is a potential mutagen, which means it can cause genetic mutations if handled inappropriately. Therefore, the use of such hazardous chemicals requires careful handling and attention to prevent their detrimental effects.

6. Recent Advancements in RNA Electrophoresis

Several innovations have been introduced into this technique for the improvement of the existing methodology. These advancements include:

1. *Automated systems:* Efforts have been made to develop a fully automated system, which would perform formaldehyde gel electrophoresis with minimum human interference. Along with higher efficiency and reliability, the automated system is expected to take less time for performing electrophoretic separation. It will be a gain in the evolving world of science and technology (Schroeder and Becker 2005).
2. *Novel staining methods:* In terms of the safety of the researcher, the most hailed innovation to the current methodology of electrophoresis is the replacement of

other less harmful dyes, such as SYBR Green or SYBR Gold, with Ethidium bromide, which is known to be mutagenic. Another important advantage of using these novel staining agents is that they have enhanced sensitivity (improved hybridization efficiency) to the target molecules even at minimal concentration. In this way, less quantity of the staining dye is required to make the RNA visible under UV light (Zhou and Fang 2011).

References

Aljanabi SM, Martinez. Universal and rapid salt-extraction of high-quality genomic DNA for PCR-based techniques. BioTechniques. 1997;22(3):434–8.
Aljanabi SM, Forget L, Dookun A. An improved and rapid protocol for the isolation of polysaccharide- and polyphenol-free sugarcane DNA. Plant Mol Biol Rep. 1999;17:1–8.
Alwine JC, Kemp DJ, Stark GR. Method for detecting specific RNAs in agarose gels by transfer to diazobenzyloxymethyl-paper and hybridization with DNA probes. Proc Natl Acad Sci. 1977;74(12):5350–4.
Anderson JA, Churchill GA, Autrique JE, Tanksley SD, Sorrells ME. Optimizing parental selection for genetic linkage maps. Genome. 1999;42(1):181–190.
Anderson NL, Polson SW. Mass spectrometry: a powerful tool for protein analysis. J Proteome. 2009;72(2):405–17.
Arakawa T, et al. Size exclusion chromatography. In: Protein purification: principles and practice. Springer; 2007. p. 71–83.
Baker M. Next-generation sequencing: adjusting to data overload. Nat Methods. 2012;9(6):333–7.
Basu A, et al. Genomic analysis of the methylome. Nat Rev Genet. 2009;10(7):514–20.
Berkhout B, van der Weijden CC. Electrophoretic mobility shift assay for RNA-protein interactions. Methods Mol Biol. 1996;30:279–89.
Blakeslee JH, Cummings PH. Electroelution: recovery of proteins from polyacrylamide gels. Anal Biochem. 1971;43(1):93–103.
Boom R, Sol CJ, Salimans MM, Jansen CL, Wertheim-van Dillen PM, van der Noordaa J. Rapid and simple method for purifying viral DNA and RNA. J Clin Microbiol. 1990;28(3):495–503.
Bradford MM. A rapid and sensitive method for the quantitation of microgram quantities of protein utilizing the principle of protein-dye binding. Anal Biochem. 1976;72(1):248–54.
Butler JM. Forensic DNA typing: biology, technology, and genetics of STR markers. Elsevier; 2005.
Carter P. Ion exchange chromatography of proteins. Methods Enzymol. 1984;104:49–72.
Cho YH, Zeman A. Microfluidic electroelution of proteins from gels. Lab Chip. 2010;10(9):1193–200.
Corman VM, et al. Detection of 2019 novel coronavirus (2019-nCoV) by real-time RT-PCR. Euro Surveill. 2020;25(3):2000045.
Domon B, Aebersold R. Options and considerations when selecting a mass spectrometry-based proteomics experiment. Nat Rev Mol Cell Biol. 2006;7(5):436–46.
Edwards JB, Patel R. Development of an automated electroelution device for protein analysis. J Biomol Tech. 2008;19(2):67–74.
Engvall E, Perlmann P. Enzyme-linked immunosorbent assay (ELISA). Quantitative assay of immunoglobulin G. Immunochemistry. 1972;8(9):871–4.
Feng G, et al. The challenges and limitations of polymerase chain reaction. BioTechniques. 2019;66(3):97–102.
Fong CW, Chua NH. Alternative splicing of plant pre-mRNAs. Trends Plant Sci. 1999;4(6):214–20.
Fu Y, et al. High-frequency off-target mutagenesis induced by CRISPR-Cas nucleases in human cells. Nat Biotechnol. 2013;31(9):822–6.

Garofalo M, et al. Cancer genomics: translating the molecular analysis of tumors into clinical practice. J Clin Oncol. 2015;33(26):3003–12. 8.
Ghosh S, Krishna N. Electroelution: a critical review of recovery efficiency and factors influencing elution. J Chromatogr Sci. 2004;42(10):457–63.
Green MR, Sambrook J. Molecular Cloning: A Laboratory Manual. 4th ed. Cold Spring Harbor (NY): Cold Spring Harbor Laboratory Press; 2012.
Ginsburg GS, et al. Genomic medicine: new frontiers in health care. JAMA. 2009;301(10):1065–72.
Görg A, et al. The current state of two-dimensional electrophoresis with immobilized pH gradients. Electrophoresis. 2004;25(10):2301–15.
Gould J, et al. Protein extraction methods. In: Protein purification protocols. Humana Press; 1991. p. 99–106.
Hames BD. Electrophoresis of proteins: a practical approach. Oxford University Press; 1990.
Hegde M, et al. High-throughput technologies for DNA analysis. BioTechniques. 2008;44(5):629–35.
Heid CA, et al. Real-time quantitative PCR. Genome Res. 1996;6(10):986–94.
Heukeshoven J, Dernick R. Simplified silver staining of proteins in polyacrylamide gels. Electrophoresis. 1985;6(2):103–12.
Hindson CM, et al. High-throughput single-copy DNA quantification with digital PCR. Nat Methods. 2011;6(4):271–3.
Ijaz S, Haq IU. Recombinant DNA technology. Cambridge Scholars Publishing; 2019.
Ijaz S, Haq IU, Razzaq HA. Mutation introduced in DDTFR10/A gene of ethylene response element-binding protein (EREBP) family through CRISPR/Cas9 genome editing confers increased Fusarium wilt tolerance in tomato. Physiol Mol Biol Plants. 2023;29:1–10. https://doi.org/10.1007/s12298-022-01273-6.
Inoue M, Marshall WS. Use of formaldehyde gel electrophoresis for separation and analysis of ribosomal RNA. Anal Biochem. 1984;139(1):196–200.
Lichter P, Boyle AL, Cremer T, Ward DC. Analysis of genes and chromosomes by non-isotopic in situ hybridization. Genet Anal Tech Appl. 1992;8(1):24–33.
Jinek M, et al. A programmable dual-RNA-guided DNA endonuclease in adaptive bacterial immunity. Science. 2012;337(6096):816–21.
Karas M, et al. Matrix-assisted laser desorption/ionization of proteins: a new technique for mass spectrometry. Anal Chem. 1988;60(20):2299–301.
Kirkland JL. Principles of protein electrophoresis. Elsevier Science; 2001.
Laemmli UK. Cleavage of structural proteins during the assembly of the head of bacteriophage T4. Nature. 1970;227(5259):680–5.
Maniatis T, Fritsch EF, Sambrook J. Molecular cloning: a laboratory manual. Cold Spring Harbor Laboratory Press; 1982.
Mardis ER. Next-generation DNA sequencing methods. Annu Rev Genomics Hum Genet. 2008;9:387–402.
Metzker ML. Sequencing technologies-the next generation. Nat Rev Genet. 2010;11(1):31–46.
Mikami A, et al. Detection of genetically modified organisms using PCR-based methods. J Biosci Bioeng. 2002;94(5):379–83.
Molinaro R, Telford WG. Optimizing electroelution conditions for protein recovery. Proteomics. 2003;3(5):752–60.
Möller M, Töpfer R, Laimer M. DNA extraction from plant tissues. In: Plant molecular biology manual; 1992. p. 1–11.
Mullis K, Faloona F. Specific synthesis of DNA in vitro via a polymerase-catalyzed chain reaction. Methods Enzymol. 1987;155:335–50.
Mullis K, et al. Specific synthesis of DNA in vitro via a polymerase-catalyzed chain reaction. Methods Enzymol. 1986;155:335–50.
Ng PC, et al. An integrated strategy for identifying molecular genetic markers of disease. Nat Rev Genet. 2010;11(7):423–34.

Nicolson GL, Poste G. Tumor cell diversity and host responses in cancer metastasis—Part II: Host immune responses and therapy of metastases. Curr Probl Cancer. 1983;7(7):1–42.
O'Farrell PH. High resolution two-dimensional electrophoresis of proteins. J Biol Chem. 1975;250(10):4007–21.
Peters T. All about albumin: biochemistry, genetics, and medical applications. Academic Press; 1996.
Peters T, et al. Ultracentrifugation and protein purification. Methods Enzymol. 1991;198:53–81.
Prowse TA, et al. Resolving DNA sequence errors using next-generation sequencing technologies. J Biomol Tech. 2019;30(1):34–45.
Porebski S, Bailey LG, Baum BR. Modification of a CTAB DNA extraction protocol for plants containing high polysaccharide and polyphenol components. Plant Mol Biol Rep. 1997;15:8–15.
Rao VB, Feiss M. The bacteriophage lambda DNA packaging motor. Annu Rev Genet. 1995;29:565–606.
Razzaq HA, Ijaz S, Haq IU, et al. Functional inhibition of *the StERF3* gene by dual targeting through CRISPR/Cas9 enhances resistance to the late blight disease in *Solanum tuberosum* L. Mol Biol Rep. 2022;49:11675–84. https://doi.org/10.1007/s11033-022-07958-1.
Righetti PG, Gianazza E, Gelfi C. Isoelectric focusing in immobilized pH gradients: theory and methodology. J Chromatogr. 1993;646:1–20.
Robinson D, et al. Nested PCR for the detection of microbial DNA in clinical specimens. J Clin Microbiol. 2000;38(4):1451–6.
Rupp B. Crystallography made crystal clear: a guide for users of macromolecular models. 3rd ed. Elsevier; 2010.
Sambrook J, Russell DW. Molecular cloning: a laboratory manual. 3rd ed. Cold Spring Harbor Laboratory Press; 2001.
Sambrook J, Fritsch EF, Maniatis T. Molecular cloning: a laboratory manual. 2nd ed. Cold Spring Harbor Laboratory Press; 1989.
Sanger F, Nicklen S, Coulson AR. DNA sequencing with chain-terminating inhibitors. Proc Natl Acad Sci. 1977;74(12):5463–7.
Schena M, et al. Quantitative monitoring of gene expression patterns with a complementary DNA microarray. Science. 1995;270(5235):467–70.
Schmidt T, Stamatakis A, Patro R. Direct PCR-based DNA extraction and next-generation sequencing. Methods Mol Biol. 2014;1123:105–13.
Schnell J, Miki K. Electroelution of antibodies for diagnostic applications. J Immunol Methods. 2006;316(1–2):12–22.
Schroeder H, Becker A. Automated systems for RNA electrophoresis. BioTechniques. 2005;38(6):1030–5.
Schwartz DC, Cantor CR. Separation of yeast chromosome-sized DNAs by pulsed field gradient gel electrophoresis. Cell. 1984;37(1):67–75.
Scopes RK. Protein purification: principles and practice. 2nd ed. Springer; 1987.
Shevchenko A, Wilm M. Mass spectrometric sequencing of proteins from silver-stained polyacrylamide gels. Anal Chem. 1996;68(7):850–8.
Smith PK, et al. Measurement of protein using bicinchoninic acid. Anal Biochem. 1985;150(1):76–85.
Southern EM. Detection of specific sequences among DNA fragments separated by gel electrophoresis. J Mol Biol. 1975;98(3):503–17.
Tobler H, Maixner F, Hoelzer D. Magnetic bead-based DNA extraction method. BioTechniques. 1999;27(5):1007–13.
Towbin H, et al. Electrophoretic transfer of proteins from polyacrylamide gels to nitrocellulose sheets: procedure and some applications. Proc Natl Acad Sci. 1979;76(9):4350–4.
Trapnell C, et al. Differential analysis of gene regulation at transcript resolution with RNA-seq. Nat Biotechnol. 2014;31(1):46–53.
Tu LH, Lee CP. Electroelution: a new method for recovering proteins from polyacrylamide gels. J Chromatogr. 1981;4, 208(1):1–8.

Turnbaugh PJ, et al. The human microbiome project. Nature. 2007;449(7164):804–10.
Tusher VG, Tibshirani R, Chu G. Significance analysis of microarrays applied to the ionizing radiation response. Proc Natl Acad Sci USA. 2001;98(9):5116–5121.
Verma RS, Babu A. Human chromosomes: structure, behavior, and function. Elsevier Science & Technology; 2007.
Vilfan D, Moores CA. RNA gel electrophoresis: challenges and limitations. Methods Enzymol. 2007;424:217–29.
Wakeley J, Finch CE. RNA quantification using formaldehyde gel electrophoresis. BioTechniques. 1992;13(2):128–34.
Wang F, et al. Native polyacrylamide gel electrophoresis: an essential tool for protein analysis. Proteomics. 2016;16(13):1913–25.
Weber K, Osborn M. The reliability of molecular weight determinations by dodecyl sulfate-polyacrylamide gel electrophoresis. J Biol Chem. 1969;244(16):4406–12.
Wilfinger WW, Mackey K, Wilfinger LE. The effect of alcohol precipitation in DNA recovery and quality. BioTechniques. 1997a;22(5):1062–6.
Wilfinger WW, Mackey K, Wilfinger L. The PCR protocols handbook. Humana Press; 1997b.
Wilkins MR, et al. Protein identification and analysis by mass spectrometry. Anal Chem. 1996;68(13):2306–13.
Williams MA, O'Leary M. Electroelution of peptides and proteins from polyacrylamide gels for mass spectrometric analysis. BioTechniques. 2000;28(5):982–9.
Wilson IG. Inhibition and facilitation of nucleic acid amplification. Appl Environ Microbiol. 1997;63(10):3741–51.
Wilson K. Preparation of genomic DNA from plant tissue. Curr Protoc Mol Biol. 2001:271–4.
Wüthrich K. NMR of proteins and nucleic acids. Wiley-Interscience; 1986.
Zhou XB, Fang Q. SYBR Green-based detection of RNA. BioTechniques. 2011;51(1):63–8.

Chapter 2
Restriction Endonucleases: Basics and Applications

Bakhtawar Nasir, Imran ul Haq, and Hayssam Mohamed Ali

2.1 Introduction

Molecular biology now depends significantly on restriction endonucleases, which are available from multiple commercial suppliers with extensive selections of products. The diversity of restriction enzymes, genomic organization, and mechanisms has been extensively studied; however, these topics continue to be researched and contribute to ongoing efforts in classification. Restriction endonucleases (REs), which selectively cleave both strands of DNA, are essential tools in molecular biology (Roberts and Halford 1993). Endonucleases were first discovered in the 1960s and commercialized in the early 1970s. It shares commonalities in structure, action, and usage, but can differ significantly. Restriction endonucleases are widely studied in molecular biology, including their method of cleavage, in vivo function, evolutionary origins, and use as a model for site-specific DNA recognition. Protein engineering and fusions of proteins to generate novel polypeptides have led to the development of new endonuclease activity, the cloning of existing enzymes, and the discovery of new natural enzymes. Although restriction endonucleases are mostly found in bacterial genomes and plasmids, they are also present in viruses, eukaryotes, and archaea. According to estimates, one out of every four microorganisms analyzed had one or more of them (Stein et al. 1995).

B. Nasir
The University of Faisalabad, Faisalabad, Pakistan

I. u. Haq (✉)
Department of Plant Pathology, University of Agriculture, Faisalabad, Pakistan
e-mail: imran_1614@uaf.edu.pk

H. Mohamed Ali
Department of Botany and Microbiology, College of Science, King Saud University, Riyadh, Saudi Arabia
e-mail: hayhassan@ksu.edu.sa

S. Ijaz et al. (eds.), *Molecular Biology Methods*,
https://doi.org/10.1007/978-981-92-2642-9_2

There are already 3500 restriction enzymes across various categories that can identify 259 distinct DNA sequences. Most of these, about 3460 enzymes that can recognize 234 DNA sequences, fall within either the conventional type II or type III subclasses. These are the standard molecular biology tools, featuring over 500 commercially available enzymes with more than 200 specificities. Furthermore, 58 homing endonucleases—so-called because they are encoded by jumping genes, self-splicing introns, or intein—were discovered, each with a distinct recognition site. There are currently 16 known site-specific nickases in total. There are already 297 restriction enzymes that have been sequenced and cloned (Ijaz and Haq 2019; Roberts et al. 2007).

Restriction endonucleases gained recognition for their ability to inhibit phage growth in bacterial cells by cleaving the invading DNA. They may function as antibacterial systems in this way. Methylase or methylases protect the host's DNA from restriction by recognizing the same sequence as the restriction enzyme and methylating a particular nucleotide on each strand of the sequence. Once methylated, the endonuclease can no longer use the host DNA as a substrate. Endonuclease cannot digest newly reproduced host DNA because it is methylated on both strands, including hemimethylated DNA (Rocha et al. 2001).

The concept of restriction enzymes as a protection system may be oversimplified. Several factors reduce the protective potential of an enzyme. Phage types for which the DNA was additionally altered at the critical bases or those lacking at least a dsDNA intermediate would not be affected. Before restriction takes place, the host may methylate a limited number of phages, allowing them to replicate protectively methylated copies of themselves. Additionally, small phage genomes typically have few recognition sites for enzymes. In a group of bacteria, restriction site avoidance is more significant than in a corresponding group of phages. If the methylase is not properly maintained, the endonucleases may be fatal to the host because they typically have a longer half-life than their corresponding methylases (Nobusato et al. 2000).

2.2 Nomenclature

According to the nomenclature system proposed by Smith and Nathans, restriction enzymes are named based on the source organism. The first letter comes from the genus, followed by the first two letters of the species name, both in italics. An extra, nonitalicized letter may be added to indicate a specific strain. Roman numerals distinguish enzymes discovered in sequence within the same organism. The observation that many enzymes recognize a limited number of distinct DNA sequences led to the idea of isoschizomers—enzymes from different sources that recognize the same DNA sequence. A subset of these, called neoschizomers, not only recognize the same sequence but also cut it at different positions (Smith and Nathans 1973).

Advances in cloning and sequence analysis have shown that methyltransferases and restriction endonucleases, even when recognizing identical DNA sequences, have little to no sequence homology. For example, the Haemophilus strains that produce the enzymes HhaII and HinfI both recognize the sequence GANTC and cut between G and A; however, their amino acid sequences are only about 19% identical (Wilson and Murray 1991). Additionally, methylation patterns and sensitivity to restriction can differ among endonuclease/methylase pairs that recognize the same DNA motif. Structural and mutational studies have identified a conserved catalytic motif, PD…D/EXK, in roughly ten endonucleases from different types (type II, IIe, IIs, IV, and intron-encoded endonucleases) (Stahl et al. 1998). Conversely, methyltransferases responsible for 6-methyladenine, 4-methylcytosine, and 5-methylcytosine modifications have characteristic conserved motifs.

The restriction endonuclease and modification enzymes together form what is known as the restriction–modification (R/M) system. Genes encoding these methylases and endonucleases are usually located next to each other on bacterial DNA, with transcriptional orientations that can be convergent, divergent, or sequential. The widespread presence of these genes has been utilized in a cloning method called the "Hungarian Trick" (Szomolányi et al. 1980). This technique involves digesting bacterial genomic DNA containing the R/M system with its specific endonuclease to create a clone library. The recognition sequence must be included on the expression vector used. When purified plasmids from the clone library are treated with the same restriction enzyme in vitro, only plasmids containing the methylase gene stay intact and resist cleavage.

To protect their own DNA from self-digestion, bacteria generally methylate their genomes before restriction endonuclease activity occurs. One strategy to prevent self-restriction is to decrease the number of recognition sites within the genome. Alternatively, bacteria can control the timing of enzyme expression, ensuring methyltransferase activity happens before the endonuclease. In some R/M systems, an upstream open reading frame encodes a regulatory "C" protein that positively regulates the endonuclease gene. This regulatory mechanism guarantees methylation is established before the restriction enzyme becomes active (Williams 2003). The presence of such C genes is especially common when methylase and endonuclease genes are organized in divergent or convergent transcriptional orientations.

2.3 Types of Restriction Endonucleases

Enzymes that undergo restriction can be classified into at least four categories (type I, II, III, and IV) according to their cleavage properties, cofactor needs, recognition sequence symmetry, and enzyme composition (Table 2.1). AdoMet acts as the donor substrate for methyl groups in all MTase actions. Enzymes of type I R-M are multienzyme complexes made up of three different subunits: R, M, and S. Specificity is determined by the S subunit for both R and M. The unmodified DNA is cleaved at a relatively random location far from the recognition site by type I R-M enzymes, which require ATP and magnesium. In addition to site-specific methylation of the target DNA, which competes with the corresponding ENase activity, type I enzymes

Table 2.1 Comparison of restriction-modification (R-M) system types

Feature	Type I	Type II	Type IIS	Type III	Type IV
Enzyme structure	Single multifunctional enzyme complex	Single enzyme (restriction + modification)	Separate restriction and modification enzymes	Single enzyme complex (2 subunits)	Separate monomeric enzymes
Subunit composition	R + M + S (3 subunits)	R: dimer, M: monomer	R: monomer, M: monomer	Two subunits (Mod + Res)	Monomeric
Cofactors required	Mg^{2+}, ATP, AdoMet	Mg^{2+}	Mg^{2+}	Mg^{2+}, AdoMet	Mg^{2+}
Recognition site	Asymmetric, bipartite	Palindromic or asymmetric	Asymmetric	Asymmetric	Asymmetric
Cleavage site	Variable distance from recognition site (either side)	Fixed distance, usually 3′ side	25–27 bp downstream of the recognition site	~14 bp downstream of the recognition site	Undefined or variable
Methylation pattern	The M subunit methylates both DNA strands	One methylase per strand	Both strands methylated	One strand methylated	Typically targets modified DNA only
Function notes	Cleaves DNA only when unmethylated; complex recognition	Most commonly used in molecular biology	Used for precise, nonpalindromic cutting	Requires two sites in opposite orientation	Targets methylated or modified DNA

hydrolyze ATP as part of their restriction activity. The R-ENase and M-MTase activities are physically distinct in the *type II* R-M system.

In contrast to M-MTases, which are monomeric enzymes, R-ENases are typically dimers with identical secondary units. All that is needed for *type II* R-ENases to have restriction activity is Mg^{2+}. Type II R-ENase and M-MTase recognize the particular nucleotide sequences. *Type II* R-ENases break DNA at the specific recognition sequence, whereas a few enzymes known as *type II* S cleave DNA at a specific distance from the recognition site (Table 2.1). The restriction activity of *type III* R-M enzymes depends on ATP and Mg^{2+}, and they are made up of two nonidentical subunits. *Type III* enzymes cut DNA at predetermined locations about the unaltered recognition sequence after recognizing brief nonpalindromic sequences. Hydrolyzing ATP is not done by *type III* enzymes. Although it promotes ENase activity, AdoMet is not necessary for restriction (Rao et al. 2014).

Between the *type II* and *type III* enzymes in the evolutionary pathway, there is another type of R-M system known as *type IV*. *Type IV* enzymes, like Eco57l, are made up of distinct R-ENase and M-MTase, but the monomeric R-ENase also has methylase activity. In vivo, the host DNA cannot be protected by the restriction-associated methylase activity. M-MTase and R-ENase are capable of identifying asymmetric sequences. R-ENase cleaves target DNAs at a specific distance from the recognition site (such as 16–14 nucleotides away with *Eco*57l) when Mg^{2+} is present. With AdoMet, the restriction activity is enhanced (Loenen et al. 2014) (Table 2.1).

The majority of *type II* endonucleases are often homodimers, with the monomeric subunit typically weighing between 25 and 35 kDa. They need magnesium, and they can cut within, partial, or interrupted palindromes. Despite having different primary sequences, *type II* endonucleases share a similar three-dimensional structure: a dimeric "U"-shaped holoenzyme with bridging domains at the bottom and recognition and catalytic domains on the sides of each of the identical subunits. Monomers are inactive on their own. Similar to the enzymes MutH and α-exonuclease, the crystal structures of *type II* enzymes, including their many subclasses, appear to share a core composed of five β-sheets with an α-helix on either side (Kovall and Matthews 1999; Kubareva et al. 2000). It is required to specify subdivisions, as many *type II* restriction endonucleases do not fit this stringent description. It has recently been suggested that these heterodox *type II* restriction endonucleases be given a new name (Table 2.2).

Type II S restriction endonucleases, such as FokI, can identify asymmetric sequences and cleave them at a specific distance. It was previously thought that these enzymes operated as monomers. However, research on *Fok*I has now demonstrated that it dimerizes on DNA, suggesting that this may be a more widespread occurrence (Bitinaite et al. 1998). The restriction endonucleases of *type II*E interact with two copies of their recognition sequence, such as *Nae*I, one of which is the target for cleavage and the other of which acts as an allosteric effector (Krüger et al. 1995).

Because *type II*F restriction endonucleases interact with two copies of their recognition sequence, they are comparable to *type II*E enzymes. (Halford et al. 1999).

Table 2.2 Nomenclature and features of type II restriction endonucleases

Subtype	Characteristic feature	Enzyme example	Recognition and cleavage sequence
Orthodox	Palindromic site; recognized by a *homodimeric enzyme*; cleavage occurs *within or adjacent* to the site	***Eco*RI**	G↓AATTC CTTAA↑G
		***Eco*RV**	GAT↓ATC CTA↑TAG
		***Bgl*I**	GCCNNNN↓NGGC CGGN↑NNNNCCG
Type IIS	Asymmetric site; cleavage at a *defined distance* from the site	*Fok*I	GGATGN$_9$↓NNNN CCTAC N$_9$↑NNNN
Type IIE	*Two recognition sites* are required; one acts as *an allosteric effector*	*Nae*I	GCG↓CGC CGC↑GCG
Type IIF	*Two recognition sites* are cleaved *simultaneously* by a *homotetrameric enzyme.*	*Ngo*MIV	G↓CCGGC CGGCC↑G
Type IIT	*Different subunits* for restriction and modification activity	*Bpu*10I	CC↓TNAGC GGTNA↑TCG
		*Bsl*I	CCNNNNNNN↓GG GG↑NNNNNNNNCC
Type IIG	*Single polypeptide* with both restriction and modification activities	*Eco*57I	CTGAAGN$_{14}$↓NN GACTTC N$_{14}$↑NN
Type IIB	Cleavage on *both sides* of the recognition site	*Bcg*I	NN↓N$_{10}$ CGA N$_6$ TGC N$_{10}$ NN↓ ↑NNN$_{10}$ GCT N$_6$ ACG N$_{10}$↑NN
		*Bpl*I	NN$_4$↓N$_8$ GAG N$_5$ CTC N$_8$ NN$_4$↓ ↑NN$_4$ N$_8$ CTC N$_5$ GAG N$_8$↑NN$_4$
Type IIM	Recognizes and cleaves *methylated DNA* (e.g., Dam methylation)	*Dpn*I	G*m*A↓TC CT↑*m*AG

↓ and ↑ Cleavage site, *N* Any nucleotide, N_4, N_6, N_8, N_9, N_{10}, N_{14} Number of unspecified bases, *m*A Methylated adenine base (methylation)

*Type II*T restriction endonucleases, such as *Bpu*10I and *Bsl*I, are made up of two distinct subunits. Recognizing an asymmetric sequence, *Bpu*10I operates as a heterodimer (αβ), with both subunits likely having a single active site. When a palindromic sequence is recognized, *Bsl*I acts as a heterotetramer ($\alpha_2\beta_2$) (Stankevicius et al. 1998). *Type II*B restriction endonucleases, such as *Bcg*I, which detects an asymmetric sequence, and *Bpl*I, which identifies a symmetric sequence, cleave DNA at both ends of the recognition sequence. These enzymes can restrict and modify substances and are made up of various subunits (*Bcg*I, $\alpha_2\beta$; *Bpl*I, αβ). For restriction, they need AdoMet to be present. The α-subunit of *Bcg*I was shown to include the target recognition domain, while the β-subunit possesses the catalytic centers for restriction and modification. AdoMet stimulates *type II* restriction endonucleases, such as *Eco57I* enzymes, which possess both restriction and modification activities within a single polypeptide chain. Methylated DNA is recognized by *type IIM* restriction endonucleases, such as *Dpn*I (Hsieh et al. 2000).

2.4 Structure, Specificity, and Mechanism

In prokaryotic species, restriction endonucleases are widely distributed. Their primary biological role is to protect the host genome from foreign DNA, especially that of bacteriophages. There is ongoing discussion regarding additional roles, including a role in transposition and recombination. (Pingoud and Jeltsch 2001).

Endonuclease and methyltransferase activity are the two components of restriction–modification (RM) systems, which are, by definition, restriction endonucleases. The host's DNA is altered at the recognition sequence by the modification enzyme, protecting it from attack by the restriction endonuclease. In contrast, the restriction enzyme's substrate is foreign DNA that is cleaved in response to specific recognition sites. Three categories of RM systems have been identified and categorized based on their mode of action, cofactor requirements, and subunit composition (Smith and Nathans 1973).

Double-stranded DNA (dsDNA) has particular, nonmethylated, bipartite, and asymmetric DNA sequences that the REase complexes first identify and attach to in order to restrict DNA (Studier and Bandyopadhyay 1988). These sequences consist of two distinct domains, each measuring 35 bp, separated by a nonspecific spacer that is characterized as being 68 bp long (Sistla and Rao 2004). The enzyme hydrolyzes −1 ATP with a k_{cat} of ~100,000 min after binding to the target sequence. At the same time, it bidirectionally translocates dsDNA in the direction of the holoenzyme linked at the recognition sequence (Bianco and Hurley 2005; Firman and Szczelkun 2000). Restrictions occur when translocation is hindered, usually thousands of base pairs away from the recognition sequence. Remarkably, these enzymes are believed to stay attached to the recently cleaved DNA for several hours, during which time they rapidly hydrolyze ATP. It means that each holoenzyme can only cleave one DNA substrate.

The mechanism of action of restriction enzymes comprises three steps: identification, cleavage, and cutting of the specific DNA sequences. Type II enzymes, which cut DNA within or close to their recognition sites, are the most often utilized restriction enzymes (Loenen and Raleigh 2014). A restriction enzyme binds to DNA and forms sequence-specific interactions with the bases when it comes into contact with the distinctive recognition site, identifying the symmetry of the sequence. After the restriction enzyme binds to the DNA, its catalytic center is activated due to a conformational change in the enzyme. The catalytic center subsequently cleaves the DNA backbone at a specified location within or close to the recognition site, releasing the enzyme. The particular restriction enzyme and the sequence of the recognition site determine the precise location of the cleavage (Pingoud and Jeltsch 2001).

Depending on the particular enzyme and the location of the cleavage, a restriction enzyme's cleavage might produce blunt or sticky ends for the DNA (Loenen and Raleigh 2014). When an enzyme cuts DNA at the same point on both strands, it produces straight, blunt ends. When the enzyme cuts the DNA at slightly different locations on the two strands, sticky ends are produced. These overhanging ends can base pair with complementary overhangs on other DNA molecules. Different kinds of DNA ends are produced by the specific recognition and cleavage of DNA at its recognition sites, which is how restriction enzymes work.

2.5 Applications of Restriction Endonucleases

REs are extensively used across various applications in the field of molecular biology and biotechnology. DNA fingerprinting utilizes restriction enzymes to generate a distinct banding pattern from DNA fragments of various individuals, which can be used to identify or establish relationships with relatives. The investigation of restriction fragment-length polymorphisms (RFLP) is a particularly interesting application for restriction mapping. This method allows for the comparison of DNA samples from different sources by analyzing the size variation of specific restriction fragments. DNA fragments from a specific region of an individual's genome can be evaluated for a unique biological marker or compared to a reference sample. RFLP analysis is now well established as a highly reliable technique for prenatal diagnosis (Di Felice et al. 2019).

The primary function of restriction enzymes, since Danna and Nathans' first investigation in 1971, has been to identify the specific DNA molecules on which they cut. Because restriction enzymes produce highly specific cleavage patterns across different DNA regions, they became invaluable tools for comparing genetic material during a time when rapid and efficient DNA sequencing technologies were not yet available. This specificity paved the way for numerous groundbreaking studies exploring the structure, function, and evolution of various gene systems (Holsinger and Jansen 1993).

It became as easy as comparing the restriction profiles of various organisms, although closely related species, to compare their DNA. In fact, cleavage and a distinct digesting pattern can be caused by a single nucleotide change that takes place at a specific restriction cutting site. The Southern blot technique, introduced in 1976, marked a breakthrough in the physical mapping of DNA through restriction enzyme digestion. Edwin Southern's technique for analyzing certain DNA restriction fragments after they had been electrophoretically sorted by size via an agarose gel slab essentially standardized the experimental setting. (Southern 1975). Restriction enzymes are now essential tools in molecular biology, especially for recombinant DNA technologies and physical DNA mapping. The exact fragmentation of DNA made possible by these enzymes, commonly referred to as molecular scissors, enables researchers to examine, work with, and reassemble genetic material. They have had a profound impact on genetics, biotechnology, and biomedicine since their discovery and application.

In 1952, Lederberg coined the term "plasmid" to describe extrachromosomal DNA fragments that can carry genetic traits. In the late 1960s, plasmids were identified as circular DNA molecules, and electron microscopy was used to define and observe their structure and characteristics fully. In a seminal experiment conducted in 1972, Cohen and his coresearchers effectively selected transformed bacteria by antibiotic screening after introducing antibiotic resistance genes into bacterial plasmids. Plasmids were first used as gene cloning vectors because of this (Cohen et al. 1972).

The DNA ligase enzyme, which links DNA fragments by creating phosphodiester bonds, had been discovered at that time, but there was still no way to cut DNA at specific locations. Following Cohen's experiments with mechanical fragmentation, the discovery of restriction enzymes in the late 1960s and early 1970s marked a significant advancement in the field of molecular biology. These enzymes, particularly the class II enzymes like *Eco*RI, which Herbert Boyer's group discovered, cut DNA in staggered chunks, producing cohesive or "sticky" ends that could be joined together regardless of the DNA's origin (Smith and Welcox 1970; Danna and Nathans 1971).

The development of gene libraries was the next important step. Due to their short recognition sequences (4–6 base pairs), these libraries involved the fragmentation of genomic DNA through digestion with restriction enzymes that cut frequently. To ensure statistically adequate coverage of the entire genome, the acquired fragments were in vitro linked to multiple plasmid molecules. Therefore, restriction segments extracted from a particular genome were dispersed among many plasmids, which together constituted the organism's entire genome. Because of these libraries, entire genomes—including the large human genome—can now be studied methodically. Overall, recombinant DNA technology and the initial attempts at DNA sequencing were made possible by the employment of restriction enzymes to break up big DNA segments into fragments with specific ends and sizes (Goeddel et al. 1979; Khan et al. 2016).

These discoveries have significantly accelerated the study of site-specific genome modification, with a focus on creating endonuclease-based instruments that can target and cleave almost any sequence. As a result, two strong systems have been developed: transcription activator-like effector nuclease (TALEN) and zinc-finger nuclease (ZFN) (Liu et al. 1997). Zinc finger nucleases (ZFNs) utilize engineered nucleases in which the C-terminal of a zinc-finger DNA-binding domain is fused to the DNA-cleavage domain of the type II restriction enzyme FokI. In this fusion protein, the zinc-finger domain specifically recognizes and binds to the target DNA sequence, while the FokI domain executes the double-stranded break (Christian et al. 2010; Miller et al. 2011; Kim and Kim 2014).

Similarly, the TALEN (transcription activator-like effector nucleases) system employs a fusion protein comprising the FokI cleavage domain and a TALE DNA-binding domain. Both ZFNs and TALENs represented significant advances in targeted genome editing, though their development and application are technically demanding and time-consuming.

The emergence of RNA-guided genome editing tools—most notably RGENs (RNA-guided engineered nucleases)—marked a major step forward in terms of efficiency, programmability, and reduced cost. Among these, CRISPR-Cas9 has rapidly become the preferred system due to its simplicity, versatility, and effectiveness in a wide range of organisms and applications. CRISPR-Cas systems are part of a bacterial adaptive immune mechanism that distinguishes between methylated self-DNA and unmethylated foreign DNA, such as that introduced by bacteriophages. Restriction endonucleases are a core component of the innate bacterial defense, recognizing and cleaving foreign, nonmethylated DNA while sparing the host genome.

In addition to restriction enzymes, bacteria employ a more sophisticated defense mechanism involving clustered regularly interspaced short palindromic repeats (CRISPR) elements () and Cas (CRISPR-associated) proteins. During phage infection, fragments of the invading genome are integrated into the bacterial genome at CRISPR loci. Upon subsequent infections, these loci are transcribed, and the resulting small RNAs—derived from the integrated phage sequences—guide Cas proteins to recognize and cleave matching exogenous DNA. Cas9, encoded near the CRISPR locus, is an RNA-guided endonuclease that mediates targeted DNA cleavage when its RNA guide base-pairs with the invading DNA (Jiang and Doudna 2017).

Type II restriction enzymes remain essential tools in genetic engineering, particularly within the *E. coli* K12 system, which underpins a wide range of molecular biology applications, including cloning, DNA library construction, sequencing, and the synthesis of recombinant proteins. Since the advent of synthetic DNA, the scope of these enzymes has broadened to include in vitro DNA packaging into phage particles, vector construction, protein stabilization, and overexpression (Loenen et al. 2014).

Since their discovery, restriction enzymes have been indispensable in foundational DNA technologies, including cloning, mapping, and manipulation. Their ability to generate specific DNA fragments revolutionized molecular biology. These enzymes continue to serve as powerful "molecular witnesses" to the evolution of genetic research, enabling increasingly complex manipulation and understanding of the DNA double helix. Remarkably, the CRISPR/Cas system is presently proceeding in a very similar direction. Overall, these findings provide a great illustration of how fundamental, unfocused research can yield a wealth of novel discoveries with significant practical implications.

References

Bianco PR, Hurley EM. The type I restriction endonuclease EcoR124I, couples ATP hydrolysis to bidirectional DNA translocation. J Mol Biol. 2005;352(4):837–59.

Bitinaite J, Wah DA, Aggarwal AK, Schildkraut I. Fok I dimerization is required for DNA cleavage. Proc Natl Acad Sci. 1998;95(18):10570–5.
Christian M, Cermak T, Doyle EL, Schmidt C, Zhang F, Hummel A, Bogdanove AJ, Voytas DF. Targeting DNA double-strand breaks with TAL effector nucleases. Genetics. 2010;186(2):757–61.
Cohen SN, Chang AC, Hsu L. Nonchromosomal antibiotic resistance in bacteria: genetic transformation of Escherichia coli by R-factor DNA. Proc Natl Acad Sci. 1972;69(8):2110–4.
Danna K, Nathans D. Specific cleavage of simian virus 40 DNA by restriction endonuclease of Hemophilus influenzae. Proc Natl Acad Sci. 1971;68(12):2913–7.
Di Felice F, Micheli G, Camilloni G. Restriction enzymes and their use in molecular biology: an overview. J Biosci. 2019;44(2):38.
Firman K, Szczelkun MD. Measuring motion on DNA by the type I restriction endonuclease EcoR124I using triplex displacement. EMBO J. 2000;19:2094.
Goeddel DV, Kleid DG, Bolivar F, Heyneker HL, Yansura DG, Crea R, Hirose T, Kraszewski A, Itakura K, Riggs AD. Expression in Escherichia coli of chemically synthesized genes for human insulin. Proc Natl Acad Sci. 1979;76(1):106–10.
Halford SE, Bilcock DT, Stanford NP, Williams SA, Milsom SE, Gormley NA, Watson MA, Bath AJ, Embleton ML, Gowers DM, Daniels LE. Restriction endonuclease reactions requiring two recognition sites. Biochem Soc Trans. 1999;27(4):696–9.
Holsinger KE, Jansen RK. Phylogenetic analysis of restriction site data. Methods Enzymol. 1993;224:439–55.
Hsieh PC, Xiao JP, O'loane D, Xu SY. Cloning, expression, and purification of a thermostable nonhomodimeric restriction enzyme, Bsl I. J Bacteriol. 2000;182(4):949–55.
Ijaz S, Haq IU. Recombinant DNA technology. Cambridge Scholars Publishing; 2019.
Jiang F, Doudna JA. CRISPR–Cas9 structures and mechanisms. Annu Rev Biophys. 2017;46(1):505–29.
Khan S, Ullah MW, Siddique R, Nabi G, Manan S, Yousaf M, Hou H. Role of recombinant DNA technology to improve life. Int J Genomics. 2016;2016(1):2405954.
Kim H, Kim JS. A guide to genome engineering with programmable nucleases. Nat Rev Genet. 2014;15(5):321–34.
Kovall RA, Matthews BW. Type II restriction endonucleases: structural, functional and evolutionary relationships. Curr Opin Chem Biol. 1999;3(5):578–83.
Krüger DH, Kupper D, Meisel A, Reuter M, Schroeder C. The significance of distance and orientation of restriction endonuclease recognition sites in viral DNA genomes. FEMS Microbiol Rev. 1995;17(1–2):177–84.
Kubareva EA, Thole H, Karyagina AS, Oretskaya TS, Pingoud A, Pingoud V. Identification of a base-specific contact between the restriction endonuclease Sso II and its recognition sequence by photocross-linking. Nucleic Acids Res. 2000;28(5):1085–91.
Liu Q, Segal DJ, Ghiara JB, Barbas CF III. Design of polydactyl zinc-finger proteins for unique addressing within complex genomes. Proc Natl Acad Sci. 1997;94(11):5525–30.
Loenen WA, Raleigh EA. The other face of restriction: modification-dependent enzymes. Nucleic Acids Res. 2014;42(1):56–69.
Loenen WA, Dryden DT, Raleigh EA, Wilson GG, Murray NE. Highlights of the DNA cutters: a short history of the restriction enzymes. Nucleic Acids Res. 2014;42(1):3–19.
Miller JC, Tan S, Qiao G, Barlow KA, Wang J, Xia DF, Meng X, Paschon DE, Leung E, Hinkley SJ a, Dulay GP. A TALE nuclease architecture for efficient genome editing. Nat Biotechnol. 2011;29(2):143–8.
Nobusato A, Uchiyama I, Kobayashi I. Diversity of restriction–modification gene homologues in Helicobacter pylori. Gene. 2000;259(1–2):89–98.
Pingoud A, Jeltsch A. Structure and function of type II restriction endonucleases. Nucleic Acids Res. 2001;29(18):3705–27.
Rao DN, Dryden DT, Bheemanaik S. Type III restriction-modification enzymes: a historical perspective. Nucleic Acids Res. 2014;42(1):45–55.

Roberts RJ, Halford SE. Type II restriction endonucleases. In: Nucleases. Cold Spring Harbor: Cold Spring Harbor Laboratory Press; 1993.

Roberts RJ, Vincze T, Posfai J a, Macelis D. REBASE—enzymes and genes for DNA restriction and modification. Nucleic Acids Res. 2007;35(suppl_1):D269–70.

Rocha EP, Danchin A, Viari A. Evolutionary role of restriction/modification systems as revealed by comparative genome analysis. Genome Res. 2001;11(6):946–58.

Sistla S, Rao DN. S-Adenosyl-L-methionine–dependent restriction enzymes. Crit Rev Biochem Mol Biol. 2004;39(1):1–19.

Smith HO, Nathans D. A suggested nomenclature for bacterial host modification and restriction systems and their enzymes. J Mol Biol. 1973;81(3):419–23.

Smith HO, Welcox KW. A restriction enzyme from Hemophilus influenzae: I. Purification and general properties. J Mol Biol. 1970;51(2):379–91.

Southern EM. Detection of specific sequences among DNA fragments separated by gel electrophoresis. J Mol Biol. 1975;98(3):503–17.

Stahl F, Wende W, Jeltsch A, Pingoud A. The mechanism of DNA cleavage by the type II. Restriction enzyme EcoRV: Asp36 is not directly involved in DNA cleavage but serves to couple indirect. Readout to catalysis. Biol Chem. 1998;379(4–5):467–73.

Stankevicius K, Lubys A, Timinskas A, Vaitkevicius D, Janulaitis A. Cloning and analysis of the four genes coding for bpu 10i restriction—modification enzymes. Nucleic Acids Res. 1998;26(4):1084–91.

Stein DC, Gunn JS, Radlinska M, Piekarowicz A. Restriction and modification systems of Neisseria gonorrhoeae. Gene. 1995;157(1–2):19–22.

Studier FW, Bandyopadhyay PK. Model for how type I restriction enzymes select cleavage sites in DNA. Proc Natl Acad Sci. 1988;85(13):4677–81.

Szomolányi É, Kiss A, Venetianer P. Cloning the modification methylase gene of Bacillus sphaericus R in Escherichia coli. Gene. 1980;10(3):219–25.

Williams RJ. Restriction endonuclease: classification, properties, and applications. Mol Biotechnol. 2003;23(3):225–43.

Wilson GG, Murray NE. Restriction and modification systems. Annu Rev Genet. 1991;25(1):585–627.

Chapter 3
Molecular Cloning

Siddra Ijaz and Muhammad Hassan Jahanzaib

3.1 Introduction

Cloning involves the use of biological substances, such as genes, cells, or even entire organisms, to create exact copies of a particular organism. There are two main in vitro methods for producing these copies, including somatic cell nuclear transfer and the splitting or twinning of embryos (Rahbaran et al. 2021). In the context of genetic engineering and biotechnology, the rise of molecular cloning has had unintended consequences, marking a significant shift in the field since the early 1970s. Although scientists had already developed the key concepts of recombinant DNA construction and replication in bacterial hosts, the initial procedures were unsuccessful and technologically limiting.

There were clumsy vector systems, and primitive restriction enzymes were impure and unpredictable, as were the sources of DNA inserts, creating bottlenecks. However, a series of breakthroughs in later decades, including the availability of standardized cloning vectors like pBR322 and pUC plasmids, the commercialization of predictable restriction enzymes and buffer cocktails, significantly increased the effectiveness and accuracy of molecular cloning. The development of technologies, including blue-white screening, the use of alkaline phosphatases to finalize vector construction, and the automation of DNA sequencing, has improved the cloning process. The stage of insert preparation was radically transformed by pioneering technologies, especially the polymerase chain reaction (PCR) and reverse transcriptase, making cloning possible on a small scale or with small amounts of template from synthetic products (RNA). Again, all this development led to the creation of modern molecular cloning, where genetic manipulation becomes highly

S. Ijaz (✉) · M. H. Jahanzaib
Centre of Agricultural Biochemistry and Biotechnology (CABB), University of Agriculture, Faisalabad, Pakistan

S. Ijaz et al. (eds.), *Molecular Biology Methods*,
https://doi.org/10.1007/978-981-92-2642-9_3

precise, thus paving the way for foundational synthetic biology (Tirabassi and Bio 2014).

The importance of molecular cloning in molecular biology and biotechnology lies in its ability to facilitate the isolation and replication of a specific stretch of DNA. As one of the most fundamental approaches, it enables various applications, ranging from basic studies of genes to more advanced therapeutic applications. Molecular cloning, a technique in clinical microbiology, is highly significant in the diagnosis of infections, particularly in cases where standard cultural methods have been ineffective, such as those involving polymicrobial or fastidious organisms.

It also enables production of recombinant antigens for sensitive diagnostic tests for HIV, HBV, HCV, CMV, and *Treponema pallidum*, which are safer and more standardized than older serological tests. The technology has allowed the development of recombinant vaccines with better safety than traditional live or attenuated vaccines. Cloning of gene probes is used in early detection of genetic diseases, forensics, and diagnostics, improving molecular test accuracy. Molecular cloning produces therapeutic peptides, antimicrobial agents, and cytokines vital for treating infections and modulating immune responses (Sharma et al. 2014).

The initial step involves extracting the DNA of interest from a biological source, which can be achieved using techniques such as mechanical homogenization, enzymatic digestion, or chemical extraction. After isolating the vectors, their preparation includes selecting an appropriate type—like a plasmid, bacteriophage, or cosmid—and modifying it to carry the foreign DNA.

The inserted DNA is next hypothesized (attached) in the constructed vector using a protein aptly called DNA ligase to assemble a recombinant DNA fragment. This recombinant DNA is inserted into a host cell, usually by a process such as transformation, which can be induced by heat shock or electroporation. After intercourse, the host cells are then nonselectively chosen using antibiotic markers on the vector. They only then are the stored cells or those that have acquired the recombinant plasmid selected. These steps, which include DNA isolation, purification into vectors, ligation, transformation, and selection, enable the processing and analysis of recombinant DNA using viable cells (Visconte 2023).

Molecular cloning is also crucial in the fields of clinical microbiology and medicine, particularly in the production of recombinant antigens used for the diagnosis of viral diseases such as HIV, HBV, and HCV. It will enable the safe manufacturing of recombinant vaccines and diagnostic gene probes, which aid in the early detection of diseases. The technique is also useful in the production of therapeutic molecules, such as antimicrobial peptides and cytokines, which enable improved treatment (Sharma et al. 2014).

Molecular cloning has revolutionized the agricultural field by creating genetically engineered crops with superior traits, such as insect resistance (e.g., Bt crops), herbicide resistance, and enhanced nutritional value, as seen in Golden Rice fortified with beta-carotene. It facilitates the effective mapping of genes, marker-assisted selection, and map-based cloning of agronomic traits, such as disease resistance and enhanced yield. Developed cloning methods also complement the molecular design breeding process, in which multiple beneficial genes can be accurately assembled and mixed into highly productive and quality crop strains. Additionally, cloning can be used to generate transgenic plants and preserve elite or endangered germplasm, ensuring the sustainability of crop enhancement (Lewis et al. 1998).

3.2 Cloning Vectors and Strategies

3.2.1 Types of Cloning Vectors

Genetic cloning is based on specialized DNA vectors. Cloning vectors that enable the insertion, replication, and retrieval of foreign DNA. The selection of the vector is primarily based on the size of the insert, the number of copies, and compatibility with the host, as well as its suitability for downstream uses. Major types of vectors are plasmids, bacteriophages, and cosmids.

Plasmids are self-replicative, extrachromosomal, circular (usually circular) DNA molecules that are commonly found in large numbers in bacteria and play a significant role in horizontal gene transfer (Couturier et al. 1988). Plasmids can be small (just a few kilobase pairs) or large (greater than several hundred kilobase pairs), and their molecular structure may include replication origins, partition systems, and accessory genes, giving them a particular characteristic to the host (Clowes 1972). These genetic factors play a significant role in the spread of antimicrobial resistance genes, particularly in the clinical environment, where selection pressure is very high (Ijaz et al. 2023; Carattoli 2013). Mediated by conjugation, transformation, or mobilization, their movement enables rapid genetic exchange between and among bacterial species, with a significant impact on microbial evolution (Smillie et al. 2010). Plasmids are functionally grouped into different incompatibility groups based on their replication and maintenance systems, which prevent the coexistence of closely related plasmids in the same host (Couturier et al. 1988; Summers 2009). In the past, plasmid research has been used to investigate bacterial adaptability and genome plasticity, with pioneering studies explaining the molecular organization and mode of replication (Ijaz 2011; Helinski 2022). In sum, plasmids are not only useful as molecular tools in biotechnology but are also important vectors of resistance determinants of clinical relevance (Carattoli 2013).

Bacteriophages, also known as phages, are viruses that have a specificity for bacteria and are one of the most widespread and diverse biological components in the biosphere (Sharma et al. 2017). Morphologically, they are nucleic acid genomes, either DNA or RNA, embedded in a protein capsid and, in a few cases, a lipid envelope, and are morphologically diverse, ranging from complex, tailed structures at one extreme to simple, filamentous structures at the other (Clokie et al. 2009).

Their genomes vary in size and composition, enabling different infection modes: lytic cycles, where the phage genome is released into the host, and lysogenic cycles, where it integrates into the bacterial chromosome as a prophage (Calendar 2006; Hatfull and Hendrix 2011). Bacteriophages significantly influenced bacterial evolution via horizontal gene transfer, increasing diversity (Hendrix 2002). Besides ecological roles, they are crucial in molecular biology as cloning vectors for gene insertion, expression, and sequencing due to their specificity and efficiency (Kutter and Sulakvelidze 2004; Clokie et al. 2009). Historically, they have been vital in developing molecular genetics, offering insights into genetic material, regulation, and recombinant DNA technology.

Cosmids are hybrid cloning vectors combining plasmid and bacteriophage lambda features, allowing cloning of large DNA segments up to 45 kb while enabling plasmid propagation (Collins and Hohn 1978). They contain a replication origin for autonomous maintenance in **Escherichia coli** and cos sites for in vitro packaging into phage heads (Hohn et al. 1988). This duality offers high cloning capacity and stable maintenance of inserted DNA (Hohn and Collins 1980). Cosmids are vital for constructing genomic libraries, especially in large genomes, as they insert large DNA fragments with minimal rearrangements (Ish-Horowicz and Burke 1981; Evans and Lewis 1989). They are used in physical genome mapping to produce contiguous sequences (Evans and Lewis 1989). Cosmids have also been adapted for eukaryotic systems, enabling transfer and expression in mammalian cells (Grosveld et al. 1982). Advances in vectors and packaging systems have expanded their use in molecular cloning, genomic analysis, and positional cloning (Hohn and Collins 1988).

BACs and YACs are artificial chromosome-based cloning vectors that overcome size limits of plasmids, cosmids, and bacteriophages, allowing stable maintenance of large DNA fragments for genomic studies (Monaco and Larin 1994). YACs support inserts up to 1–2 Mb by incorporating yeast chromosomal elements like telomeres and centromeres, making them useful for mapping and cloning (Monaco and Larin 1994; Giraldo and Montoliu 2001). However, YACs are unstable, chimeric, and prone to rearrangements, complicating their use (Cocchia et al. 2000; Zeng et al. 2001). Conversely, BACs, based on *E. coli* F factor, carry 100–300 kb inserts and are more stable for accurate propagation (Kaufer and Tischer 2013). Large-insert BAC libraries have been vital for creating physical maps and sequencing complex genomes of plants and animals (Krzywinski et al. 2004). BACs are also used in transgenics, comparative genomics, and functional studies, as they carry large genomic regions with regulatory elements (Hokanson 2001). Using YACs and BACs together has advanced large-scale genome analysis, cloning, mapping, and sequencing of higher eukaryotes (Anderson 1993).

3.2.2 Key Features of Vectors

Cloning vectors are an integral component of molecular biology, and their effectiveness largely relies on the peculiarities of their structure and function, which predetermine their use in various projects. Although numerous forms of vectors have been engineered to support inserts of varying sizes, including plasmids, bacteriophages, cosmids, and artificial chromosomes, they possess several universally important features that guarantee the effective replication, selection, and expression of foreign DNA in host cells. These are the main characteristics one must be aware of when selecting the right type of vector system for use in genetic engineering, genomic mapping, and recombinant DNA technology.

An essential characteristic of any cloning vector is the origin of replication (ori), which will guarantee the autonomous replication of the vector in the host cell. In prokaryotic organisms, the ori specifies the nucleotide sequence at which the initiation point of DNA replication occurs and where initiator proteins congregate to form the replication complex (Worning et al. 2006). Replication origins are fundamental control points found in both prokaryotic and eukaryotic chromosomes, used to coordinate DNA synthesis, regulate cell cycles, and control gene expression (Taylor 1984).

The origins of replication vary across the three domains of life, reflecting diversity in replication machinery (Robinson and Bell 2005). In eukaryotes, replication is defined by specific sequences and chromatin states that control activation during S-phase (Leonard and Méchali 2013). These origins have mapped initiation points within genome regions, forming domains that limit DNA synthesis (Guilbaud et al. 2011). Computational analysis identifies motifs in yeast origins believed to regulate initiation (Dao et al. 2019). Oligomeric sequence skewness helps predict origins in other genomes through bioinformatics (Salzberg et al. 1998). Recent research shows origins are dynamic, influenced by chromatin and epigenetic changes, affecting genome stability (Prioleau and MacAlpine 2016). Overall, replication origins serve as structural and regulatory hubs vital for genome integrity and cell division (Ekundayo and Bleichert 2019).

Selectable markers are essential in cloning vectors for identifying successful transformations, especially in plant and microbial systems where they confer resistance to herbicides or antibiotics. These markers allow only transformed cells to grow under selective conditions (Miki and McHugh 2004). Advances in molecular biology have introduced new marker genes, including those for metabolic and stress tolerance, improving transformation efficiency and biosafety (Sundar and Sakthivel 2008). A key development is designing markers for removal posttransformation to improve transgene stability and meet regulatory standards (Hare and Chua 2002). Early tests show that kanamycin resistance markers are safe for plants and humans (Flavell et al. 1992).

In addition to plants, engineering has developed special markers adapted to specific host conditions, such as selectable markers for hyperthermophiles (Brouns et al. 2005). Cereal research has shown that the consistent incorporation and dependable expression of selectable markers play a major role in generating genetically stable transformants (Hauptmann et al. 1988). Equally, marker assessment in rice systems assisted in optimization of transformation protocols and expanded the range of their applications in monocot species (Dekeyser et al. 1989). Selectable marker recycling schemes have been applied to the chloroplast genome in organelle transformation, allowing genetic engineering to be repeated multiple times without the accumulation of marker genes (Fischer et al. 1996). Together, the selectable markers are essential in designing vectors, and newer developments favor marker-free or recyclable systems to enhance safety and regulatory tolerability.

The multiple cloning site (MCS), also known as a polylinker, is a short piece of DNA designed into cloning vectors that provides a set of distinct recognition sites for restriction enzymes, allowing the insertion of foreign DNA fragments. The initial advances of MCS-containing plasmids significantly enhanced recombinant DNA technology, as they allowed the construction of recombinant viruses and plasmids to express genes in a rapid and efficient manner (Boyle et al. 1985). Multiple restriction sites can be introduced into a single region to offer flexibility in cloning strategies, enabling the efficient ligation of DNA fragments with various restriction ends into the vector (Jobling and Holmes 1990). There were also sophisticated designs that have even provided duplication in MCS regions to increase the cloning efficiency and increase the variety of restriction enzymes that can be used to prepare the insert (Cocea 1997). High-capacity specialized expression vectors, expressed with versatile MCS regions, were later invented to support fusion protein and transgene constructions (Koerner et al. 1991; Wang et al. 1992). In some applications, however, the existence of MCS regions has been observed to disrupt gene regulatory experiments, especially when placed adjacent to promoter components in mammalian expression systems (Beliveau et al. 1999). To overcome the limitations of traditional designs, re-engineered MCS sequences have been developed to be more functionally versatile, thereby simplifying the complexity of cloning and the design of constructs (Crook et al. 2011). Other innovations have also developed simplified cloning methods, combining universal MCS designs and in vivo bacteria assembly systems that offer efficient and cost-effective plasmid assembly (Chen et al. 2021). Also, lentiviral vectors containing new MCS elements have enabled long-term and stable expression of transgenes in vitro and in vivo, which has significantly increased therapeutic and research uses of the technology (Santhosh et al. 2008). The MCS is a significant part of cloning vectors that is flexible, precise, and adaptable in molecular cloning research.

3.2.3 Cloning Strategies

Molecular biology relies on manipulating DNA via vectors and cloning strategies to insert, maintain, and express foreign DNA in host cells. Cloning methods have evolved from classical recombination techniques to advanced recombination- and assembly-based methods, each suited for different sizes, complexities, and applications. Studying these principles helps choose the best method for molecular cloning.

Restriction enzyme-based cloning involves cutting DNA with restriction endonucleases and ligating into vectors, foundational in genetic engineering (Bertero et al. 2017). It requires selecting restriction sites for compatible ends (Tóth et al. 2014), but limits flexibility when internal sites exist. To overcome this, restriction enzyme-free methods like the Univector system and RF cloning have been developed, enabling manipulation without predefined cut sites (Tillett and Neilan 1999;

Tseng 1999; Liu et al. 1998; Van Den Ent and Löwe 2006). These methods broaden cloning options in genetic engineering.

TA cloning (Thymidine–Adenine cloning) is a common cloning method for PCR-amplified DNA fragments, which takes advantage of the capacity of some DNA polymerases, including Taq polymerase, to nick a single deoxyadenosine (A) overhang at the 3′ ends of PCR products. These overhangs can be ligated directly into vectors containing complementary 3-end thymidine (T) overhangs, eliminating the need for restriction enzyme digestion, which makes the cloning process significantly easier (Zhou and Gomez-Sanchez 2000). Small and flexible, TA cloning has become one of the most accessible methods for use in routine molecular biology, particularly in cases where the insert DNA lacks convenient restriction sites (Green and Sambrook 2021).

The TA cloning methodology has been modified and refined over time to become more efficient and useful. Particularly, modified TA cloning or TA has been used to investigate the production of entry clones to be subject to recombination cloning systems, and thus, making them applicable individually, as opposed to fragment insertion (Chen et al. 2006a, b). Universal TA cloning systems have incorporated the most recent advancements, which have enriched functional genomic research, making it easier to obtain a wide range of diverse DNA sequences (Yao et al. 2016). Additionally, the innovations have introduced new forms of vector designs, including those with high efficiencies that support TA and blunt-end ligation, thereby enabling improved adaptability in generating a variety of PCR products (Liu et al. 2018; Motohashi 2019).

TA cloning does not operate within the elitism of elementary life cloning applications; in fact, it can also be used as an important tool in a wide range of biological research. The mixed-sequence analysis system for genetically unanticipated malaria infections in birds is a more pertinent instance, in which its capability to establish the diversity of infected agents is confirmed in the avian model (Perez-Tris and Bensch 2005). Similarly, TA based cloning systems have been utilized in the development of effective transgenic tools in model organisms, such as *C. elegans*, where they facilitate RNAi-based genetic research (Gengyo-Ando et al. 2006). More recently, the protocols of universal TA cloning have been reconsidered and optimized in modern molecular biology toolkits, further cementing the applicability of these in current cloning procedures (Zhou and Gomez-Sanchez 2023).

Blunt-end cloning is one of the cornerstone cloning methods, which involves the direct ligation of bluntly ended DNA fragments produced by certain DNA polymerases or restriction enzymes into linearized vectors with compatible blunt termini (Green and Sambrook 2020). Blunt-end cloning, as opposed to sticky-end cloning, does not require cohesive ends, making it suitable for use with a wide variety of DNA fragments. The lack of overhangs, however, typically results in decreased ligation efficiency, which has traditionally been considered a drawback of this technique (Bhat et al. 1991).

To address this problem, better blunt-end cloning techniques and vectors have been established. Initial discoveries, such as Turbo Cloning, have shown that PCR products and other blunt-ended DNA fragments can be easily transformed into plasmids under favorable conditions, significantly reducing the cloning time (Boyd 1993). Similarly, altered vectors, such as the pGEM-T system, also supported the use of blunt-ended fragments by integrating TA cloning concepts with blunt-end compatibility, thereby increasing the range of applications in molecular biology workflows (Kobs 1995).

Subsequent improvements made directional blunt-end cloning possible, in which the orientation of inserts could be directed by the use of specially designed vectors and ligation methods, increasing its applicability in gene expression experiments (Liu et al. 2000). Recent advances in the creation of new high-efficiency vectors have further expanded the capabilities of blunt-end cloning, providing compatibility with both TA and blunt-end methods to enable the use of a wide range of PCR products (Motohashi 2019). Moreover, recent implementations have applied blunt-end cloning to state-of-the-art systems. Furthermore, CRISPR/Cas9-based non-homologous end-joining enables in vivo blunt-end cloning, making it a powerful tool for directing genome editing (Geisinger et al. 2016).

Overall, blunt-end cloning is a flexible and efficient cloning technique, particularly when restriction sites or overhangs are not present in the DNA fragments being cloned. Although blunt-end cloning has relatively lower efficiency than cohesive-end ligation, technological advances and dedicated systems of vectors, as well as other specialized tools, have enabled the use of blunt-end cloning as an important aspect of the modern molecular cloning process.

One of the promising new molecular methodologies is ligation-independent cloning (LIC), which does not require DNA ligase but instead utilizes the exonuclease activity of DNA polymerase to tick complementary single-stranded tips overhangs of the insert DNA into the job of the system (Aslanidis and De Jong 1990). Contrary to traditional approaches to cloning, LIC is significantly more efficient and versatile, as it is not based on the recognition sites of restriction enzymes or the activity of ligase.

Since its introduction, this has been followed by several variations of LIC to enhance versatility and throughput. One such application is that LIC-PCR was shown to enable the direct cloning of PCR products without restriction site limitations (Li and Evans 1997), and quick and clean cloning (QC cloning) provided a way to selectively isolate specific PCR products from a complex mixture (Thieme et al. 2011). Particular applications have also been created where LIC is used to screen protein expression constructs at high throughput (Berrow et al. 2007) and for plant functional genomics using specialized vector systems (De Rybel et al. 2011).

More innovations were made in the form of sequence- and ligation-independent cloning (SLIC), which utilizes controlled exonuclease digestion to generate recombination-compatible ends on DNA fragments, providing a more flexible and efficient means of assembling DNA fragments (Li and Elledge 2012). One-step SLIC has been particularly helpful in functional genomics, where construction generation must be both fast and precise (Jeong et al. 2012). Equally, high-throughput LIC-based approaches have been extended to assembly of complex synthetic constructs, including transcription activator-like effector (TALE) genes (Schmid-Burgk et al. 2013).

Comparisons of LIC and its derivatives have shown that they are practical and efficient in various cloning applications, offering benefits over more traditional methods based on ligation (Stevenson et al. 2013). Combined, these developments underscore the flexibility of LIC, which is what makes it one of the most important methods of modern molecular cloning in both basic research and practical biotechnology.

Recombination-based cloning utilizes the natural recombination mechanism to assemble DNA fragments or to assemble them in a more efficient and precise manner. Compared to traditional restriction-ligation methods, the strategy allows easy insertion of DNA sequences into vectors without the need to use compatible restriction sites and is thus a viable alternative to modern molecular biology applications (Celie et al. 2016).

One of the most common techniques is called gateway cloning, which entails the utilisation of bacteriophage four-site-specific recombination between att sites to enable the directional transfer of DNA fragments between entry and destination vectors (Hartley et al. 2000). This system has been optimized to serve a wide range of applications, including functional genomics and high-throughput screening, as it is simple and highly faithful (Reece-Hoyes and Walhout 2018).

The other significant advancement is ET-cloning (Enzyme Targeting cloning), a development that focuses on homologous recombination to achieve DNA manipulation, offering more flexibility than enzyme-based approaches (Muyrers et al. 2000). Based on this rule, recombineering has become a flexible approach to genetic engineering in *Escherichia coli*, which utilizes bacteriophage-coded recombination proteins to introduce specific genetic modifications (Sharan et al. 2009). It was followed by later developments known as recombinogenic engineering, which made even greater use of recombination-based methods in the large-scale manipulation of DNA (Muyrers et al. 2001).

Yeast systems have been useful in cloning through recombination. Yeast homologous recombination is versatile because it can easily be used to assemble several overlapping DNA fragments into a functional construct, providing a universal, ligation-free cloning system (Joska et al. 2014). This approach has been effectively used to assemble expression vectors for fungal pathogens, including Zymoseptoria tritici, demonstrating its versatility in other organisms as well (Kilaru and Steinberg 2015).

3.3 Gateway® Cloning System

The Gateway cloning system is a site-specific recombination method enabling straightforward and directional transfer of DNA fragments among different vector backbones. It serves as a universal alternative to traditional restriction enzyme cloning, functioning through recombination at att sites to generate entry and expression clones with consistent accuracy and efficiency (Katzen 2007).

3.3.1 Principle and Mechanism

The Gateway cloning system uses the site-specific recombination of bacteriophage λ to insert DNA into *E. coli*. It involves specialized sites and proteins that facilitate DNA transfer between vectors through two steps: BP and LR reactions. In the BP step, an attB-flanked DNA joins a donor vector with attP, forming an entry clone with attL sites. This clone is then inserted into a destination vector with attR through the LR reaction, creating an expression clone with attB sites for host expression. The process is directional, avoids restriction enzymes, and is useful for protein expression, plant transformation, and genomics applications.

The system exploits the recombination system of bacteriophage λ, known as the Gateway cloning system, enabling precise DNA exchange between the phage genome and *Escherichia coli* chromosome. It is controlled by short, asymmetric DNA recognition sequences called att sites and specific recombinase enzymes (Hartley et al. 2000).

Four att types are key to Gateway cloning recombination: attB (bacterial), attP (phage), attL (left), and attR (right). attB-flanked DNA fragments recombine with attP in the BP reaction with donor vectors, creating entry clones with attL sites. In the LR reaction, attL sites recombine with attR sites on target vectors, forming expression clones with attB sites on both ends (Reece-Hoyes and Walhout 2018). Integration host factor (IHF) and excisionase (Xis) catalyze recombination, making the process specific and efficient (Wall et al. 2013). Detecting att sites prevents unwanted rearrangements and guides directional movement of genetic elements across vectors. This principle underpins Gateway cloning and its variants like multiSite Gateway, where multiple att site versions enable parallel assembly of DNA sequences into a single product (Petersen and Stowers 2011). Att-site recombination is powerful and flexible, making Gateway cloning popular in plants, microbes, and mammalian cells (Gottschamel et al. 2013).

The Gateway cloning system employs BP and LR reactions for specific DNA insertion into vectors, using bacterial phage integrase that eliminates restriction enzymes or ligases, enabling bidirectional allelic exchange between att sites. Under BP recombination, DNA with attB sites is recombined with a donor vector with attP sites. Integrationase and IHF catalyze this, creating an entry clone with attL sites. The exit clone is a stable, recombinable middle phase for further cloning (Untergasser 2006; Liang et al. 2013).

The LR reaction transfers the gene of interest from the entry clone to the destination. During this process, attL sites are replicated with attR sites using 3C-L integrase, excisionase (Xis), and IHF. The resulting expression clone is flanked by attB sites (Matsuyama and Yoshida 2009; Chee and Chin 2015). The combination of BP and LR reactions is central to Gateway cloning, enabling directional, high-throughput DNA transfer across various vectors. Performing BP/LR reactions simultaneously enhances speed and capacity (Gruet et al. 2012; Liang et al. 2013).

3.3.2 Components of the Gateway System

The Gateway cloning system uses specialized vectors for directional DNA transfer via site-specific recombination. These components improve adaptability, modularity, and accuracy in cloning workflows (Katzen 2007; Reece-Hoyes and Walhout 2018).

The entry vector is a versatile cloning plasmid where the DNA fragment of interest is inserted via the BP recombination reaction. Its attP sites flank the vector and recombine with the attB sites of the gene to create an attL site in the resulting entry clone. Additionally, this vector can be used repeatedly to insert the same sequence into different destination vectors because of its robustness and reusability (Xu and Qingshun 2008; Matsuyama and Yoshida 2009).

The destination vector contains attR sites, replicated by attL sites of the entry clone during the LR reaction to produce an expression clone. These vectors are highly adaptable, featuring various promoters, markers, and fusion tags, suitable for bacteria, yeast, plants, and animal cells (Karimi et al. 2007; Esposito et al. 2009).

After BP recombination, the donor vector and gene of interest form the entry clone. LR recombination produces an expression clone that expresses functional genes in the host. Some expression clones are engineered to produce proteins, tagged, or for high-throughput analysis (Petersen and Stowers 2011).

3.3.3 Advantages of Gateway Cloning

The Gateway technology has several advantages over traditional restriction enzyme technology-based cloning systems, and it has become the technique of choice in both fundamental research and biotechnology applications. This design allows it to be extremely efficient, directional, and scalable, which is essential in large-scale functional genomics and protein expression experiments (Katzen 2007; Reece-Hoyes and Walhout 2018). The system is virtually 100 percent efficient in cloning because site-specific recombination is done using att sites, and no restriction digestion and ligation are required. This efficiency saves time during experiments, reduces errors, and offers fidelity of insert transfer (Esposito et al. 2009; Chee and Chin 2015).

In contrast to the conventional techniques, Gateway 2 cloning provides the appropriate positioning of the DNA insert by directional recombination between specific att sites. It addresses the issue of reversed insert orientation, a common problem in standard cloning, thereby making it more reliable in downstream applications (Dubin et al. 2008; Reece-Hoyes and Walhout 2018). Cloning Gateway-based cloning can be particularly useful when a large number of functional genomics studies are required, such as ORFeome efforts, protein expression screening, and establishing large mutant libraries. Its modular platform enables transference of a single entry clone into many destination vectors, and is thus able to perform concurrent analysis with various expression systems (Xu and Qingshun 2008; Vemanna et al. 2013).

3.3.4 Limitations and Considerations

Gateway 2 cloning, despite its efficiency, has significant limitations. Applications that need smooth junctions may be compromised by the existence of att sites scars, especially when using protein expression or fusion constructs (Walhout et al. 2000; Hartley 2002). It is also costly and less flexible because it depends on proprietary enzymes and vectors, unlike open-source techniques (Carl et al. 2013). Moreover, it proves to be more effective in standard cloning methods; nevertheless, it is less versatile compared to scarless assembly methodologies or more complex applications of synthetic biology, with options like Golden Gate offering more flexibility (Pryor et al. 2020).

3.4 Golden Gate Cloning

Golden Gate cloning is a flexible DNA assembly method based on type(II) restriction endonucleases and T4 DNA ligase, enabling seamless, directional fusion of multiple fragments in a single reaction. This technique prevents extra sequences at junctions and dominates synthetic biology due to its effectiveness and scalability for modular assembly and high-throughput experimentation.

3.4.1 Concept and Principle

Golden Gate cloning is based on the selective cleavage of DNA to the exterior of type IIS restriction sites by restriction enzymes, thereby forming predefined overhangs that can be manipulated to facilitate multiple fragment assembly (Engler and Marillonnet 2013). Unlike in traditional cloning, where the restriction sites are placed at the end of the fragment involved in the assembly, the recognition sites of type IIS enzymes are removed in the assembly reaction, thus producing scarless constructs (Bird et al. 2022). The reaction occurs in a single pot, where both the digestion process and the ligation take place under thermal cycling conditions. The gene pool is then reduced to DNA fragments through the action of the type IIS enzyme, followed by recombining the pieces with the DNA ligase chosen in the desired order (Laborda-Mansilla and García-Ruiz 2024). Due to the removal of the sites of recognition during the process of ligation, well-assembled probes are not presented before the restriction enzyme anymore, and indefinitely repeated steps of digestion and re-ligation are taken to achieve the desired outcome, as a consequence of the removal of the recognition sites (Terfrüchte et al. 2014). It is a directional, highly efficient, and sequential digestion and ligation process that enables the assembly of multiple fragments simultaneously. The method has been seen in other applications (e.g., the genetic manipulation of plants (Lampropoulos et al. 2013), fungi systems (Terfrüchte et al. 2014), microbial systems (Behrendt et al. 2022), or even when constructing DNA nanostructures (Zhang et al. 2023). This flexibility has made Golden Gate cloning stand out as one of the strategic approaches in modular cloning pipelines and the broader field of synthetic sciences.

3.4.2 Vector Design and Assembly

Golden Gate cloning's success lies in carefully designed vectors and overhang sequences. Type II-I enzymes, with four-nucleotide overhangs, enable precise DNA fragment creation (Marillonnet and Grützner 2020). Overhangs must be compatible to ensure efficient assembly and avoid errors (Sorida and Bonasio 2023). To simplify construct assembly, Golden Gate is integrated into layered pipelines like Modular Cloning (MoClo), which uses common elements for recursive, multilevel assembly, including multigene pathways (Werner et al. 2012). The system's flexibility is boosted by vector libraries such as BEVA, adaptable to various expression systems (Geddes et al. 2019). Combined with rational overhang design and modular structures, Golden Gate enables scalable applications in plant engineering, microbial expression, and protein engineering (Weber et al. 2011).

3.4.3 Advantages of Golden Gate Cloning

Golden Gate cloning is an efficient method for assembling DNA by using type IIS restriction enzymes to cut near recognition sites, enabling seamless, scarless integration of multiple fragments in a single reaction (Engler and Marillonnet 2013; Bird et al. 2022). It can incorporate random sequences, increasing versatility for synthetic biology. The method allows precise control over fragment arrangement, making it a crucial tool for DNA assembly.

Golden Gate assembly's inherent scalability and modularity offer benefits over classical cloning, enabling the simultaneous connection of many fragments and reducing time and effort (Marillonnet and Grützner 2020; Liu 2025). It is especially useful in high-throughput workflows, library building, and large-scale genetic engineering, where efficiency and reproducibility are key (Laborda-Mansilla and García-Ruiz 2024).

Golden Gate cloning is versatile, used across bacterial, fungal, plant, and mammalian systems (Terfrüchte et al. 2014; Chiasson et al. 2019). Recent improvements have expanded its use in complex cloning projects due to better compatibility with many vectors and restriction sites (Sorida and Bonasio 2023). These strengths make it a valuable tool for precision, speed, and scalability in modern molecular biology.

Comparison with Traditional Cloning and Gateway Cloning
Golden Gate cloning offers advantages over traditional methods using restriction enzymes and Gateway system. Unlike classical cloning that relies on restriction sites and ligation, Golden Gate uses type IIS restriction enzymes to produce non-palindromic overhangs, enabling the assembly of multiple fragments in one reaction without scar sequences. It shortens cloning time and avoids extraneous sequences. Compared to Gateway, which uses site-specific recombination and can leave scars, Golden Gate is more efficient for assembling many DNA fragments, especially for large libraries and modular cloning due to its directional and scarless nature.

Nevertheless, Gateway cloning retains its prowess in flexibility and convenience, particularly in the study of protein expression and functional genomics, where recombinational cloning provides a simple transfer of inserts to various destination vectors (Dubin et al. 2008). Generally, Golden Gate cloning is faster and more precise than the traditional method, complementing Gateway cloning, which excels in high-throughput and complex assembly applications. Both methods still have critical roles in molecular biology, and their use depends on the research purpose.

Cloned Gene Expression
Gene cloning involves transcribing and translating cloned sequences into RNA or proteins for large-scale protein production, gene regulation, and applications like therapeutics and industry (Roberts et al. 1979; Cullen 1987). Expression cloning studies reveal gene functions, pathways, and diseases (Soares 1997; Hu et al. 2000). Systems in prokaryotic and eukaryotic hosts, especially *E. coli* vectors, enhance productivity in cloning, sequencing, and protein production (Wang and Kushner 1991). Engineered microbes like lactic streptococci serve industrial and therapeutic ends (De Vos 1987). High-throughput systems boost recombinant protein production, supporting screening and functional analysis (Dieckman et al. 2002). Overall, gene expression links DNA cloning to practical uses in medicine, biotech, and microbiology.

Expression Vectors
Cloning vectors designed for expression are used to efficiently transcribe and translate genes, focusing on optimizing gene expression. The key component is the promoter, which controls gene expression and can be either constitutive or inducible. Constitutive promoters provide constant expression, while inducible ones can be regulated in response to stimuli (Brosius 1989; Moss et al. 1990). The choice depends on whether regulated expression is needed, especially when overexpression of recombinant proteins could be lethal. These vectors may also include tags like His-tags or green fluorescent protein (GFP), aiding in protein purification, detection, and localization. His-tags facilitate purification via affinity chromatography, and GFP enables real-time visualization of protein expression and localization (Sheffield et al. 1999).

Vectors can create fusion proteins with stabilizing or functional partners, enhancing solubility, stability, and applications (Brosius 1989). Different systems for expression cloning exist: baculovirus vectors are used in eukaryotic systems for high protein production and posttranslational modifications (Miller 1988; Kidd and Emery 1993). Poxvirus vectors efficiently express proteins in mammalian cells (Carroll and Moss 1997), and bacterial systems offer engineered vectors with increased versatility (Luckow and Summers 1988; O'Reilly et al. 1994). Overall, expression vectors are essential for customizing gene expression in research and industry.

Host Systems

The choice of host system is a highly important factor that determines the success of gene expression cloning, as each has its own distinct benefits regarding yield, posttranslational changes, scalability, and cost. Bacterial systems, especially *Escherichia coli*, are the most commonly used hosts for expressing recombinant proteins due to their rapid growth, well-characterized genetics, and ability to produce large amounts of proteins economically. Nevertheless, their applications to some eukaryotic proteins may be limited by the absence of complicated posttranslational modifications (Twyman et al. 2003). Yeast expression systems aim to leverage the ease of microbial growth, enabling the execution of certain eukaryotic posttranslational modifications. They are especially applicable in the production of recombinant proteins, which may require glycosylation; however, variations in glycan structure relative to mammalian cells can affect protein functionality (Twyman et al. 2003). Higher-order eukaryotic folding and modification capabilities are found in insect cell systems, usually with the help of a baculovirus, and are suitable for more complex proteins and large-scale production. The ability to produce functional multi-subunit proteins has made them useful in both research and industrial biotechnology (Twyman et al. 2003). Mammalian systems are the new gold standard recombinant protein expression system when authentic folding, glycosylation, and biological activity are required. They are also commonly used in the production of therapeutic proteins, such as monoclonal antibodies and hormones, but are more expensive and technologically challenging than microbial or insect systems (Twyman et al. 2003).

Factors Affecting Gene Expression

The process of efficient gene expression is determined by several molecular and cellular conditions that dictate the processes of transcription, translation, and protein stability. Codon optimization is a critical parameter, as it enables the codons of a transgene to be compatible with the host's desired codon usage, thereby maximizing translation rates and protein yield (Mellor et al. 1985; Pastinen et al. 2004). Vector-host compatibility is also important, as conflicts between regulatory elements in the expression vector and the host's transcriptional machinery may lead to decreased or fluctuating expression of the genes. It has been demonstrated that genetic variation and epigenetic regulation play a key role in determining expression outcomes in various organisms and experimental systems (Pascussi et al. 2003; Williams et al. 2007). The ability to solubilize proteins and maintain proper folding is central to defining the functional competency of the protein being expressed. Controlled folding can result in aggregation or degradation, making protein recovery uncontrolled. Environmental and physiological factors, such as cellular stress responses and chaperone systems, usually affect solubility and stability (Mantikou et al. 2016; Ricciardi et al. 2018). All of this, together with the optimization of codon usage, host-vegetable compatibility, and sophisticated protein folding plans, is essential to the creation of robust and reliable gene expression in the recombinant system.

Applications of Cloned Gene Expression, Future Perspectives, and Emerging Techniques

Genetic cloning has widespread use in biotechnology and industry. Cloned genes enable mass production of stable, selective enzymes used in food, textiles, and biofuels (Lodge et al. 2007). In medicine, cloned gene expression produces therapeutic proteins like insulin, growth hormones, and monoclonal antibodies for genetic and metabolic diseases (Tefferi 2006). Vaccines based on cloned genes have revolutionized infectious disease control with safe, specific recombinant subunit vaccines with low pathogenicity (Sharma et al. 2014). Overall, cloned gene expression is a cornerstone of modern molecular biotechnology with vital clinical and industrial roles.

Molecular cloning evolves with new tools and methods, opening new research and biotech opportunities. CRISPR offers precise DNA editing, surpassing traditional techniques, enabling high-throughput genome engineering (Samiec 2022). Synthetic biology tools like modular DNA assemblies streamline gene construction, aiding functional genomics and aquaculture (Watson and García-Nafría 2019). Automation and AI are transforming cloning, increasing speed, accuracy, and scalability for industrial and clinical uses (Tirabassi and Bio 2014). These trends make molecular cloning more focused, automated, and application-driven, shaping its future in medicine, agriculture, and biotech.

References

Anderson C. BAC to the future? Science. 1993;259:1686.

Aslanidis C, De Jong PJ. Ligation-independent cloning of PCR products (LIC-PCR). Nucleic Acids Res. 1990;18:6069–74.

Behrendt G, Frohwitter J, Vlachonikolou M, Klamt S, Bettenbrock K. Zymo-parts: a golden gate modular cloning toolbox for heterologous gene expression in Zymomonas mobilis. ACS Synth Biol. 2022;11:3855–64.

Beliveau A, Leclerc S, Rouleau M, Guerin S. Multiple cloning sites from mammalian expression vectors interfere with gene promoter studies in vitro. Eur J Biochem. 1999;261:585–90.

Berrow NS, Alderton D, Sainsbury S, Nettleship J, Assenberg R, Rahman N, Stuart DI, Owens RJ. A versatile ligation-independent cloning method suitable for high-throughput expression screening applications. Nucleic Acids Res. 2007;35:e45.

Bertero A, Brown S, Vallier L. Methods of cloning. In: Basic science methods for clinical researchers. Elsevier, Amsterdam, Netherlands; 2017. p. 19–39.

Bhat GJ, Lodes MJ, Myler PJ, Stuart KD. A simple method for cloning blunt ended DNA fragments. Nucleic Acids Res. 1991;19:398.
Bird JE, Marles-Wright J, Giachino A. A user's guide to golden gate cloning methods and standards. ACS Synth Biol. 2022;11:3551–63.
Boyd AC. Turbo cloning: a fast, efficient method for cloning PCR products and other blunt-ended DNA fragments into plasmids. Nucleic Acids Res. 1993;21:817–21.
Boyle DB, Coupar BE, Both GW. Multiple-cloning-site plasmids for the rapid construction of recombinant poxviruses. Gene. 1985;35:169–77.
Brosius J. Superpolylinkers in cloning and expression vectors. DNA. 1989;8:759–77.
Brouns SJ, Wu H, Akerboom J, Turnbull AP, De Vos WM, Van Der Oost J. Engineering a selectable marker for hyperthermophiles. J Biol Chem. 2005;280:11422–31.
Calendar R. The bacteriophages. Oxford University Press, New York, USA; 2006.
Carattoli A. Plasmids and the spread of resistance. Int J Med Microbiol. 2013;303:298–304.
Carl UD, Batz L, Schuchardt I, Germeroth L, Schmidt TG. StarGate®: a high-capacity expression cloning system to speed-up biopharmaceutical development. In: Modern biopharmaceuticals: recent success stories; Springer, Berlin/Heidelberg, Germany; 2013. p. 147–64.
Carroll MW, Moss B. Poxviruses as expression vectors. Curr Opin Biotechnol. 1997;8:573–7.
Celie PH, Parret AH, Perrakis A. Recombinant cloning strategies for protein expression. Curr Opin Struct Biol. 2016;38:145–54.
Chee J, Chin C. Gateway cloning technology: advantages and drawbacks. Clon Transgenes. 2015;4:138.
Chen Q-J, Zhou H-M, Chen J, Wang X-C. A Gateway-based platform for multigene plant transformation. Plant Mol Biol. 2006a;62:927–36.
Chen Q-J, Zhou H-M, Chen J, Wang X-C. Using a modified TA cloning method to create entry clones. Anal Biochem. 2006b;358:120–5.
Chen F, Li Y-Y, Yu Y-L, Dai J, Huang J-L, Lin J. Simplified plasmid cloning with a universal MCS design and bacterial in vivo assembly. BMC Biotechnol. 2021;21:24.
Chiasson D, Giménez-Oya V, Bircheneder M, Bachmaier S, Studtrucker T, Ryan J, Sollweck K, Leonhardt H, Boshart M, Dietrich P. A unified multi-kingdom Golden Gate cloning platform. Sci Rep. 2019;9:10131.
Clokie MR, Kropinski AM, Lavigne R. Bacteriophages. Springer, New York, USA; 2009.
Clowes RC. Molecular structure of bacterial plasmids. Bacteriol Rev. 1972;36:361–405.
Cocchia M, Kouprina N, Kim S-J, Larionov V, Schlessinger D, Nagaraja R. Recovery and potential utility of YACs as circular YACs/BACs. Nucleic Acids Res. 2000;28:e81.
Cocea L. Duplication of a region in the multiple cloning site of a plasmid vector to enhance cloning-mediated addition of restriction sites to a DNA fragment. BioTechniques. 1997;23:814–6.
Collins J, Hohn B. Cosmids: a type of plasmid gene-cloning vector that is packageable in vitro in bacteriophage lambda heads. Proc Natl Acad Sci. 1978;75:4242–6.
Couturier M, Bex F, Bergquist PL, Maas WK. Identification and classification of bacterial plasmids. Microbiol Rev. 1988;52:375–95.
Crook NC, Freeman ES, Alper HS. Re-engineering multicloning sites for function and convenience. Nucleic Acids Res. 2011;39:e92.
Cullen BR. Use of eukaryotic expression technology in the functional analysis of cloned genes. Methods Enzymol. 1987;(United States), 152:684.
Dao F-Y, Lv H, Wang F, Feng C-Q, Ding H, Chen W, Lin H. Identify origin of replication in Saccharomyces cerevisiae using two-step feature selection technique. Bioinformatics. 2019;35:2075–83.
De Rybel B, van Den Berg W, Lokerse AS, Liao C-Y, Van Mourik H, Möller B, Llavata-Peris CI, Weijers D. A versatile set of ligation-independent cloning vectors for functional studies in plants. Plant Physiol. 2011;156:1292–9.
De Vos WM. Gene cloning and expression in lactic streptococci. FEMS Microbiol Rev. 1987;3:281–95.

Dekeyser R, Claes B, Marichal M, Van Montagu M, Caplan A. Evaluation of selectable markers for rice transformation. Plant Physiol. 1989;90:217–23.

Dieckman L, Gu M, Stols L, Donnelly MI, Collart FR. High throughput methods for gene cloning and expression. Protein Expr Purif. 2002;25:1–7.

Dubin MJ, Bowler C, Benvenuto G. A modified Gateway cloning strategy for overexpressing tagged proteins in plants. Plant Methods. 2008;4:3.

Ekundayo B, Bleichert F. Origins of DNA replication. PLoS Genet. 2019;15:e1008320.

Engler C, Marillonnet S. Golden gate cloning. In: DNA cloning and assembly methods. Springer, New York, USA; 2013. p. 119–31.

Esposito D, Garvey LA, Chakiath CS. Gateway cloning for protein expression. In: High throughput protein expression and purification: methods and protocols. Springer, New York, USA; 2009. p. 31–54.

Evans G, Lewis K. Physical mapping of complex genomes by cosmid multiplex analysis. Proc Natl Acad Sci. 1989;86:5030–4.

Fischer N, Stampacchia O, Redding K, Rochaix J-D. Selectable marker recycling in the chloroplast. Mol Gen Genet. 1996;251:373–80.

Flavell RB, Dart E, Fuchs RL, Fraley RT. Selectable marker genes: safe for plants? Bio/Technology. 1992;10:141–4.

Geddes BA, Mendoza-Suárez MA, Poole PS. A bacterial expression vector archive (BEVA) for flexible modular assembly of golden gate-compatible vectors. Front Microbiol. 2019;9:3345.

Geisinger JM, Turan S, Hernandez S, Spector LP, Calos MP. In vivo blunt-end cloning through CRISPR/Cas9-facilitated non-homologous end-joining. Nucleic Acids Res. 2016;44:e76.

Gengyo-Ando K, Yoshina S, Inoue H, Mitani S. An efficient transgenic system by TA cloning vectors and RNAi for C. elegans. Biochem Biophys Res Commun. 2006;349:1345–50.

Giraldo P, Montoliu L. Size matters: use of YACs, BACs and PACs in transgenic animals. Transgenic Res. 2001;10:83–103.

Gottschamel J, Waheed MT, Clarke JL, Lössl AG. A novel chloroplast transformation vector compatible with the Gateway® recombination cloning technology. Transgenic Res. 2013;22:1273–8.

Green MR, Sambrook J. Cloning in plasmid vectors: blunt-end cloning. Cold Spring Harbor Protocols, 2020: pdb. prot101246; 2020.

Green MR, Sambrook J. Cloning polymerase chain reaction (PCR) products: TA cloning. Cold Spring Harbor Protocols, 2021: pdb. prot101303; 2021.

Grosveld F, Lund T, Murray E, Mellor A, Dahl H, Flavell R. The construction of cosmid libraries which can be used to transform eukaryotic cells. Nucleic Acids Res. 1982;10:6715–32.

Gruet A, Longhi S, Bignon C. One-step generation of error-prone PCR libraries using Gateway® technology. Microb Cell Factories. 2012;11:14.

Guilbaud G, Rappailles A, Baker A, Chen C-L, Arneodo A, Goldar A, d'Aubenton-Carafa Y, Thermes C, Audit B, Hyrien O. Evidence for sequential and increasing activation of replication origins along replication timing gradients in the human genome. PLoS Comput Biol. 2011;7:e1002322.

Hare PD, Chua N-H. Excision of selectable marker genes from transgenic plants. Nat Biotechnol. 2002;20:575–80.

Hartley JL. Use of the gateway system for protein expression in multiple hosts. Curr Protoc Protein Sci. 2002;30:5.17. 11–15.17. 10.

Hartley JL, Temple GF, Brasch MA. DNA cloning using in vitro site-specific recombination. Genome Res. 2000;10:1788–95.

Hatfull GF, Hendrix RW. Bacteriophages and their genomes. Curr Opin Virol. 2011;1:298–303.

Hauptmann R, Vasil V, Ozias-Akins P, Tabaeizadeh Z, Rogers S, Fraley R, Horsch RB, Vasil IK. Evaluation of selectable markers for obtaining stable transformants in the Gramineae. Plant Physiol. 1988;86:602–6.

Helinski DR. A brief history of plasmids. EcoSal Plus. 2022;10:eESP-0028-2021.

Hendrix RW. Bacteriophages: evolution of the majority. Theor Popul Biol. 2002;61:471–80.

Hohn B, Collins J. A small cosmid for efficient cloning of large DNA fragments. Gene. 1980;11:291–8.
Hohn B, Collins J. Ten years of cosmids. Trends Biotechnol. 1988;6:293–8.
Hohn B, Koukolikova-Nicola Z, Lindenmaier W, Collins J. Cosmids. In: Vectors; 1988. p. 113–27.
Hokanson SC. SNiPs, chips, BACs, and YACs: are small fruits part of the party mix? HortScience. 2001;36:859–71.
Hu R-M, Han Z-G, Song H-D, Peng Y-D, Huang Q-H, Ren S-X, Gu Y-J, Huang C-H, Li Y-B, Jiang C-L. Gene expression profiling in the human hypothalamus-pituitary-adrenal axis and full-length cDNA cloning. Proc Natl Acad Sci. 2000;97:9543–8.
Ijaz S. Mitochondrial plasmid: could this be the next hit for genetic engineering? Genet Mol Res. 2011;10(1):391–2.
Ijaz S, Ul Haq I, Razzaq HA, Nasir B, Ali HM, Kaur S. In silico structural-functional characterization of three differentially expressed resistance gene analogs identified in Dalbergia sissoo against dieback disease reveals their role in immune response regulation. Front Plant Sci. 2023;14:1134806.
Ish-Horowicz D, Burke J. Rapid and efficient cosmid cloning. Nucleic Acids Res. 1981;9:2989–898.
Jeong J-Y, Yim H-S, Ryu J-Y, Lee HS, Lee J-H, Seen D-S, Kang SG. One-step sequence-and ligation-independent cloning as a rapid and versatile cloning method for functional genomics studies. Appl Environ Microbiol. 2012;78:5440–3.
Jobling MG, Holmes RK. Construction of vectors with the p15a replicon, kanamycin resistance, inducible lacZ alpha and pUC18 or pUC19 multiple cloning sites. Nucleic Acids Res. 1990;18:5315.
Joska TM, Mashruwala A, Boyd JM, Belden WJ. A universal cloning method based on yeast homologous recombination that is simple, efficient, and versatile. J Microbiol Methods. 2014;100:46–51.
Karimi M, Depicker A, Hilson P. Recombinational cloning with plant gateway vectors. Plant Physiol. 2007;145:1144–54.
Katzen F. Gateway® recombinational cloning: a biological operating system. Expert Opin Drug Discov. 2007;2:571–89.
Kaufer B, Tischer B. BACs (bacterial artificial chromosomes). In: Brenner's Encyclopedia of genetics: Springer,York, USA;251; 2013.
Kidd IM, Emery VC. The use of baculoviruses as expression vectors. Appl Biochem Biotechnol. 1993;42:137–59.
Kilaru S, Steinberg G. Yeast recombination-based cloning as an efficient way of constructing vectors for Zymoseptoria tritici. Fungal Genet Biol. 2015;79:76–83.
Kobs G. pGEM®-T Vector: cloning of modified blunt-ended DNA fragments. Promega Notes. 1995;55:28–9.
Koerner T, Hill JE, Myers AM, Tzagoloff A. [33] High-expression vectors with multiple cloning sites for construction of trpE fusion genes: pATH vectors. In: Methods in enzymology. Elsevier, Oxford University Press (for Nucleic Acids Research), Oxford, UK; 1991. p. 477–90.
Krzywinski M, Wallis J, Gösele C, Bosdet I, Chiu R, Graves T, Hummel O, Layman D, Mathewson C, Wye N. Integrated and sequence-ordered BAC-and YAC-based physical maps for the rat genome. Genome Res. 2004;14:766–79.
Kutter E, Sulakvelidze A. Bacteriophages: biology and applications. CRC Press; 2004.
Laborda-Mansilla J, García-Ruiz E. Advancements in Golden Gate cloning: a comprehensive review. In: Golden Gate cloning: methods and protocols; 2024. p. 481–500.
Lampropoulos A, Sutikovic Z, Wenzl C, Maegele I, Lohmann JU, Forner J. GreenGate-a novel, versatile, and efficient cloning system for plant transgenesis. PLoS One. 2013;8:e83043.
Leonard AC, Méchali M. DNA replication origins. Cold Spring Harb Perspect Biol. 2013;5:a010116.
Lewis I, Peura T, Trounson A. Large-scale applications of cloning technologies for agriculture: an industry perspective. Reprod Fertil Dev. 1998;10:677–82.

Li MZ, Elledge SJ. SLIC: a method for sequence-and ligation-independent cloning. In: Gene synthesis: methods and protocols. Springer; 2012. p. 51–9.
Li C, Evans RM. Ligation independent cloning irrespective of restriction site compatibility. Nucleic Acids Res. 1997;25:4165–6.
Liang X, Peng L, Baek C-H, Katzen F. Single step BP/LR combined Gateway reactions. BioTechniques. 2013;55:265–8.
Liu Z. Comparative analysis of Golden Gate and classical cloning techniques in E. coli: a study in molecular cloning efficiency. Asia Pacific J Clin Med Res. 2025;1:412.
Liu Q, Li MZ, Leibham D, Cortez D, Elledge SJ. The univector plasmid-fusion system, a method for rapid construction of recombinant DNA without restriction enzymes. Curr Biol. 1998;8:S1.
Liu VW, Nagley P, Ngan HY. One-step directional cloning of blunt-ended polymerase chain reaction products. Anal Biochem. 2000;287:349–52.
Liu Q, Dang H-J, Wu Y-H, Li M, Chen Y-H, Niu X-L, Li K-M, Luo L-J. pXST, a novel vector for TA cloning and blunt-end cloning. BMC Biotechnol. 2018;18:44.
Lodge J, Lund P, Minchin S. Gene cloning. Taylor & Francis (CRC Press) Boca Raton, Florida, USA; 2007.
Luckow VA, Summers MD. Trends in the development of baculovirus expression vectors. Bio/Technology. 1988;6:47–55.
Mantikou E, Jonker MJ, Wong KM, van Montfoort AP, De Jong M, Breit T, Repping S, Mastenbroek S. Factors affecting the gene expression of in vitro cultured human preimplantation embryos. Hum Reprod. 2016;31:298–311.
Marillonnet S, Grützner R. Synthetic DNA assembly using golden gate cloning and the hierarchical modular cloning pipeline. Curr Protoc Mol Biol. 2020;130:e115.
Matsuyama A, Yoshida M. Systematic cloning of an ORFeome using the gateway system. In: Reverse chemical genetics: methods and protocols. Springer, New York, USA; 2009. p. 11–24.
Mellor J, Dobson M, Roberts N, Kingsman A, Kingsman S. Factors affecting heterologous gene expression in Saccharomyces cerevisiae. Gene. 1985;33:215–26.
Miki B, McHugh S. Selectable marker genes in transgenic plants: applications, alternatives and biosafety. J Biotechnol. 2004;107:193–232.
Miller LK. Baculoviruses as gene expression vectors. Annu Rev Microbiol. 1988;42:177.
Monaco AP, Larin Z. YACs, BACs, PACs and MACs: artificial chromosomes as research tools. Trends Biotechnol. 1994;12:280–6.
Moss B, Elroy-Stein O, Mizukami T, Alexander W, Fuerst T. New mammalian expression vectors. Nature. 1990;348:91–2.
Motohashi K. A novel series of high-efficiency vectors for TA cloning and blunt-end cloning of PCR products. Sci Rep. 2019;9:6417.
Muyrers JP, Zhang Y, Stewart AF. ET-cloning: think recombination first. In: Genetic engineering: principles and methods; Oxford University Press (for Nucleic Acids Research), Oxford, UK; 2000. p. 77–98.
Muyrers JP, Zhang Y, Stewart AF. Techniques: recombinogenic engineering–new options for cloning and manipulating DNA. Trends Biochem Sci. 2001;26:325–31.
O'Reilly DR, Miller LK, Luckow VA. Baculovirus expression vectors: a laboratory manual. Oxford University Press; 1994.
Pascussi JM, Dvorák Z, Gerbal-Chaloin S, Assenat E, Maurel P, Vilarem MJ. Pathophysiological factors affecting CAR gene expression. Drug Metab Rev. 2003;35:255–68.
Pastinen T, Sladek R, Gurd S, Sammak BG, Lepage P, Lavergne K, Villeneuve A, Gaudin T, Brändström H. A survey of genetic and epigenetic variation affecting human gene expression. Physiol Genomics. 2004;16:184.
Perez-Tris J, Bensch S. Diagnosing genetically diverse avian malarial infections using mixed-sequence analysis and TA-cloning. Parasitology. 2005;131:15–23.
Petersen LK, Stowers RS. A Gateway MultiSite recombination cloning toolkit. PLoS One. 2011;6:e24531.
Prioleau M-N, MacAlpine DM. DNA replication origins—where do we begin? Genes Dev. 2016;30:1683–97.

Pryor JM, Potapov V, Kucera RB, Bilotti K, Cantor EJ, Lohman GJ. Enabling one-pot Golden Gate assemblies of unprecedented complexity using data-optimized assembly design. PLoS One. 2020;15:e0238592.

Rahbaran M, Razeghian E, Maashi MS, Jalil AT, Widjaja G, Thangavelu L, Kuznetsova MY, Nasirmoghadas P, Heidari F, Marofi F. Cloning and embryo splitting in mammalians: brief history, methods, and achievements. Stem Cells Int. 2021;2021:2347506.

Reece-Hoyes JS, Walhout AJ. Gateway recombinational cloning. Cold Spring Harbor Protocols; 2018. 2018: pdb. top094912.

Ricciardi A, Ianniello RG, Parente E, Zotta T. Factors affecting gene expression and activity of heme-and manganese-dependent catalases in Lactobacillus casei strains. Int J Food Microbiol. 2018;280:66–77.

Roberts TM, Kacich R, Ptashne M. A general method for maximizing the expression of a cloned gene. Proc Natl Acad Sci. 1979;76:760–4.

Robinson NP, Bell SD. Origins of DNA replication in the three domains of life. FEBS J. 2005;272:3757–66.

Salzberg SL, Salzberg AJ, Kerlavage AR, Tomb J-F. Skewed oligomers and origins of replication. Gene. 1998;217:57–67.

Samiec M. Molecular mechanism and application of somatic cell cloning in mammals—past, present and future. Int J Mol Sci. 2022;23:13786.

Santhosh C, Tamhane MC, Kamat RH, Patel VV, Mukhopadhyaya R. A lentiviral vector with novel multiple cloning sites: stable transgene expression in vitro and in vivo. Biochem Biophys Res Commun. 2008;371:546–50.

Schmid-Burgk JL, Schmidt T, Kaiser V, Höning K, Hornung V. A ligation-independent cloning technique for high-throughput assembly of transcription activator–like effector genes. Nat Biotechnol. 2013;31:76–81.

Sharan SK, Thomason LC, Kuznetsov SG, Court DL. Recombineering: a homologous recombination-based method of genetic engineering. Nat Protoc. 2009;4:206–23.

Sharma K, Mishra AK, Mehraj V, Duraisamy GS. Advances and applications of molecular cloning in clinical microbiology. Biotechnol Genet Eng Rev. 2014;30:65–78.

Sharma S, Chatterjee S, Datta S, Prasad R, Dubey D, Prasad RK, Vairale MG. Bacteriophages and its applications: an overview. Folia Microbiol. 2017;62:17–55.

Sheffield P, Garrard S, Derewenda Z. Overcoming expression and purification problems of RhoGDI using a family of "parallel" expression vectors. Protein Expr Purif. 1999;15:34–9.

Smillie C, Garcillán-Barcia MP, Francia MV, Rocha EP, de la Cruz F. Mobility of plasmids. Microbiol Mol Biol Rev. 2010;74:434–52.

Soares MB. Identification and cloning of differentially expressed genes. Curr Opin Biotechnol. 1997;8:542–6.

Sorida M, Bonasio R. An efficient cloning method to expand vector and restriction site compatibility of Golden Gate Assembly. Cell Rep Methods. 2023;3:100564.

Stevenson J, Krycer JR, Phan L, Brown AJ. A practical comparison of ligation-independent cloning techniques. PLoS One. 2013;8:e83888.

Summers D. The biology of plasmids. Wiley, New York, USA; 2009.

Sundar IK, Sakthivel N. Advances in selectable marker genes for plant transformation. J Plant Physiol. 2008;165:1698–716.

Taylor JH. Origins of replication and gene regulation. Mol Cell Biochem. 1984;61:99–109.

Tefferi A. Genomics basics: DNA structure, gene expression, cloning, genetic mapping, and molecular tests. In: Seminars in cardiothoracic and vascular anesthesia. Thousand Oaks: Sage Publications Sage CA; 2006. p. 282–90.

Terfrüchte M, Joehnk B, Fajardo-Somera R, Braus GH, Riquelme M, Schipper K, Feldbrügge M. Establishing a versatile Golden Gate cloning system for genetic engineering in fungi. Fungal Genet Biol. 2014;62:1–10.

Thieme F, Engler C, Kandzia R, Marillonnet S. Quick and clean cloning: a ligation-independent cloning strategy for selective cloning of specific PCR products from non-specific mixes. PLoS One. 2011;6:e20556.

Tillett D, Neilan BA. Enzyme-free cloning: a rapid method to clone PCR products independent of vector restriction enzyme sites. Nucleic Acids Res. 1999;27:e26–8.

Tirabassi R, Bio B. Foundations of molecular cloning-past, present and future. NEB, New England Bio Labs Inc; 2014. www.neb.com/tools-and-resources/featurearticles/foundations-of-molecular-cloning-past-present-and-future

Tóth E, Huszár K, Bencsura P, Kulcsár PI, Vodicska B, Nyeste A, Welker Z, Tóth S, Welker E. Restriction enzyme body doubles and PCR cloning: on the general use of type IIs restriction enzymes for cloning. PLoS One. 2014;9:e90896.

Tseng H. DNA cloning without restriction enzyme and ligase. BioTechniques. 1999;27:1240–4.

Twyman RM, Stoger E, Schillberg S, Christou P, Fischer R. Molecular farming in plants: host systems and expression technology. Trends Biotechnol. 2003;21:570–8.

Untergasser A. Cloning–Gateway BP-reaction II. Untergasser's Lab. Published online; 2006.

Van Den Ent F, Löwe J. RF cloning: a restriction-free method for inserting target genes into plasmids. J Biochem Biophys Methods. 2006;67:67–74.

Vemanna RS, Chandrashekar BK, Hanumantha Rao H, Sathyanarayanagupta SK, Sarangi K, Nataraja KN, Udayakumar M. A modified multisite gateway cloning strategy for consolidation of genes in plants. Mol Biotechnol. 2013;53:129–38.

Visconte S. Molecular cloning: the process and its applications. J Glycobiol. 2023;12(2):1–4. https://www.longdom.org/open-access/molecular-cloning-the-process-and-its-applications-98732.html

Walhout AJ, Temple GF, Brasch MA, Hartley JL, Lorson MA, van den Heuvel S, Vidal M. [34] GATEWAY recombinational cloning: application to the cloning of large numbers of open reading frames or ORFeomes. In: Methods in enzymology. Elsevier: 575-IN577; 2000.

Wall VE, Garvey LA, Mehalko JL, Procter LV, Esposito D. Combinatorial assembly of clone libraries using site-specific recombination. In: DNA cloning and assembly methods. Springer; 2013. p. 193–208.

Wang RF, Kushner SR. Construction of versatile low-copy-number vectors for cloning, sequencing and gene expression in Escherichia coli. Gene. 1991;100:195–9.

Wang Q, Maher VM, McCormick JJ. Mammalian expression vectors with modulatable promoters and two multiple cloning sites. Gene. 1992;119:155–61.

Watson JF, García-Nafría J. In vivo DNA assembly using common laboratory bacteria: a re-emerging tool to simplify molecular cloning. J Biol Chem. 2019;294:15271–81.

Weber E, Gruetzner R, Werner S, Engler C, Marillonnet S. Assembly of designer TAL effectors by Golden Gate cloning. PLoS One. 2011;6:e19722.

Werner S, Engler C, Weber E, Gruetzner R, Marillonnet S. Fast track assembly of multigene constructs using Golden Gate cloning and the MoClo system. Bioengineered. 2012;3:38–43.

Williams RB, Chan EK, Cowley MJ, Little PF. The influence of genetic variation on gene expression. Genome Res. 2007;17:1707–16.

Worning P, Jensen LJ, Hallin PF, Stærfeldt HH, Ussery DW. Origin of replication in circular prokaryotic chromosomes. Environ Microbiol. 2006;8:353–61.

Xu R, Qingshun LQ. Protocol: streamline cloning of genes into binary vectors in agrobacterium via the Gateway® TOPO vector system. Plant Methods. 2008;4:4.

Yao S, Hart DJ, An Y. Recent advances in universal TA cloning methods for use in function studies. Protein Eng Des Sel. 2016;29:551–6.

Zeng C, Kouprina N, Zhu B, Cairo A, Hoek M, Cross G, Osoegawa K, Larionov V, de Jong P. Large-insert BAC/YAC libraries for selective re-isolation of genomic regions by homologous recombination in yeast. Genomics. 2001;77:27–34.

Zhang W, Wu H, Wang W, Wei B. Golden Gate assembly of DNA nanostructures. ACS Mater Lett. 2023;5:3316–20.

Zhou M-Y, Gomez-Sanchez CE. Universal TA cloning. Curr Issues Mol Biol. 2000;2:1–7.

Zhou M-Y, Gomez-Sanchez CE. Universal TA cloning. In: Gene cloning and analysis. Garland Science; 2023. p. 165–78.

Chapter 4
Enzymes in Molecular Cloning: Properties and Applications

Maria Babar and Hayssam Mohamed Ali

4.1 Introduction

The genetic code and the structure of DNA were elucidated in the 1950s and early 1960s. However, it was not until the 1970s that experimental methods advanced enough for a thorough examination of the gene. Over the past 40 years, there have been remarkable advancements in DNA modification and cloning, commonly referred to as recombinant DNA technology. This technology has consistently delivered by generating a wealth of biological data. Understanding of the relationship between a gene's structure and function has improved thanks to advances in cloning, sequencing, gene engineering, and mutagenesis. Experiments across various branches of biology are now conducted differently because of the ability to introduce genes and specific mutations into a wide range of organisms, from higher mammals to plants. An understanding of an organism's function can be gained by analyzing the sequencing data derived from a cloned gene (Eidne 1996).

Molecular cloning is a fundamental technique in molecular biology that involves copying specific DNA sequences. Central to this process are enzymes that manipulate, insert, and replicate DNA fragments. These enzymes perform crucial functions, such as cutting, joining, and amplifying DNA, which enables their insertion into suitable vectors for growth in host organisms (Ijaz 2011; Green and Sambrook 2012). Laboratory suppliers have shown great interest in the rapid advancement of cloning methods and applications in recent years. As a result, numerous enzyme

M. Babar
Centre of Agricultural Biochemistry and Biotechnology (CABB), University of Agriculture, Faisalabad, Pakistan

H. Mohamed Ali (✉)
Department of Botany and Microbiology, College of Science, King Saud University, Riyadh, Saudi Arabia
e-mail: hayhassan@ksu.edu.sa

S. Ijaz et al. (eds.), *Molecular Biology Methods*,
https://doi.org/10.1007/978-981-92-2642-9_4

sources are now available, and beginners might find it challenging to choose the right enzyme for a particular task (Rittié and Perbal 2008).

Natural or engineered nucleic acids, which can range from just a few to several thousand nucleotides, are used in molecular cloning. It is possible to significantly modify nucleic acids to give them specific traits and qualities. This review highlights the main enzymes in each category, along with their characteristics and modes of action. The goal is to enhance the reader's understanding of the fundamental enzymatic processes involved in molecular cloning (Rittié and Perbal 2008).

4.2 DNA-Dependent DNA Polymerase

DNA polymerases are enzymes that facilitate the formation of many structural units, called deoxyribonucleoside triphosphates (dNTPs), which are used to build DNA polymers. So far, no DNA polymerase has been discovered that can initiate the synthesis of a polynucleotide from individual nucleotides—this process, known as de novo synthesis, is not controlled by any enzyme. DNA polymerases only add nucleotides to the 3′-OH end of an existing primer that has a 5′-phosphate group. Primers are short RNA sequences that pair with approximately ten DNA bases at the 5′ end to start replication. Primase, an RNA polymerase, produces these primers. Most polymerases employed in molecular biology are derived from bacteria or their infecting viruses, such as phages or bacteriophages.

In bacterial DNA replication, three DNA polymerases—Pol I, Pol II, and Pol III—work together. When primers and dNTPs are available, all three facilitate DNA strand extension in the 5′ → 3′ direction, though their speeds vary. DNA Pol III, the main enzyme complex for bacterial replication, was identified in 1970. Each polymerase has a 3′ → 5′ exonuclease activity, called proofreading activity, which corrects incorrectly incorporated bases during replication. This proofreading slows the process but improves fidelity. DNA Pol I also possesses a 5′ → 3′ exonuclease activity, which is involved in excision-repair and the removal of RNA primers from the DNA's 5′ ends. Additionally, it shares polymerase and proofreading functions with the others. In laboratory conditions, Pol I's 5′ → 3′ exonuclease activity is used in nick-translation, a method that replaces nucleotides with labeled analogues in a DNA sequence.

4.3 T4 DNA Polymerase

The two roles of bacterophage T4 polymerase depend on whether a template and primer are present. In the absence of dNTPs, it functions as a 3′ → 5′ (reverse) exonuclease. When dNTPs are available, it acts as a 5′ → 3′ polymerase. T4 DNA Pol lacks 5′ → 3′ exonuclease activity, unlike *Escherichia coli* DNA Pol I (Englund 1971). Consequently, T4 DNA Pol can substitute for Klenow to label the 3′ ends of double-stranded DNA, fill in 5′ overhangs of DNA fragments, and repair nicks in DNA. When conditions are controlled, it can synthesize DNA at a rate of 15,000 nucleotides per minute.

4.4 Klenow Fragment and Pol I

Before adding molecular linkers, native DNA Pol I has been effectively utilized to fill cohesive ends in the presence of dNTPs or to remove 3′ protruding DNA ends when dNTPs are absent. However, its 5′ → 3′ exonuclease activity limits its use in applications that only require polymerization. Fortunately, digestion of *E. coli* DNA Pol I (109 kDa) with proteolysis creates two fragments (76 and 36 kDa). The functional quality of these fragments has since been significantly improved through recombinant production of Klenow, which removes contaminants caused by residual natural enzymes in proteolytically treated samples.

4.5 Modified Bacteriophage T7 DNA Polymerase

Richardson et al. identified a chemically modified phage polymerase as an ideal tool for DNA sequencing (Tabor and Richardson 1987). This complex, known as modified bacteriophage Pol, comprises the 12 kDa *E. coli* thioredoxin and the 84 kDa T7 gene 5 product (Mark and Richardson 1976). The thioredoxin binds the T7 gene 5 protein to a primer-template, enabling the polymerization of thousands of nucleotides without dissociation and increasing T7 polymerase efficiency. The T7 gene 5 protein provides the complex's catalytic activity. As a result, modified Pol is over 70 times faster than avian myeloblastosis virus (AMV) reverse transcriptase, with a polymerization rate exceeding 300 nucleotides per second. This enzyme complex can amplify large DNA fragments and produce radioactive probes. Further research showed that the T7 DNA Pol's 3′ → 5′ exonuclease activity depends on a 28-amino acid sequence (residues 118–145). In vitro mutagenesis of the corresponding nucleotides in T7 gene 5 completely abolished this exonuclease activity, which increased spontaneous mutation rates by 14 times and improved polymerase efficiency by 9 times (Tabor and Richardson 1989).

4.6 Terminal Deoxynucleotidyl Transferase

Terminal deoxynucleotidyl transferase (TdT), first isolated from calf thymus, is a unique DNA polymerase that adds homopolymeric tails to the 3′-OH ends of DNA without requiring a template. This enzyme is widely used in molecular biology applications, including primer extension, DNA sequencing, and labeling of DNA 3′ ends with radiolabeled nucleotides. In the presence of magnesium ions (Mg^{2+}), TdT requires a single-stranded DNA primer of at least three nucleotides for activity. However, cobalt ions (Co^{2+}) significantly alter TdT's substrate specificity; in their presence, the enzyme can also utilize double-stranded DNA as a primer (Roychoudhury et al. 1976). This is likely due to the ability of Co^{2+} to relax the DNA helix, allowing extension from internal nicks—an effect much less pronounced with Mg^{2+}, which typically results in shorter tail lengths. Therefore, careful optimization of incubation conditions, particularly the choice of divalent cation, is essential to control TdT's specificity, activity, and product characteristics.

4.7 Thermostable DNA Polymerases Lacking Proofreading Activity

Although isolated in the early 1970s, thermostable DNA polymerases gained major attention for molecular cloning once PCR was developed, as there was a need for enzymes that could function at high temperatures for DNA synthesis. These thermostable polymerases are typically derived from thermophilic organisms because they work effectively at elevated temperatures.

4.7.1 Bst Polymerase

Bacillus stearothermophilus, a thermophilic bacterium, produces the thermostable DNA polymerase I known as Bst polymerase (Stenesh et al. 1968). Its optimal activity occurs at 65 °C, but it loses activity after 15 min at 75 °C. Bst Pol I requires a high Mg^{2+} concentration to function effectively and exhibits 5′ → 3′ exonuclease activity. Digestion by Bst DNA Pol proteases yields two protein fragments. The large fragment of Bst DNA Pol I, being thermostable, can be used to prevent hairpin formation during sequencing at 65 °C. Like Klenow, the large fragment's strand displacement activity is faster than that of the full enzyme. Epicentre Biotechnologies now offers recombinant Bst DNA Pol I and its large fragment under the brand IsoThermTM (Rittié and Perbal 2008).

4.7.2 Taq Polymerase

In 1976, Chien and colleagues successfully isolated *Taq* polymerase from *Thermus aquaticus*, a thermophilic bacterium first discovered in 1968 in the Great Fountain region of Yellowstone National Park (Brock and Freeze 1969).

T. aquaticus thrives at high temperatures, with an optimal growth temperature of approximately 70 °C and a thermal tolerance of up to 80 °C. The thermostability of *Taq* polymerase is key to its utility in the polymerase chain reaction (PCR), as it retains enzymatic activity at elevated temperatures—exhibiting a half-life of 40 min at 95 °C and approximately 5 min at 100 °C. For optimal function, the enzyme requires Mg^{2+} ions and operates most efficiently at incubation temperatures around 80 °C.

Although *Taq* polymerase is widely used in PCR, it is often classified as a low-fidelity enzyme due to its lack of 3′ → 5′ exonuclease (proofreading) activity. Its error rate typically ranges from 1×10^{-4} to 2×10^{-5} errors per base pair, depending on reaction conditions. Nonetheless, the enzyme can incorporate roughly 45,000 nucleotides before introducing an error, demonstrating a relatively high level of overall accuracy despite the absence of proofreading. Additionally, *Taq* polymerase exhibits terminal deoxynucleotidyl transferase activity, which enables it to add a few adenine residues to the 3′ ends of PCR products—an attribute common among DNA polymerases that lack proofreading functionality (Ijaz and Haq 2019).

4.8 DNA Polymerases That Have Proofreading Activity

Thermophilic organism DNA polymerases are more accurate because of their 3′ → 5′ exonuclease (proofreading) activity. Taq polymerase's error rate is approximately ten times higher (1.6×10^{-6}) than that of *Pyrococcus furiosus* (Pfu) DNA polymerase. Pow polymerase, derived from *Pyrococcus woesei*, has a low error rate of 7.4×10^{-7} and a half-life exceeding 2 h at 100 °C, though it performs best with templates smaller than 3.5 kb. Vent polymerase, from *Thermococcus litoralis* (also known as Tli polymerase), exhibits an error rate similar to Taq and Pfu, with a half-life of 7 h at 95 °C. *Pyrococcus abyssi* produces two DNA polymerases (Pol I and II) with 3′ → 5′ exonuclease activity, as described by Gueguen et al. (2001). *P. abyssi* Pol I has an error rate of 0.66×10^{-6}, comparable to Pfu, and is stable for about 5 h at 100 °C, making it highly thermostable for high-temperature PCR.

Most reverse transcriptases and DNA polymerases are available as "Hot Start" or similar formulations. Hot Start aims to prevent non-specific priming, such as template/primer hybridization or primer dimer formation, during PCR setup, which can produce unwanted secondary products in later cycles. This technology adds a barrier—such as an oligonucleotide, antibody, or reversible chemical modification—between the DNA-binding site and secondary structures. This barrier is removed during the first denaturation cycle, restoring activity. These modifications improve PCR setup and specificity by reducing polymerase activity at room temperature (Rittié and Perbal 2008).

4.9 RNA-Dependent DNA Polymerase

For many years, it was believed that genetic information flowed exclusively from DNA to RNA. This classical view persisted until the late 1950s (Crick 1958). However, groundbreaking research in the 1970s revealed a surprising reversal of this flow through the discovery of reverse transcriptase (RT) enzymes. These enzymes, identified independently in the Rauscher Leukemia Virus (RLV) (Baltimore 1970) and the Rous Sarcoma Virus (RSV) (Temin and Mizutani 1970), demonstrated that RNA could serve as a template to synthesize complementary DNA (cDNA). This process fundamentally challenged the central dogma by showing that RNA-dependent DNA polymerization occurs naturally.

Reverse transcriptase, an enzyme also known as RNA-dependent DNA polymerase, exists in both viral and nonviral forms. In retroviruses like RSV and RLV, RT facilitates the reverse transcription of the viral RNA genome into double-stranded DNA, which then integrates into the host genome as a provirus. This integrated DNA acts as a template for transcription, ultimately producing new infectious viral particles. Many viruses employ reverse transcription at various stages of their life cycle.

Beyond retroviruses, Metaviridae—a family of retrotransposons found in eukaryotic genomes—also rely on reverse transcription. These mobile genetic elements propagate by generating an RNA intermediate, which is then reverse transcribed into DNA and integrated at new genomic locations. Examples include the Ty3 element in *Saccharomyces cerevisiae*, the gipsy retrotransposon in *Drosophila melanogaster*, and the Tas element in *Ascaris lumbricoides*.

Members of the Pseudoviridae family, such as the Ty1 element in *S. cerevisiae* and the copia element in *D. melanogaster*, have segmented single-stranded RNA genomes. Meanwhile, the Hepadnaviridae family—exemplified by hepatitis B virus (HBV)—possesses a unique partially double-stranded, circular DNA genome with one complete and one incomplete strand. HBV's viral polymerase exhibits multifunctionality, combining RNA- and DNA-dependent DNA polymerase activities, as well as RNase H and protein priming functions. Upon infecting a host cell, the relaxed circular DNA (rcDNA) is converted into covalently closed circular DNA, which serves as a template for transcription by host RNA polymerases. This transcription produces various viral RNAs, including the pregenomic RNA, which the viral reverse transcriptase subsequently converts back into DNA (cDNA) to generate the viral genome (Seeger and Mason 2000).

4.10 DNA-Dependent RNA Polymerases

Transcription, the synthesis of RNA from a DNA template, is catalyzed by DNA-dependent RNA polymerases. These enzymes resemble DNA polymerases but differ notably by lacking proofreading exonuclease activity, allowing them to add an extra nucleotide at the transcript's end. Among the most widely used RNA polymerases in molecular biology are the bacteriophage-derived SP6 and T7 RNA polymerases. T7 RNA polymerase, a 98 kDa protein produced by the T7 bacteriophage, and SP6 RNA polymerase, a 96 kDa protein derived from *Salmonella typhimurium* infected with SP6 bacteriophage, are valuable tools for generating RNA transcripts in vitro.

Both enzymes require a double-stranded DNA template and Mg^{2+} ions for activity, with dithiothreitol commonly included to maintain enzyme stability, especially for T7 polymerase. SP6 and T7 RNA polymerases exhibit high promoter specificity, transcribing RNA exclusively from DNA sequences located downstream of their respective promoters. Interestingly, these polymerases can transcribe through polyadenine (poly(A)) regions, allowing RNA synthesis to continue around circular templates multiple times before enzyme dissociation. To ensure controlled and complete transcription, DNA templates are typically linearized before use, often by creating blunt ends or 5′ overhangs. The RNA products generated are functional and spliceable, suitable for a variety of molecular biology applications, including antisense RNA synthesis, labeled probe preparation, and RNase protection assays (Melton et al. 1984; Schenborn and Mierendorf 1985).

4.11 Ligase Enzymes

4.11.1 DNA Ligases

DNA ligases are essential enzymes that repair breaks in double-stranded DNA by forming phosphodiester bonds between adjacent 3′-hydroxyl and 5′-phosphate groups. This activity is crucial during DNA replication, where ligases join Okazaki fragments on the lagging strand, as well as in various DNA repair pathways. In molecular biology, DNA ligases are widely used to seal nicks in DNA molecules, attach linkers or adaptors, and join DNA fragments with either blunt or cohesive (sticky) ends generated by restriction enzymes.

The ligation process typically occurs through a three-step enzymatic mechanism. First, a conserved lysine residue within the ligase reacts with adenosine triphosphate (ATP) or NAD^+ to form a ligase-adenylate intermediate. Next, the adenosine monophosphate (AMP) group is transferred to the 5′-phosphate at the site of the nick, creating a DNA-adenylate (AppDNA) intermediate. Finally, the free 3′-OH group attacks this adenylated intermediate, resulting in the formation of a new phosphodiester bond and release of AMP. Notably, eukaryotic, viral, and bacteriophage ligases generally utilize ATP as the adenylation cofactor, whereas bacterial ligases prefer NAD^+; some ligases can accept both cofactors but often differ in catalytic efficiency depending on which is used.

4.11.2 Thermostable DNA Ligases

Thermostable DNA ligases are derived from thermophilic organisms and maintain activity over a broad temperature range, typically between 45 and 80 °C. These enzymes efficiently recognize and repair nicks in double-stranded DNA, exhibiting high substrate specificity that makes them particularly suited for applications requiring stringent ligation conditions. Examples of such ligases originate from bacteria like *Thermus thermophilus*, *Bacillus stearothermophilus*, *Thermus scotoductus*, and *Rhodothermus marinus*. While thermostable ligases typically complement rather than replace commonly used enzymes, such as T4 DNA ligase, they are invaluable in specialized techniques, including the ligase chain reaction (LCR). LCR leverages the ligase's high specificity to detect single-base mutations by ligating fragments only when they are perfectly matched to a template, enabling sensitive mutation detection and subsequent PCR amplification.

4.11.3 Thermolabile DNA Ligases

Eukaryotic DNA ligases are characterized by structural features including zinc finger motifs and BRCT domains (related to the BRCA-1 protein), as well as localization signals targeting the nucleus or mitochondria. In contrast, ligases encoded by bacteriophage T7 and some algae, such as *Chlorella*, contain primarily an oligonucleotide/oligosaccharide-binding (OB) fold and a nucleotidyl-transferase domain. Among the most commonly employed ligases in molecular biology is the T4 DNA ligase—a 68 kDa monomeric enzyme that requires ATP and Mg^{2+} as cofactors. T4 DNA ligase efficiently repairs single-stranded nicks and ligates both cohesive and blunt-ended DNA fragments.

The efficiency of DNA ligation is influenced by multiple factors, including temperature, DNA fragment concentration, the nature of DNA ends (blunt or sticky), the length and sequence of sticky ends, and hydrogen-bonding interactions between overhangs. Consequently, determining optimal ligation conditions often requires empirical testing. For instance, ligation of DNA fragments cut with HindIII is significantly faster—by 10–40 times—than ligation of fragments generated by SalI, despite both producing sticky ends. To standardize activity measurements, many manufacturers define one unit of ligase as the amount of enzyme needed to catalyze the ligation of 50% of HindIII-digested lambda DNA fragments within 30 min at 16 °C. Protocols typically recommend overnight incubation at 16 °C; however, ligation times can range from 10 min at room temperature to 24 h at 4 °C, depending on the DNA concentration and fragment type. Regular testing of ligase activity is advisable, particularly after long-term storage at −20 °C. For quality control, ligation of HindIII-digested lambda DNA is commonly used to assess sticky-end ligation, while blunt-end ligation is often tested using HaeIII-digested DNA.

4.11.4 RNA Ligases

RNA ligases catalyze the formation of phosphodiester bonds between the 5′-phosphate and 3′-hydroxyl termini of single-stranded RNA or DNA molecules in an ATP-dependent manner. Their primary biological function is to repair and process damaged RNA by sealing breaks or joining RNA fragments. Like DNA ligases, RNA ligases operate via a conserved three-step mechanism: first, the enzyme forms a covalent intermediate with AMP attached to a lysine residue, releasing pyrophosphate. Next, the AMP is transferred to the 5′-phosphate of the RNA substrate, generating an adenylated RNA intermediate (AppRNA). Finally, the 3′-OH group of an adjacent RNA strand attacks the adenylated phosphate, forming a phosphodiester bond and completing the ligation.

T4 RNA ligase 2 (T4 Rnl2, also known as gene product 24.1 or gp24.1) is particularly effective at ligating both intra- and intermolecular RNA fragments. Unlike T4 RNA ligase 1, which favors joining single-stranded RNA ends, T4 Rnl2 exhibits superior efficiency in sealing nicks within double-stranded RNA molecules. It recognizes RNA nicks with adjacent 3′-OH and 5′-phosphate termini. It is also capable of catalyzing the ligation of RNA to DNA strands in a double-stranded context, specifically joining the 3′-OH of RNA to the 5′-phosphate of DNA. This versatility makes T4 Rnl2 a valuable tool for diverse RNA manipulation techniques.

4.12 Addition and Removal of Phosphate

4.12.1 Alkaline Phosphatase

Alkaline phosphatase, derived from sources such as *Escherichia coli* or higher eukaryotes, is widely utilized in molecular biology to remove 5′-phosphate groups from nucleic acids. This enzymatic activity is crucial during cloning procedures to prevent vector self-ligation by dephosphorylating DNA ends, thereby facilitating the insertion of foreign DNA fragments. Notably, alkaline phosphatase targets terminal phosphate groups without cleaving internal phosphodiester bonds. Although *E. coli* and calf intestinal alkaline phosphatases differ in sequence and structure, both require divalent metal ions, such as zinc and magnesium, to function optimally. Their activity can be inhibited by chelating agents such as EGTA or by the presence of inorganic phosphate, which must be carefully removed after restriction enzyme digestion, typically during gel purification or dialysis of DNA fragments. For successful 5′ end-labeling of DNA, alkaline phosphatase treatment should precede polynucleotide kinase incubation in the presence of nucleoside triphosphates.

4.12.2 T4 Polynucleotide Kinase

T4 Polynucleotide Kinase is a multifunctional enzyme that catalyzes the transfer of phosphate groups to nucleic acids, including phosphorylation of 5′-hydroxyl termini using ATP, as well as the exchange of 5′-phosphate groups. This homotetrameric enzyme, encoded by the *pseT* gene of bacteriophage T4, has a subunit molecular weight of approximately 33 kDa. It is commonly employed for radiolabeling DNA or RNA at their 5′ ends with γ-^{32}P-ATP, enabling applications such as sequencing, restriction mapping, and nucleic acid footprinting. Optimal kinase activity occurs at pH 7.6 and 37 °C, with a molar ATP-to-substrate ratio around 5:1, in the presence of Mg^{2+} and reducing agents like β-mercaptoethanol. Additionally, T4 PNK possesses 3′-phosphatase activity, which removes phosphate groups from 3′ ends of nucleic acids and is most active at a lower pH (5.0–6.0), allowing experimental manipulation of substrates. A mutant form lacking 3′-phosphatase but retaining kinase activity has also been characterized.

4.12.3 Nucleases

Nucleases are enzymes that cleave the phosphodiester bonds within nucleic acids. They are broadly classified into endonucleases, which cut within nucleic acid chains, and exonucleases, which degrade nucleic acids from their ends. Both DNases and RNases fall into these categories, playing critical roles in nucleic acid manipulation for cloning and analytical purposes.

4.12.4 DNase I

Deoxyribonuclease I (DNase I) is an endonuclease capable of cleaving both single- and double-stranded DNA, whether free in solution or chromatin-bound. Commonly sourced from bovine pancreas, it consists of a mixture of isoenzymes and has a molecular weight of approximately 31 kDa. DNase I preferentially cleaves phosphodiester bonds adjacent to pyrimidines, generating fragments with 3′-hydroxyl and 5′-phosphate ends. The enzyme requires divalent cations such as Mg^{2+} and Ca^{2+} for activity, and its cleavage pattern can be modulated by the presence of Mn^{2+}, which promotes double-stranded cuts with blunt or short overhanging ends. DNase I is widely used to remove DNA contamination in RNA or protein samples, generate DNA fragments for sequencing, and perform footprinting assays. It can be inactivated by heat treatment in the presence of EGTA.

4.12.5 Exonuclease III

Exonuclease III, a 28 kDa enzyme derived from *E. coli*, exhibits 3′-to-5′ exonuclease activity specific to double-stranded DNA. It also possesses phosphatase and RNase H activities. Unique among exonucleases, Exonuclease III cannot digest single-stranded DNA and produces 5′ overhangs that block further degradation. It is employed for generating nested deletions in DNA sequencing and, in combination with other enzymes, to prepare substrates for various molecular biology applications. Optimal activity occurs at pH 7.6–8.5, with Mg^{2+} or Mn^{2+} as cofactors, while Zn^{2+} acts as an inhibitor.

4.12.6 Exonuclease VII

Exonuclease VII, an 88 kDa polypeptide isolated from *E. coli* K-12, specifically degrades single-stranded DNA from either the 5′ or 3′ ends. Unlike exonucleases I and III, it remains active in the presence of EDTA and produces oligonucleotide fragments. This property makes it valuable for removing single-stranded primers after PCR and processing single-stranded DNA regions.

4.12.7 RNase A

Ribonuclease A (RNase A), also known as pancreatic RNase, cleaves single-stranded RNA at the 3′-phosphodiester bonds, producing 3′-phosphorylated nucleotides via a 2′,3′-cyclic phosphate intermediate. It is frequently used to eliminate RNA contamination in nucleic acid preparations and to map mutations via mismatch cleavage. RNase A is exceptionally stable and difficult to inactivate, so stringent precautions are necessary to avoid contamination. Inhibitors such as human placenta RNase inhibitor, vanadyl-ribonucleoside complexes, and chemical treatments like DEPC can reduce its activity.

4.12.8 RNase H

RNase H specifically degrades the RNA strand of RNA-DNA hybrids, facilitating removal of RNA primers or probes. The enzyme is active over a pH range of 7–9, requires Mg^{2+} ions (which can be partially substituted by Mn^{2+}), and tolerates moderate ionic strength. It is inhibited by sulfhydryl-reactive compounds such as N-ethylmaleimide.

4.13 Other Ribonucleases for RNA Sequencing

Various ribonucleases with distinct cleavage specificities have been utilized in RNA sequencing. RNase Phy I (from *Physarum polycephalum*) cleaves G, A, and U residues; RNase CL3 (from chicken liver) preferentially cleaves RNA adjacent to cytidine residues; RNase T1 (from *Aspergillus oryzae*) targets guanine residues; and RNase U2 (from *Ustilago sphaerogena*) cleaves adjacent to purines, complementing RNase T1. Endoribonucleases such as Cereus ribonuclease and Phy M also exhibit specific cleavage patterns useful for RNA analysis.

4.13.1 S1 Nuclease

S1 nuclease, purified from *Aspergillus oryzae*, selectively degrades single-stranded nucleic acids without affecting duplex DNA or RNA-DNA hybrids. This Zn^{2+}-dependent enzyme operates optimally at acidic pH (4.0–4.4) and is inhibited by chelators like EDTA and phosphate ions. It is thermostable and resistant to denaturants, making it ideal for mapping single-stranded regions and processing nucleic acid ends.

4.13.2 Mung Bean Nuclease

Similar to S1 nuclease, mung bean nuclease is derived from mung bean sprouts and specifically targets single-stranded DNA and RNA regions. It generates mononucleotides and oligonucleotides with 5′-phosphate ends. While it primarily removes single-stranded overhangs, excessive enzyme concentrations or extended incubation can degrade double-stranded DNA. Optimal activity requires Zn^{2+} and reducing agents such as cysteine, while high salt concentrations inhibit its function. The enzyme is stabilized by low concentrations of detergents like Triton X-100 to prevent surface adsorption during reactions.

4.13.3 Methylases

Infectious agents, such as bacteriophages, introduce foreign DNA, which is then degraded using restriction endonucleases. To protect their DNA, bacteria add methyl groups—a process known as methylation—to prevent restriction enzymes from degrading it. This modification typically targets only a few specific nucleotides on each DNA strand and must not disrupt base pairing (Wilson 1991). Although there are some exceptions (e.g.,), DNA methylation typically prevents restriction enzymes from recognizing their target sequences (Sistla and Rao 2004).

This occurs in two main ways:

(i) Certain restriction endonucleases—such as PvuI, ApyI, Sau3AI, BstNI, BamHI, and BglII—fail to cleave DNA if Dam or Dcm methylases methylate their recognition sites.
(ii) Conversely, some bacterial restriction systems specifically target and cleave methylated DNA, helping the host discriminate between its own unmethylated genome and potentially harmful, methylated phage DNA.

In molecular cloning, methylases are frequently employed to protect DNA from restriction digestion, particularly when constructing cDNA or genomic libraries. For example, when cloning a fragment into a BamHI site of a vector, BamHI methylase is used to methylate internal BamHI sites within the vector. This protects them from cleavage when BamHI linkers are added and the DNA is subsequently digested with BamHI endonuclease (Rittié and Perbal 2008).

Methylase can block restriction endonucleases from several RM systems; for example, TaqI methylase inhibits BamHI. The impact of methylation on endonuclease cleavage specificity and activity varies depending on the degree of overlap between the restriction and methylation sequences. There are two types of overlaps: in one, methylation alters cleavage selectivity, and the methylation sequence fully overlaps with the restriction enzyme's recognition sequence. In the other, nucleotide methylation changes the recognition sequence itself, providing resistance to endonuclease activity, with the methylation sequence partially overlapping the restriction site (Rittié and Perbal 2008).

Two types of overlaps between methylase activity and restriction enzyme recognition can affect DNA digestion.

The first type occurs when a methylase modifies only one of several possible recognition sequences for a restriction endonuclease, leading to selective protection. For example, the recognition sequence for HincII is GTPyPuAC, allowing it to cleave DNA at four possible sites: GTCGAC, GTCAAC, GTTGAC, and GTTAAC. The TaqI methylase (M.TaqI) specifically methylates the adenine in the TCGA sequence. As a result, only TCGA-containing sites are methylated and thus protected from HincII cleavage, while the other three variants remain susceptible. This kind of selective methylation can effectively redefine the specificity of certain restriction enzymes within duplex DNA.

The second type of overlap involves partial sequence overlap between a methylase recognition site and a restriction enzyme recognition site. A notable example is the overlap between the BamHI recognition sequence (GGATCC) and the methylation site of M.MspI (CCGG). The G of the BamHI site falls within the CCGG sequence recognized by M.MspI. When M.MspI methylates the internal cytosine of this site—transferring a methyl group to the 5′ cytosine residue—it renders the overlapping BamHI site resistant to cleavage (Nelson et al. 1984).

4.14 Additional Enzymes

4.14.1 Polynucleotide Adenylyl-Transferase

The 58 kDa polyA polymerase from *E. coli* synthesizes adenylate residues at the 3′ end of polyribonucleotides without a template. It is used for in vitro RNA 3′ extension for cDNA synthesis or RNA purification with poly-dT. Depending on conditions, it can produce polyA stretches from 20 to 2000 nucleotides. *E. coli* PolyA polymerase adds adenosine monophosphates (AMPs) using ATP. CTP and UTP are also substrates, but at a rate less than 5% of ATP's rate; ADP, dATP, and GTP are not substrates. The enzyme is inhibited by phosphate, pyrophosphate, and aurin tricarboxylic acid, which blocks RNA polymerase and mRNA binding. It requires high monovalent cation concentrations (such as 400 mM NaCl) for activity, is stimulated by Mn^{2+}, and is resistant to antibiotics like streptolydigin and rifampicin, which inhibit transcription. Unlike polynucleotide phosphorylase, polyA polymerase does not degrade its product. It can use various single-stranded RNAs as primers, though polyUG, polyC, and double-stranded RNA are poor primers, and DNA cannot serve as a primer. Most poly (A) polymerases share this characteristic, except for plant enzymes from maize, which can use both natural and synthetic oligo- and poly-dNTPs.

4.14.2 Guanylyl Transferase

The guanylyl transferase complex derived from vaccinia virus functions as a multifunctional RNA capping enzyme. It catalyzes the addition of a cap structure to the 5′ ends of RNA molecules bearing either diphosphate or triphosphate termini, as well as to 5′ ends from which the terminal 7-methylguanosine (m^7G) cap has been removed. This enzyme complex exhibits three distinct catalytic activities: (i) RNA 5′-triphosphatase, which removes the γ-phosphate from 5′ triphosphorylated RNA; (ii) guanylyltransferase, which transfers GMP from GTP to the diphosphorylated RNA end; and (iii) RNA (guanine-N7)-methyltransferase, which uses S-adenosylmethionine to methylate the guanine at the N7 position, forming the mature m^7G cap.

Notably, the guanylyl transferase complex does not act on RNA substrates with a monophosphate at the 5′ end, ensuring that only intact or decapped 5′ ends with diphosphate or triphosphate groups are substrates. This specificity allows for selective labeling of damaged or nicked RNA cap structures.

Enzyme dosage should be optimized empirically for each experimental context to ensure efficient capping. Furthermore, to confirm that labeling occurs specifically at the cap structure, the capped RNA can be treated with tobacco acid pyrophosphatase. Following digestion, successful labeling will be verified by the release of labeled 5′ GMP, confirming incorporation into a genuine cap structure (Rittié and Perbal 2008).

4.14.3 Topoisomerase I

To aid in relaxing supercoiled DNA, topoisomerase I, a 105 kDa enzyme (Liu and Miller 1981), transiently cuts one strand of the DNA sugar-phosphate backbone before religating the ends. This nuclear enzyme is crucial for cellular DNA activities, including recombination, replication, and DNA-protein interactions. The process results in increased supercoiling upstream of the unwinding point (Champoux 2001). In vitro, topoisomerase I was first utilized to generate circular DNA (Martin et al. 1983), study DNA tertiary conformation (Peck and Wang 1981), and facilitate nucleosome assembly. Later research revealed its ability to catalyze the covalent transfer of a single-stranded DNA (donor) to a different DNA molecule (acceptor) (Halligan et al. 1982). The enzyme forms a covalent complex with DNA when it binds single-stranded DNA, allowing ligation of the single-stranded fragment to the acceptor's 5′-OH end (Halligan et al. 1982). 0.2% SDS halts topoisomerase activity but remains active at pH 7.5 in the presence of 50–200 mM NaCl. Furthermore, 5–10 mM Mg^{2+} enhances its covalent attachment to DNA.

New cloning vectors, known as TOPO® Cloning Vectors, have been created by leveraging the dual function of topoisomerase I as both a ligase and a nuclease. This enzyme, derived from vaccinia virus and used in the TOPO® cloning method, specifically cleaves double-stranded DNA at the end of a (C/T)CCTT sequence and forms a covalent bond with the 3′-phosphate of the thymidine in the generated DNA. An acceptor DNA with compatible ends can then be joined through topoisomerase I's ligase activity. Invitrogen's TOPO® vectors are linear duplex DNA molecules, featuring a GTGG overhang on one strand and a covalently attached topoisomerase I to the 3′-phosphate on the other strand. PCR products prepared with a forward primer containing four extra bases (CACC) at the 5′ end are suitable for cloning. The topoisomerase I enzyme then ligates the PCR product in the proper orientation once the overhang (GTGG) in the cloning vector hybridizes to the 5′ end of the PCR product and anneals to the additional bases. Furthermore, the blunt 3′ end of the PCR product is also ligated by topoisomerase I.

4.14.4 Topoisomerase II

The purified enzyme topoisomerase II from M. Luteus facilitates the breakage and rejoining of both duplex DNA strands, altering the total number of supercoiled DNA forms by two (Brown et al. 1979; Liu and Miller 1981). Moreover, undamaged circular DNA can be reversibly knotted by topoisomerase II (Brown et al.

1979; Liu and Miller 1981). These multimeric proteins, known as type II topoisomerases, require ATP to function effectively. Etoposide, a chemotherapeutic agent that slows the proliferation of rapidly dividing cancer cells, inhibits type II topoisomerase activity (Baldwin and Osheroff 2005).

References

Baldwin EL, Osheroff N. Etoposide, topoisomerase II and cancer. Curr Med Chem Anticancer Agents. 2005;5:363–72. https://doi.org/10.2174/1568011054222364.

Baltimore D. RNA-dependent DNA polymerase in virions of RNA tumour viruses. Nature. 1970;226:1209–11. https://doi.org/10.1038/2261209a0.

Brock TD, Freeze H. Thermus aquaticus gen. n. and sp. n., a nonsporulating extreme thermophile. J Bacteriol. 1969;98:289–97. https://doi.org/10.1002/path.1710980409.

Brown PO, Peebles CL, Cozzarelli NR. A topoisomerase from Escherichia coli related to DNA gyrase. Proc Natl Acad Sci USA. 1979;76:6110–4. https://doi.org/10.1073/pnas.76.12.6110.

Champoux JJ. DNA topoisomerases: structure, function, and mechanism. Annu Rev Biochem. 2001;70:369–413. https://doi.org/10.1146/annurev.biochem.70.1.369.

Crick FH. On protein synthesis. Symp Soc Exp Biol. 1958;12:138–63.

Eidne KA. Molecular cloning. In: Principles of medical biology, vol. 5. Elsevier; 1996. p. 289–305.

Englund PT. Analysis of nucleotide sequences at 3′ termini of duplex deoxyribonucleic acid with the use of the T4 deoxyribonucleic acid polymerase. J Biol Chem. 1971;246:3269–76.

Gellert M. DNA topoisomerases. Annu Rev Biochem. 1981;50:879–910. https://doi.org/10.1146/annurev.bi.50.070181.004311.

Green MR, Sambrook J. Molecular cloning: a laboratory manual. 4th ed. Cold Spring Harbor Laboratory Press; 2012.

Gueguen Y, Rolland JL, Lecompte O, Azam P, Le Romancer G, al Flament D e. Characterization of two DNA polymerases from the hyperthermophilic euryarchaeon Pyrococcus abyssi. Eur J Biochem. 2001;268:5961–9. https://doi.org/10.1046/j.0014-2956.2001.02550.x.

Halligan BD, Davis JL, Edwards KA, Liu LF. Intra- and intermolecular strand transfer by HeLa DNA topoisomerase I. J Biol Chem. 1982;257:3995–4000.

Ijaz S. Mitochondrial plasmid: could this be the next hit for genetic engineering? Genet Mol Res. 2011;10(1):391–2.

Ijaz S, Haq IU. Recombinant DNA technology. Cambridge Scholars Publishing; 2019.

Liu LF, Miller KG. Eukaryotic DNA topoisomerases: two forms of type I DNA topoisomerases from HeLa cell nuclei. Proc Natl Acad Sci USA. 1981;78:3487–91. https://doi.org/10.1073/pnas.78.6.3487.

Mark DF, Richardson CC. Escherichia coli thioredoxin: a subunit of bacteriophage T7 DNA polymerase. Proc Natl Acad Sci USA. 1976;73:780–4. https://doi.org/10.1073/pnas.73.3.780.

Martin SR, McCoubrey WK Jr, McConaughy BL, Young LS, Been MD, Brewer BJ, et al. Multiple forms of rat liver type I topoisomerase. Methods Enzymol. 1983;100:137–44. https://doi.org/10.1016/0076-6879(83)00050-6.

Martins A, Shuman S. Characterization of a baculovirus enzyme with RNA ligase, polynucleotide 5′-kinase, and polynucleotide 3′-phosphatase activities. J Biol Chem. 2004a;279:18220–31. https://doi.org/10.1074/jbc.M313386200.

Martins A, Shuman S. An RNA ligase from Deinococcus radiodurans. J Biol Chem. 2004b;279:50654–61. https://doi.org/10.1074/jbc.M407657200.

Melton DA, Krieg PA, Rebagliati MR, Maniatis T, Zinn K, Green MR. Efficient in vitro synthesis of biologically active RNA and RNA hybridization probes from plasmids containing a bacteriophage SP6 promoter. Nucleic Acids Res. 1984;12:7035–56. https://doi.org/10.1093/nar/12.18.7035.

Nelson M, Christ C, Schildkraut I. Alteration of apparent restriction endonuclease recognition specificities by DNA methylases. Nucleic Acids Res. 1984;12:5165–73. https://doi.org/10.1093/nar/12.13.5165.
Peck LJ, Wang JC. Sequence dependence of the helical repeat of DNA in solution. Nature. 1981;292:375–8. https://doi.org/10.1038/292375a0.
Reddy R, Henning D, Epstein P, Busch H. Primary and secondary structure of U2 snRNA. Nucleic Acids Res. 1981;9:5645–58. https://doi.org/10.1093/nar/9.21.5645.
Rittié L, Perbal B. Enzymes used in molecular biology: a useful guide. J Cell Commun Signal. 2008;2:25–45.
Roychoudhury R, Jay E, Wu R. Terminal labeling and addition of homopolymer tracts to duplex DNA fragments by terminal deoxynucleotidyl transferase. Nucleic Acids Res. 1976;3:863–77.
Schenborn ET, Mierendorf RC Jr. A novel transcription property of SP6 and T7 RNA polymerases: dependence on template structure. Nucleic Acids Res. 1985;13:6223–36. https://doi.org/10.1093/nar/13.17.6223.
Seeger C, Mason WS. Hepatitis B virus biology Microbiol Mol Biol Rev. 2000;64:51–68. https://doi.org/10.1128/MMBR.64.1.51-68.2000.
Sistla S, Rao DN. S-Adenosyl-L-methionine-dependent restriction enzymes. Crit Rev Biochem Mol Biol. 2004;39:1–19. https://doi.org/10.1080/10409230490440532.
Stenesh J, Roe BA, Snyder TL. Studies of the deoxyribonucleic acid from mesophilic and thermophilic bacteria. Biochim Biophys Acta. 1968;161:442–54.
Tabor S, Richardson CC. DNA sequence analysis with a modified bacteriophage T7 DNA polymerase. Proc Natl Acad Sci USA. 1987;84:4767–71. https://doi.org/10.1073/pnas.84.14.4767.
Tabor S, Richardson CC. Selective inactivation of the exonuclease activity of bacteriophage T7 DNA polymerase by in vitro mutagenesis. J Biol Chem. 1989;264:6447–58.
Tabor S, Huber HE, Richardson CC. Escherichia coli thioredoxin confers processivity on the DNA polymerase activity of the gene 5 protein of bacteriophage T7. J Biol Chem. 1987;262:16212–23.
Temin HM, Mizutani S. RNA-dependent DNA polymerase in virions of Rous sarcoma virus. Nature. 1970;226:1211–3. https://doi.org/10.1038/2261211a0.
Wilson GG. Organization of restriction-modification systems. Nucleic Acids Res. 1991;19:2539–66. https://doi.org/10.1093/nar/19.10.2539.

Chapter 5
Genome Modification: Methods and Applications

Amjad Abbas, Muhammad Usman, Muhammad Jabran, Javeria Yasin, Muhammad Ehetisham-Ul-Haq, Muhammad Huzaifa Tanveer, Hafiza Arooj Razzaq, and Muhammad Amjad Ali

5.1 Introduction

Genetic modification of crops was first subjected to criticism, with claims that it was unnatural or even dangerous, and was dubbed "Frankenstein crops." At that time, the word "Frankenstein" symbolized things that are unnatural or tampered with. However, at the same time, ancient humans were working to alter the genome to produce crops that were qualitatively and quantitatively better. Genome alteration was first associated with basic selective breeding techniques, but more recently, it has been sped up through modern methods that now speed up the process in an accurate and regulated manner. These modern techniques that enhance genomic modifications are crucial to meet global crop demands, as Tilman et al. (2011) speculate that, with an increasing population, the need for crop production might double by 2050. While the annual growth rate of major crops recorded in 2013 was wheat 0.9%, rice 1.0%, soybean 1.3%, and maize 1.6% (Ray et al. 2013). Factors such as reduced arable land, water scarcity, climate change, and rising biofuel consumption may hinder future crop production. Apart from crop yield, the development of new, better-performing varieties, good in terms of traits,

A. Abbas · M. Usman · J. Yasin · M. H. Tanveer · M. A. Ali (✉)
Department of Plant Pathology, University of Agriculture, Faisalabad, Pakistan
e-mail: amjad.ali@uaf.edu.pk

M. Jabran
State Key Laboratory for Biology of Plant Disease and Insect Pests, Institute of Plant Protection, Chinese Academy of Agricultural Sciences, Beijing, China

M. Ehetisham-Ul-Haq
Oilseeds Research Institute, Ayub Agricultural Research Institute, Faisalabad, Pakistan

H. A. Razzaq
Center of Agricultural Biochemistry and Biotechnology, University of Agriculture, Faisalabad, Pakistan

S. Ijaz et al. (eds.), *Molecular Biology Methods*,
https://doi.org/10.1007/978-981-92-2642-9_5

nutrition, stress tolerance, disease resistance, and with less input requirements is also the choice of the era. However, traditional breeding methods to improve crops mainly rely on spontaneous genetic variation or on chemical and physical mutagenesis agents, and may use T-DNA insertional mutagenesis or transposon tagging techniques (Songstad et al. 2017) as natural genetic modification and artificial mutagenesis are limited by their random nature, low efficacy, and time- and labor-intensive nature. While gene targeting via homologous recombination allows precise mutation, its integration frequency is low and limited to a few crops (Puchta and Fauser 2013). In addition, genome editing techniques are revolutionizing crop improvement with high efficiency. The edited crops are not only transgene-free but also indistinguishable from those obtained from traditional techniques (Songstad et al. 2017).

Transgenesis, Cisgenesis, and Intragenesis Transgenes are a new genetic improvement strategy for breeders or geneticists, commonly employed in genetically modified (GM) plants. They are made up of foreign genes or portions of foreign genes (Joshi et al. 2011). A wide range of foodstuffs and other materials can be obtained through GM plants. However, GM crops pose risks under strict genetically modified organism (GMO) legislation and are not widely accepted by consumers. The major concern with transgenic crops is the mixing of genetic material from two species that do not hybridize naturally or are sexually incompatible. To address this concern, the concept of incorporating changes through the transformation of genetic material from the same species, from different genotypes of the same species, or from related species that can undergo hybridization or are sexually compatible (Holme et al. 2013) emerged. This concept is called cisgenesis or intragenesis (Holme et al. 2013). Conventional breeding strategies are commonly adopted, but gene delivery via Agrobacterium-mediated or biolistic methods is also used. Cisgenesis or intragenesis reduced selection step duration and linkage drag with generally unchanged genotype and phenotype (Jogdand et al. 2017). The risks associated with cisgenic crops are similar to conventional breeding (EFSA Panel on Genetically Modified Organisms (GMO) 2012). Even these are more acceptable to consumers than transgenic crops (Delwaide et al. 2015). The genetic approaches used to improve different traits in different crops are given in Table 5.1.

Agrobacterium-Mediated Transformation This story began with a detailed study of crown gall-producing bacteria, *Agrobacterium tumefaciens* (Smith and Towsend 1907), and its oncogenic behavior in different plants, especially in the rose family, i.e., peach, pear, cherry, apple, raspberry, almond, and roses. Thanks to Binns and Thomashaw (1988), who explain the ability of *A. tumefaciens* to transfer a DNA segment, namely transfer DNA (T-DNA), from its tumor-inducing (Ti) plasmid into the nucleus of the infected cell. This T-DNA integrates into the host genome. This study paved the way toward genetic engineering. Scientists use this natural ability of *A. tumefaciens* to get desired results (Opabode 2006).

Table 5.1 Genetic approaches to improve different traits in different crops

Crop species	Trait(s)	Gene(s)	Approach(s)	Reference(s)
Alfalfa	Lignin content	Comt	Intragenesis Transgenesis	Weeks et al. (2008)
	Water deficit stress	Mn superoxide dismutase	Transgenesis	McKersie et al. (1996)
	Salt and drought stress tolerance	GsZFP1	Transgenesis	Tang et al. (2013)
Apple	Scab resistance	HcrVf1, HcrVf2, Rvi6	Cisgenesis	Joshi et al. (2011), Vanblaere et al. (2014) and Würdig et al. (2015)
	Fire blight resistance	FB_MR5		Kost et al. (2015)
Barley	Grain Phytase activity	HvPAPhy_a	Cisgenesis	Holme et al. (2012)
	Frost tolerance	TaDREB3	Transgenesis	Kovalchuk et al. (2013)
Durum wheat	Baking quality	1Dy10	Transgenesis	Gadaleta et al. (2008)
	Salt and drought tolerance	*Td*PIP2;1	Transgenesis	Ayadi et al. (2019)
Ryegrass	Drought tolerance	Lpvp1	Intragenesis, Cisgenesis	Bajaj et al. (2008), Sathish et al. (2007)
	Salt tolerance	Vacuolar Membrane Na^+/H^+ antiporter gene	Transgenesis	Wu et al. (2005)
Pear	Fire blight resistance	Attacin E	Transgenesis	Reynoird et al. (1999)
Poplar	Plant growth and stature, wood properties	PtGa20ox7, PtGa2ox2, PtRGL1_1, PtRGL1_2, PtGA11	Cisgenesis	Han et al. (2011)
	Environmental stresses (drought, waterlogging, salinity, and leaf beetle)	*vgb*, *SacB*, *JERF36*, *BtCry3A*, and *OC-I*	Transgenesis	Su et al. (2011)
Potato	Late blight resistance	Rpi-sto1, Rpi-vnt1.1, GBSS	Cisgenesis Intragenesis	Jo et al. (2014) de Vetten et al. (2003)
	Blackspot	Polyphenol oxidase	Intragenesis	Rommens (2004)
	Accumulation of reducing sugars after cold storage and acrylamide after high temperature processing	R1, PhL, StAs1, StAS2	Intragenesis	Chawla et al. (2012), Rommens et al. (2004) and Rommens et al. (2006)
	Blackleg	Msr A1	Transgenesis	Osusky et al. (2000)

(Continued)

Table 5.1 (Continued)

Crop species	Trait(s)	Gene(s)	Approach(s)	Reference(s)
Strawberry	Gray mold resistance	PGIP	Intragenesis	Schaart (2004)
	Wilting	pcht28	Tansgenesis	Chalavi et al. (2003)
Tomato	Bio fortification	BCAAs	Cisgenesis	Vazquez-Vilar et al. (2023)
	Salt tolerance	BADH	Transgenesis	Goel et al. (2010)

5.2 Plant Breeding to Enhance Plant Disease Resistance

The demand for food is increasing in terms of quantity and quality. The term quality emphasizes nutritional balance and healthy diet. Thus, improvement in any plant type seems simple if we think about a single trait, but several traits need to be addressed at the same time. The expertise and knowledge behind this intervention is a combined effort as a multidisciplinary advancement by targeting several traits improvement (Bueren 2016). To respond to the ever-increasing demand for food, our understanding and background knowledge become more important.

The aim of crop improvement is to find and explore the genetic variations present behind a trait or phenotypic variation (Haussmann et al. 2004). With high confidence after dealing with such variations, the next step is to select desired plant types with a more profound countenance, with potential to be stable in the stand under biotic and abiotic stresses, and then finally probability toward multiplication for commercial growers and adaptation (Caligari and Forster 2001).

In self-pollinated crops, the purity of a cultivar is important and pure lines fulfilling the optimum performance criterion of target traits and objectives of a breeding program are used as commercial varieties after all testing and approvals. Developing hybrids in wheat and rice are also common. To exploit natural variations is an important task during the breeding of a self-pollinated crop. Selection of promising plant types is practiced, which is further diversified through introduction and then selection process (Begna 2021).

For the introduction of germplasm, it is important to choose the plant types with high heritability, and the program is further strengthened by conducting adaptability trials before release of a variety based on selection, because local adaptation is the key indicator for the success of this genetic intervention (Lyon et al. 2020). The selection process requires expertise to execute because it becomes difficult to reject similar plant types with minor differences in target traits. Selection is continuous in the generations until the plant types with the desired and uniform make up are achieved. The varieties that are developed in the self-pollinated crops through the selection procedure are more durable, and the genetic purity is more stable compared to hybrid concepts in self-pollinated crops. Additionally, such varieties are well adapted to local conditions, and they tend to have comparatively higher resistance to insects and diseases. This process is simple, cost-effective, and does not involve complex breeding techniques and expertise (Hussain et al. 2021). In the

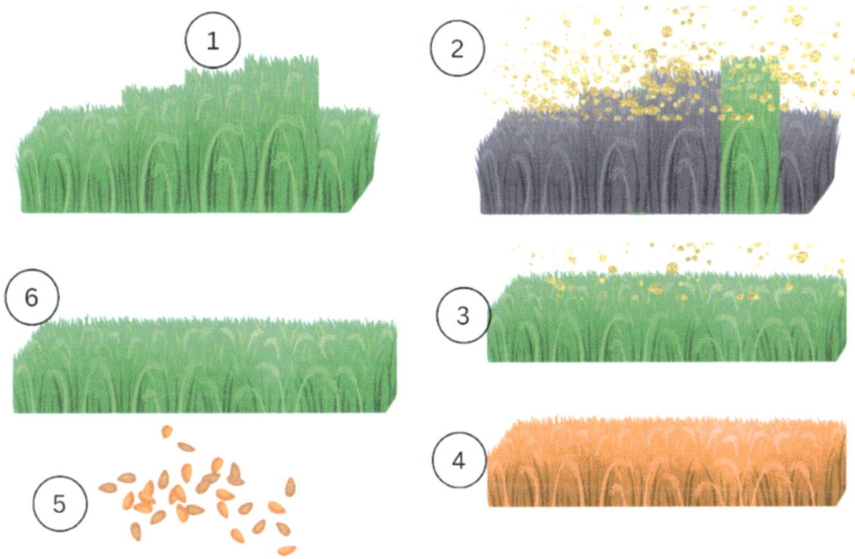

Fig. 5.1 Screening of wheat germplasm with an objective to find the rust resistance selection where 1. Shows planting diverse germplasm; 2. Application of high inoculum pressure of rust spores followed by selection of surviving genotypes of wheat; 3. Second round of screening followed by; 4. Crop maturity; 5. Harvesting of seed; and 6. Sowing of candidate rust resistant line which could be used in release of cultivar after meeting all other requirements

scenario of disease resistance, for instance, in case of leaf rust and wheat plant-pathogen system, this practice is common to screen the germplasm under high pressure of rust inoculum through artificial applications and by using susceptible disease spreader lines. The disease screening is shown in Fig. 5.1. The surviving lines are kept after disease screening and could be used in the development of a variety (Thakur et al. 2007).

The exposure of target germplasm to leaf rust results in the availability of candidate resistant lines. But the evolution of rusts is faster and more dynamic, which results in new pathogenic races. Under such conditions, it becomes important to combine different resistant genes in a high-yielding cultivar, which is compromised with disease susceptibility. The resistant gene is fixed in the high-yielding through crossing. The F1 is backcrossed to the recurrent parent for several generations to revive the productive cultivar, which is now equipped with resistant genes from a donor parent. It is shown in Fig. 5.2.

There are a number of ways to improve the genetics of plants through self- and cross-pollination methods as described in this section. These methods mostly involve parents, which are of the same genus and species, and further selection, which depends on the ability of breeders. These human efforts resulted in the improvement of yield contributing attributes of the plant, enhanced defense to diseases and insect pests, and responsive plant types to mechanized farming.

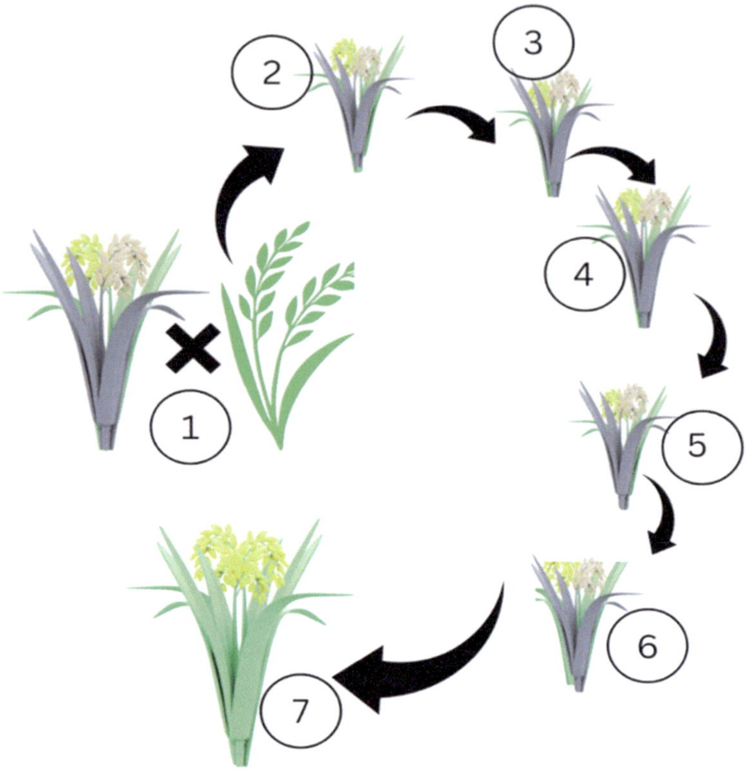

Fig. 5.2 The resistant gene is fixed in the high-yielding through crossing. The F1 is backcrossed to the recurrent parent for several generations to revive the productive cultivar, which is now equipped with resistant genes from a donor parent

In cross-pollinated crops, open-pollinated varieties are common, which are improved through recurrent selection and are maintained by open pollination. The genetic base of these Open-Pollinated Varieties (OPVs) is broad, and maintenance in the farmers' field is possible. Usually, such varieties are moderate in production, and they need regular improvement. Since the genetic background of these varieties is diverse, they show higher adaptability under biotic and abiotic stress. On the other hand, hybrids exploit the concept of heterosis or hybrid vigor and exceed in the performance of traits than both the parents (Bar-Zvi et al. 2017). The vigor of hybrids is attributed to the phenomenon of heterosis. Hybrids are F1 generation seeds which are given to sow in farmer fields, and their performance changes in the segregating generations. So, in each cropping season, there is a need to plant fresh seeds of hybrids. This also necessitates maintaining the parental inbred lines and performing the seed development under controlled fields (Afzal et al. 2019). Parents, which are inbred lines should be pure for the desired traits for the development of hybrids. The plant stand of hybrids is more uniform, and they are exceptional in performance. Inbred lines are developed through repeated self-pollination until genetic purity is achieved. One line is used as the male parent, while the other is used as the female parent. The female parent bears the F1 seed after crossing with the male parent. The female is kept pure by removing the male part to prevent self-pollination by the process of detasseling, and designated male parent is used for fertilization, and the ultimate seed will be F1 or hybrid seed (Yadav et al. 2022).

5.3 Use of Marker-Assisted Selection

The female parent is the inbred line on which the hybrid seed is produced. Its tassels (male parts) are removed through a process called detasseling to prevent seed formation through self-pollination. The exposed silks (female flowers) then receive pollen from the selected male parent. The seeds formed on these female plants are the F_1 hybrid seeds, which are later harvested and sold to farmers. Conventional breeding relies on the phenotypic data and the selection process. For some traits, such selection could be misleading, for instance, in case of selection for disease resistance, there could be a chance of selection of a plant that has escaped disease conditions or could not be confronted with disease conditions (Xu and Crouch 2008).

In this scenario and for fast results, instead of entirely depending on visual selection and physiological markers, some molecular tools like marker-assisted selection (MAS) are included in the breeding program, where molecular markers are used to assist the selection process. These markers are PCR-based and are easy to use in the selection process. In the MAS, the selection process could be performed much earlier and more accurately. Findings of MAS of the germplasm are correlated with field performance for some important traits like disease resistance. This approach is also helpful for the pyramiding of resistance or R-genes in a single line for more profound resistance against several pathogenic races of a particular pathogen.

When conventional sources like plant breeding are not useful due to lack of variation or heritability in certain plant traits, then there is a dire need to fix or enhance such traits for optimum financial benefits in plants; then a transgenic approach could be an option. Knowledge of the problem and possible solutions is very important during the improvement plan for plants. From the literature and previous research on the issue under consideration, the transgenic approach becomes applicable for the improvement of plant traits. For instance, if we want to introduce disease resistance in a rice plant, which lacks the natural potential of resistance in the available germplasm, then selection of a donor plant having an established edge in performance is chosen for transforming and genetic manipulation from a distantly related plant species like tomato. The process begins with the selection of a suitable donor organism from which a gene of interest, such as one conferring disease resistance or a specific trait, is to be isolated.

The genomic DNA of the donor plant is isolated, and specific disease resistance gene or gene of interest is amplified through PCR by using gene-specific primers. Confirmation of the PCR product is important before cloning into the binary vector, and gel electrophoresis and sequencing are sufficient for this confirmation. The binary vector and newly amplified gene of interest are double-digested with the endonuclease restriction enzymes. These enzymes produce either sticky or blunt end, depending on the restriction site of the enzymes. These fragments are joined end-to-end by using T4 DNA ligase enzyme. The newly synthesized construct is transformed into the competent chemical cells of *E. coli* for multiplication of the construct. The selection of transformed *E. coli* cells is performed by using drug-resistant markers like kanamycin.

The gene of interest is placed between the left and right T-DNA borders of the disarmed vector, and it is coupled with crucial regulatory sequences, including a promoter, often the Cauliflower Mosaic Virus 35S promoter for constitutive expression, and a terminator sequence. Additionally, the vector is equipped with a selectable marker gene.

This constructed plasmid vector is then introduced into competent cells of *Agrobacterium tumefaciens*, a soil bacterium naturally capable of transferring DNA into plant genomes, through a method like electroporation or freeze-thaw. The transformed *Agrobacterium* cells are selected on antibiotic-containing media to ensure only bacteria carrying the plasmid are grown and can be confirmed through PCR. For the transformation itself, healthy recipient plants are grown until they produce numerous flowers. The entire aerial parts of these plants, or just the developing floral buds, are submerged in a liquid culture of the engineered *Agrobacterium* for a few minutes. This method, known as the floral dip technique, relies on the bacterium to transfer the T-DNA region containing the gene of interest from the plasmid into the genome of the plant's egg cells or early embryos.

After dipping, the plants are kept in a humid environment for a day before being returned to normal growth conditions to allow seed development. The harvested seeds, representing the first generation (T1), are then sown on a selective medium, typically agar plates containing the appropriate antibiotic, like kanamycin. Seeds that have not integrated the T-DNA will die, while the transgenic seedlings, which express the selectable marker and thus survive, are identified by their healthy green appearance and transferred to soil. These selected T1 plants are grown to maturity, and their progeny (T2 generation) are analyzed to confirm the stable integration and inheritance of the transgene. This confirmation is achieved through PCR to detect the presence of the gene and other methods like Southern blotting to verify the integration event, ultimately resulting in a population of transgenic plants for further phenotypic and molecular analysis, as shown in Fig. 5.3.

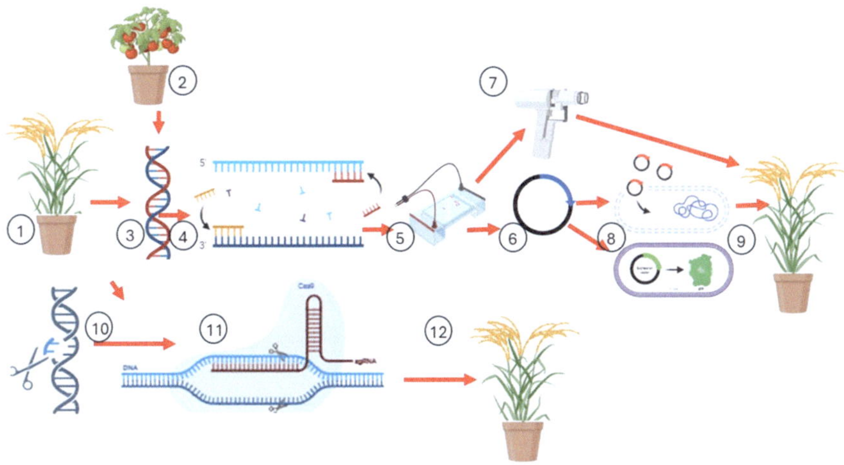

Fig. 5.3 The steps involved in the production of genetically modified disease resistant plants, where 1. is rice plant facing issue of bacterial invasion; 2. doner plant with resistant gene; 3. DNA isolated from the doner plant; 4. resistant gene is amplified through PCR; 5. PCR product is confirmed through agarose gel electrophoresis; 6. gene is cloned in the binary vector; 7. gene is transferred through gene gun to the recipient plant; 8. the cloned plasmid is multiplied in the *E. coli* bacteria and then transformed in the Agrobacterium; 9. Agrobacterium mediated transformation of the gene into plant; 10. targeting gene by CRISPER Cas 9 and fixed in the rice plant

5.4 Advances in CRISPR/Cas-Based Genome Editing

The CRISPR/Cas system has quickly established itself as one of the most cutting-edge tools in modern plant biotechnology. It was discovered as part of the adaptive immune system in bacteria and serves as protection against viral diseases by cutting the viral DNA. In plants, the system has been developed into a powerful molecular scissor that edits any gene. The CRISPR mechanism employs a Cas endonuclease (Cas9 or Cas12a) guided by a small RNA (sgRNA) that binds to a matching segment of DNA, resulting in double-stranded breaks in the DNA. The plant utilizes its inherent DNA repair processes to mend the breaks through predetermined methods, including Non-Homologous End Joining (NHEJ), which usually results in a knock-out of the gene or Homology-Directed Repair (HDR) which allows for the insertion of a precise sequence edit or correction to give the desired trait.

CRISPR/Cas evolved from a mechanism of microbial defense to a programmable genome-editing tool when scientists demonstrated a Cas endonuclease could be directed by short RNA to cut DNA at user-defined sites. That modularity of RNA → protein allowed rapid deployment of CRISPR as a general editing system (Jinek et al. 2012). The immediate challenge was to adapt the system for eukaryotic cells, and work developing single-guide RNA (sgRNA) structures and editing in mammalian cells made CRISPR viable for complex organisms and multiplexed targeting. Those early adaptations established workflows for designing guides, delivering nucleases, and leveraging cellular repair pathways (NHEJ and HDR) to create knockouts and targeted insertions (Cong et al. 2013; Mali et al. 2013).

Expanding the menu of effectors beyond the typical SpCas9 was a significant milestone. Cas12a (Cpf1) was described as a single-RNA class-2 endonuclease that recognizes T-rich PAMs and makes staggered cuts, creating an alternate window for targeting and newer outcomes in repair that are beneficial for multiple genomes. The Cas12 family also made the guide architecture simpler through the elimination of a tracrRNA in some of the systems (Zetsche et al. 2015). A complementary advance was the discovery and engineering of RNA-targeting CRISPR effectors (Cas13 family), which allowed for programmable RNA knockdown and RNA editing without the alteration of genomic DNA. RNA targeting provided a transient and reversible means to change gene expression and combat RNA viruses in plants and animals (Abudayyeh et al. 2016).

The concern of reducing off-target activity grew to an urgent translational application need. Structure-guided protein engineering yielded high-fidelity Cas9 variants that greatly reduced nonspecific cleavage while preserving on-target activity. These engineered enzymes (SpCas9-HF1, eSpCas9 variants) are now recommended for sensitive applications and clinical pipelines (Kleinstiver et al. 2016; Slaymaker et al. 2016). In parallel to the growing number of engineered nucleases, empirical genome-wide off-target detection methods were developed (GUIDE-seq), which permit unbiased mapping of double-strand breaks in living cells. These validation metrics are now the gold standards for evaluation of new nucleases, guide designs, and therapeutic candidates (Tsai et al. 2015).

Guide-RNA design itself has improved tremendously with the development of large-scale screening datasets and machine learning models. Now using deep learning or ensemble models trained on tens of thousands of guides, on-target efficiencies and increasingly off-target risk can be easily predicted, thereby reducing failed experimental trials and fueling an increased level of high-throughput functional screens (Lee 2023; Wang et al. 2019). Improvements in multiplexing capacity, editing multiple loci at once, through more efficient array designs and multiplexed guide expression cassettes allow us to engineer complex traits (stacking resistance alleles, rewiring combinatorial pathways) in a single generation. This is particularly salient in polyploid crops, as these often require multi-allelic edits (Cong et al. 2013; Liang et al. 2017).

Discovery of hypercompact effectors was driven by size-constrained delivery, bacteriophage-derived CasΦ (Cas-phi) and other small Cas12-like proteins deliver crRNA processing and target DNA cleavage, in a single, small protein, simplifying packaging into viral vectors or nanocarriers, while expanding use cases for in vivo as well as direct plant delivery. Delivery is also expanded in multi-allelic editing because of expanded PAM compatibility and reduced delivery barriers by hypercompact systems (Pausch et al. 2020). Delivery technologies have evolved concurrently; furthermore, DNA-free technology (direct delivery of Cas9–sgRNA ribonucleoprotein [RNP] complexes) are inherently less risky for integration and persistence off targets and are successfully demonstrated in crops, including wheat and also exploit nanomaterials (carbon nanotubes, carbon dots, and other nanocarriers) for delivering pared down ribonucleoproteins information from intact plant tissue without integrating into the genome, and thus providing relatively practical and regulatory friendly paths to transgene-free editing (Demirer et al. 2019; Liang et al. 2017).

Studies of structure (cryo-EM, X-ray) and mechanism have provided atomic-level information on PAM recognition, guide binding, and conformational activation of nucleases, allowing for rational redesign of PAM specificity, catalytic rates, and fidelity. These structure-function maps can explain why newer generations of editors have improved specificity and have altered targeting windows (Zhu et al. 2019; Zhu and Huang 2019). CRISPR as a technology has become a multifaceted precision platform by integrating CRISPR approaches with other adaptations, base editors for direct base conversions, prime editors for "search-and-replace" edits, and emerging AI-guided guide design approaches. These new approaches allow for organelle-based and genome-based editing applications. Continued research is aimed at breaking endogenous byproducts, increasing edit types, and improving delivery, incrementally advancing the technology toward safe clinical and large-scale agricultural use (Kleinstiver et al. 2016; Xiang et al. 2021).

The CRISPR/Cas9 platform has effectively been applied to improve rice resistance against important pathogens by editing several multigene target susceptibility genes, including Pi21, Bsr-d1, and ERF922, simultaneously. The mutant lines demonstrated sustainable resistance to blast and bacterial blight disease without adverse effects on yield or morphological traits, marking a breakthrough in crop protection while using targeted genomic approaches (Zhou et al. 2022). A temperature-tolerant version of the CRISPR/Cas12a system (Mb2Cas12a) has been used in cotton to generate over 90% editing efficiency across multiple genes under field-temperatures (22–32 °C). The system has also been applied to multiplex editing by editing up to eight loci from a single crRNA cassette, extending the modifications of this tool in complex crop genomes (Hui et al. 2024).

5.5 Multiplexed Orthologous-Based Editors (MODEs)

The Multiplexed Orthologous Based Editor (MODE) concept is centered around employing multiple discrete Cas orthologs or variants together in one transformation, with each recognizing different PAMs and guide-RNA structures. This orthogonality allows for parallel editing of different targets (a different gene, pathway, and allele) without any guide cross-talk and without the burden of traditional promoters, increasing the overall editing capacity and speed of the breeding cycle (McCarty et al. 2020). One-way MODEs could be implemented and by commercial breeders particularly, would be with crops with polygenic traits where dozens of genes are involved in yield, architecture, or stress tolerance. By deploying Cas9 for one gene set, Cas12a for another, and possibly even Cas12f for a third, a single concoction may knockout, modify, or regulate multiple gene families all at once. In rice, one paper explored 12 multiplex systems with Cas12a and achieved essentially 100%

biallelic edits at 16 different loci in T_0 plants, showcasing the strength of ortholog-based multiplexed editing in plants (Zhang et al. 2021).

MODE/ortholog development builds upon finding ways to design and use guide-RNA arrays, with efforts to quickly construct and process guide-RNA arrays. Methods for GUIDE array construction could include tRNA-based arrays, HH/HDV ribozymes, or self-processing Cas12a arrays. In addition, when using the ortholog route, each Cas-block will have distinct scaffolds of guide RNA and distinct promoter types, helping to minimize the cross-talk between each editing module for independent control of each editing module employed (McCarty et al. 2020). Researchers developed a toolkit for cereal crops (specifically, wheat and barley), which utilized optimized variants of Cas9 and Cas12a to perform multiplex editing across subgenomes. The toolkitAQ8 was able to target three genes at the same time. Mutagenesis using Cas9, T_0 wheat plants resulted in mutations across all three subgenomes (90%). Thus, multiplex editing of orthologs is clearly an achievable capability in a complex genome (Lawrenson et al. 2024).

MODEs provide support for stacked trait development in a single generation. For example, the editing construct is engineered to have Cas9 target genes for disease-resistance, Cas12a target gene promoters for drought tolerance, and Cas12f to edit loci for oil biosynthesis. Ultimately, the results provide evidence that one can improve a cultivar by developing multiple trait categories at the same time instead of using the more conventional breeding generations with sequential breeding approaches (Khan and Sallard 2023). The opportunity for utilizing ortholog diversity further reduces guide RNA saturation risk and reduces the risk of guide RNA overlap at the off-target or on-target site because the Cas enzymes have different PAMs and structural domains. The rice call showed that while SaCas9 and other orthologs failed, Mb3Cas12a successfully performed editing in the rice material, showing the benefit of choice among multiple orthologs for targets and efficiency (Illa-Berenguer et al. 2023).

The application of MODEs in commercial breeding pipelines will require efficient cloning and delivery strategies. However, the development of new techniques like single-vector systems that encode multiple Cas orthologs and independent guide arrays, as well as modular cloning kit (GoldenGate, MoClo), allow for compatibility and higher throughput assembly of these complex constructs and reduces experimental burden to breeders (McCarty et al. 2020). Although the developments have high throughput possibilities designated models (MODEs) provide high potency, there are motivations related to phenotypic expression balance across different Cas orthologs, limiting potential interactions, restoring developmental adequacy in which the MODE is required to reach the germline and confirmation of off-target or undesired activity in multiplex scenarios. Nevertheless, to combine design rules and validation workflows enable MODEs to become a new normal for scale plant improvement and alteration in regards to multiple traits (Zhang et al. 2023).

References

Abudayyeh OO, Gootenberg JS, Konermann S, Joung J, Slaymaker IM, Cox DB, Shmakov S, Makarova KS, Semenova E, Minakhin L, Severinov K, Regev A, Lander ES, Koonin EV, Zhang F. C2c2 is a single-component programmable RNA-guided RNA-targeting CRISPR effector. Science. 2016;353(6299):aaf5573. https://doi.org/10.1126/science.aaf5573.

Afzal I, Shabir R, Rauf S. Seed production technologies of some major field crops. In: Agronomic crops: volume 1: production technologies. Singapore: Springer; 2019. p. 655–78.

Ayadi M, Brini F, Masmoudi K. Overexpression of a wheat aquaporin gene, Td PIP2; 1, enhances salt and drought tolerance in transgenic durum wheat cv. Maali. Int J Mol Sci. 2019;20(10):2389.

Bajaj S, Puthigae S, Templeton K, Bryant C, Gill G, Lomba P, Zhang H, Altpeter F, Hanley Z. Towards engineering drought tolerance in perennial ryegrass using its own genome 6th Canadian plant genomics workshop. Abstracts. 2008;62 https://doi.org/10.1111/pbi.12055.

Bar-Zvi D, Lupo O, Levy AA, Barkai N. Hybrid vigor: the best of both parents, or a genomic clash? Curr Opin Syst Biol. 2017;6:22–7.

Begna T. Conventional breeding methods widely used to improve self-pollinated crops. Int J Res. 2021;7(1):1–16.

Binns AN, Thomashaw MF. Cell biology of agrobacterium infection and transformation of plants. Ann Rev Microbiol. 1988;42:575–606.

Bueren EV. Enhancing resilience through plant breeding requires an integrated and interdisciplinary approach. In: Proceedings of the 8th organic seed growers conference; 2016. p. 133–4.

Caligari PD, Forster BP. Plant breeding and crop improvement. In: Encyclopedia of life sciences. Wiley; 2001. p. 1–9.

Chalavi V, Tabaeizadeh Z, Thibodeau P. Enhanced resistance to Verticillium dahliae in transgenic strawberry plants expressing a Lycopersicon chilense chitinase gene. J Am Soc Hortic Sci. 2003;128(5):747–53.

Chawla R, Shakya R, Rommens CM. Tuber-specific silencing of asparagine synthetase-1 reduces the acrylamide-forming potential of potatoes grown in the field without affecting tuber shape and yield. Plant Biotechnol J. 2012;10:913–24.

Cong L, Ran FA, Cox D, Lin S, Barretto R, Habib N, Hsu PD, Wu X, Jiang W, Marraffini LA, Zhang F. Multiplex genome engineering using CRISPR/Cas systems. Science. 2013;339(6121):819–23. https://doi.org/10.1126/science.1231143.

de Vetten N, Wolters AM, Raemakers K, van der Meer I, ter Stege R, Heeres E, Heeres P, Visser R. A transformation method for obtaining marker-free plants of a cross-pollinating and vegetatively propagated crop. Nat Biotechnol. 2003;21:439–42.

Delwaide AC, Nalley LL, Dixon BL, Danforth DM, Nayga RM Jr, Van Loo EJ, Verbeke W. Revisiting GMOs: are there differences in European consumers' acceptance and valuation for cisgenically vs transgenically bred rice? PLoS One. 2015;10:e0126060.

Demirer GS, Zhang H, Matos JL, Goh NS, Cunningham FJ, Sung Y, Chang R, Aditham AJ, Chio L, Cho M-J, Staskawicz B, Landry MP. High aspect ratio nanomaterials enable delivery of functional genetic material without DNA integration in mature plants. Nat Nanotechnol. 2019;14(5):456–64. https://doi.org/10.1038/s41565-019-0382-5.

EFSA Panel on Genetically Modified Organisms (GMO). Scientific opinion addressing the safety assessment of plants developed through cisgenesis and intragenesis. EFSA J. 2012;10:2561.

Gadaleta A, Giancaspro A, Blechl AE, Blanco A. A transgenic durum wheat line that is free of marker genes and expresses 1Dy10. J Cereal Sci. 2008;48:439–45.

Goel D, Singh AK, Yadav V, Babbar SB, Bansal KC. Overexpression of osmotin gene confers tolerance to salt and drought stresses in transgenic tomato (Solanum lycopersicum L.). Protoplasma. 2010;245(1):133–41.

Han KM, Dharmawardhana P, Arias RS, Ma C, Busov V, Strauss SH. Gibberellin-associated cisgenes modify growth, stature and wood properties in Populus. Plant Biotechnol J. 2011;9:162–78.

Haussmann BIG, Parzies HK, Presterl T, Sušić Z, Miedaner T. Plant genetic resources in crop improvement. Plant Genet Resour. 2004;2(1):3–21.

Holme IB, Dionisio G, Brinch-Pedersen H, Wendt T, Madsen CK, Vincze E, Holm PB. Cisgenic barley with improved phytase activity. Plant Biotechnol J. 2012;10:237–47.

Holme IB, Wendt T, Holm PB. Intragenesis and cisgenesis as alternatives to transgenic crop development. Plant Biotechnol J. 2013;11(4):395–407.

Hui F, Tang X, Li B, Alariqi M, Xu Z, Meng Q, Hu Y, Wang G, Zhang Y, Zhang X, Jin S. Robust CRISPR/Mb2Cas12a genome editing tools in cotton plants. iMeta. 2024;3(3):e209. https://doi.org/10.1002/imt2.209.

Hussain A, Arshad K, Abdullah J, Aslam A, Azam A, Bilal M, et al. A comprehensive review on breeding technologies and selection methods of self-pollinated and cross-pollinated crops. Asian J Biotechnol Genet Eng. 2021;4(3):35–47.

Illa-Berenguer E, LaFayette PR, Parrott WA. Editing efficiencies with Cas9 orthologs, Cas12a endonucleases and temperature in rice. Front Genome Ed. 2023;5:1074641. https://doi.org/10.3389/fgeed.2023.1074641.

Jinek M, Chylinski K, Fonfara I, Hauer M, Doudna JA, Charpentier E. A programmable dual-RNA–guided DNA endonuclease in adaptive bacterial immunity. Science. 2012;337(6096):816–21. https://doi.org/10.1126/science.1225829.

Jo KR, Kim CJ, Kim SJ, Kim TY, Bergervoet M, Jongsma MA, Visser RG, Jacobsen E, Vossen JH. Development of late blight resistant potatoes by cisgene stacking. BMC Biotechnol. 2014;14:50.

Jogdand SM, Bhat S, Misra KK, Lal RL, Preet MS. The cisgenesis new tool of breeding: a review. Chem Sci Rev Lett. 2017;22:690–703.

Joshi SG, Schaart JG, Groenwold R, Jacobsen E, Schouten HJ, Krens FA. Functional analysis and expression profiling of HcrVf1 and HcrVf2 for development of scab resistant cisgenic and intragenic apples. Plant Mol Biol. 2011;75(6):579–91.

Khan S, Sallard E. Current and prospective applications of CRISPR-Cas12a in pluricellular organisms. Mol Biotechnol. 2023;65(2):196–205. https://doi.org/10.1007/s12033-022-00538-5.

Kleinstiver BP, Pattanayak V, Prew MS, Tsai SQ, Nguyen NT, Zheng Z, Joung JK. High-fidelity CRISPR–Cas9 nucleases with no detectable genome-wide off-target effects. Nature. 2016;529(7587):490–5. https://doi.org/10.1038/nature16526.

Kost TD, Gessler C, Jänsch M, Flachowsky H, Patocchi A, Broggini GAL. Development of the first cisgenic apple with increased resistance to fire blight. PLoS One. 2015;10(12):e0143980. https://doi.org/10.1371/journal.pone.0143980.

Kovalchuk N, Jia W, Eini O, Morran S, Pyvovarenko T, Fletcher S, Bazanova N, Harris J, Beck-Oldach K, Shavrukov Y, Langridge P. Optimization of TaDREB3 gene expression in transgenic barley using cold-inducible promoters. Plant Biotechnol J. 2013;11(6):659–70.

Lawrenson T, Clarke M, Kirby R, Forner M, Steuernagel B, Brown JKM, Harwood W. An optimised CRISPR Cas9 and Cas12a mutagenesis toolkit for Barley and Wheat. Plant Methods. 2024;20(1):123. https://doi.org/10.1186/s13007-024-01234-y.

Lee M. Deep learning in CRISPR-Cas systems, a review of recent studies. Front Bioeng Biotechnol. 2023;11:1226182. https://doi.org/10.3389/fbioe.2023.1226182.

Liang Z, Chen K, Li T, Zhang Y, Wang Y, Zhao Q, Liu J, Zhang H, Liu C, Ran Y, Gao C. Efficient DNA-free genome editing of bread wheat using CRISPR/Cas9 ribonucleoprotein complexes. Nat Commun. 2017;8(1):14261. https://doi.org/10.1038/ncomms14261.

Lyon A, Tracy W, Colley M, Culbert P, Mazourek M, Myers J, et al. Adaptability analysis in a participatory variety trial of organic vegetable crops. Renewable Agric Food Syst. 2020;35(3):296–312.

Mali P, Yang L, Esvelt KM, Aach J, Guell M, DiCarlo JE, Norville JE, Church GM. RNA-guided human genome engineering via Cas9. Science. 2013;339(6121):823–6. https://doi.org/10.1126/science.1232033.

McCarty NS, Graham AE, Studená L, Ledesma-Amaro R. Multiplexed CRISPR technologies for gene editing and transcriptional regulation. Nat Commun. 2020;11(1):1281. https://doi.org/10.1038/s41467-020-15053-x.

McKersie BD, Bowley SR, Harjanto E, Leprince O. Water-deficit tolerance and field performance of transgenic alfalfa overexpressing superoxide dismutase. Plant Physiol. 1996;111(4):1177–81.
Opabode JT. Agrobacterium-mediated transformation of plants: emerging factors that influence efficiency. Biotechnol Mol Biol Rev. 2006;1(1):12–20.
Osusky M, Zhou G, Osuska L, Hancock RE, Kay WW, Misra S. Transgenic plants expressing cationic peptide chimeras exhibit broad-spectrum resistance to phytopathogens. Nat Biotechnol. 2000;18:1162–6.
Pausch P, Al-Shayeb B, Bisom-Rapp E, Tsuchida CA, Li Z, Cress BF, Knott GJ, Jacobsen SE, Banfield JF, Doudna JA. CRISPR-CasΦ from huge phages is a hypercompact genome editor. Science. 2020;369(6501):333–7. https://doi.org/10.1126/science.abb1400.
Puchta H, Fauser F. Gene targeting in plants: 25 years later. Int J Dev Biol. 2013;57:629–37.
Ray DK, Mueller ND, West PC, Foley JA. Yield trends are insufficient to double global crop production by 2050. PLoS One. 2013;8:e66428.
Reynoird JP, Mourgues F, Norelli J, Aldwinckle HS, Brisset MN, Chevreau E. First evidence for improved resistance to fire blight in transgenic pear expressing the attacin E gene from Hyalophora cecropia. Plant Sci. 1999;149(1):23–31.
Rommens CM. All-native DNA transformation: a new approach to plant genetic engineering. Trends Plant Sci. 2004;9:457–64.
Rommens CM, Humara JM, Ye J, Yan H, Richael C, Zhang L, Perry R, Swords K. Crop improvement through modification of the plant's own genome. Plant Physiol. 2004;135:421–31.
Rommens CM, Ye JS, Richael C, Swords K. Improving potato storage and processing characteristics through all-native DNA transformation. J Agric Food Chem. 2006;54:9882–7.
Sathish P, Bajaj S, Bryant C, Templeton K, Gill G, Hanley Z. Perennial ryegrass (Lolium perenne L.) improvement through cisgenics®. In: Proceedings of the XXVIITH Eucarpia symposium on improvement of fodder crops and amenity grasses; 2007. p. 19–23.
Schaart JG. Towards consumer-friendly cisgenic strawberries which are less susceptible to Botrytis cinerea. PhD Wageningen UR, Wageningen; 2004.
Slaymaker IM, Gao L, Zetsche B, Scott DA, Yan WX, Zhang F. Rationally engineered Cas9 nucleases with improved specificity. Science. 2016;351(6268):84–8. https://doi.org/10.1126/science.aad5227.
Smith EF, Towsend CO. A plant tumour of bacterial origin. Science. 1907;25:671–3.
Songstad DD, Petolino JF, Voytas DF, Reichert NA. Genome editing of plants. CRC Crit Rev Plant Sci. 2017;36:1–23.
Su X, Chu Y, Li H, Hou Y, Zhang B, Huang Q, Hu Z, Huang R, Tian Y. Expression of multiple resistance genes enhances tolerance to environmental stressors in transgenic poplar (Populus× euramericana 'Guariento'). PLoS One. 2011;6(9):e24614.
Tang L, Cai H, Ji W, Luo X, Wang Z, Wu J, Wang X, Cui L, Wang Y, Zhu Y, Bai X. Overexpression of GsZFP1 enhances salt and drought tolerance in transgenic alfalfa (Medicago sativa L.). Plant Physiol Biochem. 2013;71:22–30.
Thakur RP, Reddy BV, Mathur K. Screening techniques for sorghum diseases, Information: bulletin no. 76. International Crops Research Institute for the Semi-Arid Tropics; 2007.
Tilman D, Balzer C, Hill J, Befort BL. Global food demand and the sustainable intensification of agriculture. Proc Natl Acad Sci USA. 2011;108:20260–4.
Tsai SQ, Zheng Z, Nguyen NT, Liebers M, Topkar VV, Thapar V, Wyvekens N, Khayter C, Iafrate AJ, Le LP, Aryee MJ, Joung JK. GUIDE-seq enables genome-wide profiling of off-target cleavage by CRISPR-Cas nucleases. Nat Biotechnol. 2015;33(2):187–97. https://doi.org/10.1038/nbt.3117.
Vanblaere T, Flachowsky H, Gessler C, Broggini GAL. Molecular characterization of cisgenic lines of apple 'Gala' carrying the Rvi6 scab resistance gene. Plant Biotechnol J. 2014;12:2–9.
Vazquez-Vilar M, Fernandez-del-Carmen A, Garcia-Carpintero V, Drapal M, Presa S, Ricci D, Diretto G, Rambla JL, Fernandez-Muñoz R, Espinosa-Ruiz A, Fraser PD. Dually biofortified cisgenic tomatoes with increased flavonoids and branched-chain amino acids content. Plant Biotechnol J. 2023;21(12):2683–97.

Wang D, Zhang C, Wang B, Li B, Wang Q, Liu D, Wang H, Zhou Y, Shi L, Lan F, Wang Y. Optimized CRISPR guide RNA design for two high-fidelity Cas9 variants by deep learning. Nat Commun. 2019;10(1):4284. https://doi.org/10.1038/s41467-019-12281-8.

Weeks JT, Ye J, Rommens CM. Development of an in planta method for transformation of alfalfa (Medicago sativa). Transgenic Res. 2008;17:587–97.

Wu YY, Chen QJ, Chen M, Chen J, Wang XC. Salt-tolerant transgenic perennial ryegrass (Lolium perenne L.) obtained by Agrobacterium tumefaciens-mediated transformation of the vacuolar Na+/H+ antiporter gene. Plant Sci. 2005;169(1):65–73.

Würdig J, Flachowsky H, Saß A, Peil A, Hanke M-V. Improving resistance of different apple cultivars using the Rvi6 scab resistance gene in a cisgenic approach based on the Flp/FRT recombinase system. Mol Breed. 2015;35:95.

Xiang X, Corsi GI, Anthon C, Qu K, Pan X, Liang X, Han P, Dong Z, Liu L, Zhong J, Ma T, Wang J, Zhang X, Jiang H, Xu F, Liu X, Xu X, Wang J, Yang H, et al. Enhancing CRISPR-Cas9 gRNA efficiency prediction by data integration and deep learning. Nat Commun. 2021;12(1):3238. https://doi.org/10.1038/s41467-021-23576-0.

Xu Y, Crouch JH. Marker-assisted selection in plant breeding: from publications to practice. Crop Sci. 2008;48(2):391–407.

Yadav RN, Kumar PR, Hussain Z, Yadav S, Lal SK, Kumar A, et al. Maintenance breeding. In: Fundamentals of field crop breeding. Singapore: Springer Nature; 2022. p. 703–44.

Zetsche B, Gootenberg JS, Abudayyeh OO, Slaymaker IM, Makarova KS, Essletzbichler P, Volz SE, Joung J, van der Oost J, Regev A, Koonin EV, Zhang F. Cpf1 is a single RNA-guided endonuclease of a class 2 CRISPR-Cas system. Cell. 2015;163(3):759–71. https://doi.org/10.1016/j.cell.2015.09.038.

Zhang Y, Ren Q, Tang X, Liu S, Malzahn AA, Zhou J, Wang J, Yin D, Pan C, Yuan M, Huang L, Yang H, Zhao Y, Fang Q, Zheng X, Tian L, Cheng Y, Le Y, McCoy B, et al. Expanding the scope of plant genome engineering with Cas12a orthologs and highly multiplexable editing systems. Nat Commun. 2021;12(1):1944. https://doi.org/10.1038/s41467-021-22330-w.

Zhang Y, Wu Y, Li G, Qi A, Zhang Y, Zhang T, Qi Y. Genome-wide investigation of multiplexed CRISPR-Cas12a-mediated editing in rice. Plant Genome. 2023;16(2):e20266. https://doi.org/10.1002/tpg2.20266.

Zhou Y, Xu S, Jiang N, Zhao X, Bai Z, Liu J, Yao W, Tang Q, Xiao G, Lv C, Wang K, Hu X, Tan J, Yang Y. Engineering of rice varieties with enhanced resistances to both blast and bacterial blight diseases via CRISPR/Cas9. Plant Biotechnol J. 2022;20(5):876–85. https://doi.org/10.1111/pbi.13766.

Zhu Y, Huang Z. Recent advances in structural studies of the CRISPR-Cas-mediated genome editing tools. Natl Sci Rev. 2019;6(3):438–51. https://doi.org/10.1093/nsr/nwy150.

Zhu X, Clarke R, Puppala AK, Chittori S, Merk A, Merrill BJ, Simonović M, Subramaniam S. Cryo-EM structures reveal coordinated domain motions that govern DNA cleavage by Cas9. Nat Struct Mol Biol. 2019;26(8):679–85. https://doi.org/10.1038/s41594-019-0258-2.

Chapter 6
Next-Generation Sequencing Techniques

Muhammad Waris, Zobia Jabeen, Syed Zulfiqar Ali, Aurang Zaib Jamali, Atta Ullah, Hina Ali Ahmed, Siddra Ijaz, and Zakia Habib

6.1 Introduction

Next-Generation Sequencing (NGS) has revolutionized genomics research, providing new perspectives on the structure, function, and complexity of genomes. This revolutionary technology has enabled researchers to dichotomize genetic information at a level of resolution that was previously impossible. Due to its low cost and the potential to produce a vast amount of data in a single run, NGS is being applied everywhere from basic biology to clinical applications. In addition to whole-genome sequencing (WGS), NGS has significantly hastened advancements in transcriptomics, epigenomics, metagenomics, and other omics-based research (Kumar et al. 2024). Since the development of these platforms based on microchip technologies and integrated optics, we have been able to sequence millions or even billions of DNA fragments in parallel, which has transformed our ability to study genetic variation, gene regulation, microbial ecologies, or epigenetic marks. These capabilities have not only improved our knowledge but also helped in the identification of disease-causing variants and potential therapeutic targets. This chapter covers

M. Waris (✉) · Z. Jabeen · S. Z. Ali
Department of Plant Pathology, Balochistan Agriculture College, Quetta, Pakistan

A. Z. Jamali
Department of Soil Science, Balochistan Agriculture College, Quetta, Pakistan

A. Ullah
Department of Plant Breeding and Genetics, Balochistan Agriculture College, Quetta, Pakistan

H. A. Ahmed
Department of Zoology, Sardar Bahadur Khan Women University, Quetta, Pakistan

S. Ijaz · Z. Habib
Centre of Agricultural Biochemistry and Biotechnology (CABB), University of Agriculture, Faisalabad, Pakistan

S. Ijaz et al. (eds.), *Molecular Biology Methods*,
https://doi.org/10.1007/978-981-92-2642-9_6

current NGS methods and describes how NGS has revolutionized agriculture research, clinical genomics, infectious disease monitoring, and microbiome investigations. (Singh et al. 2025).

The Pioneering Era of Nucleic Acid Sequencing Since the discovery of the double helix structure of DNA in 1953 and the identification of the genetic code as nature's plan for life in 1964, people have dreamed of sequencing DNA. Regardless, it was a quarter of a century ago that DNA sequencing methods were first perfected. Sanger sequencing is a "sequencing by synthesis" method of uncovering the nucleic acid sequence, one nucleotide at a time, for any length DNA template (Shook et al. 2023).

6.2 NGS Systematic Workflows

The workflow of next-generation sequencing (NGS) is a well-orchestrated process that integrates molecular biology, chemistry, and bioinformatics to generate high-quality genomic information (Fig. 6.1). Although specific steps may vary depending on the platform (e.g., Illumina, PacBio, Oxford Nanopore), the general pipeline follows a systematic progression from sample preparation to data interpretation (Head et al. 2014).

Sample Preparation and DNA/RNA Extraction The process begins with careful isolation of high-quality nucleic acids. For genome sequencing, genomic DNA is purified, while transcriptome studies require RNA extraction and conversion into complementary DNA (cDNA). The integrity and purity of the nucleic acids are critical, as they directly influence the accuracy of downstream analysis (Hess et al. 2021).

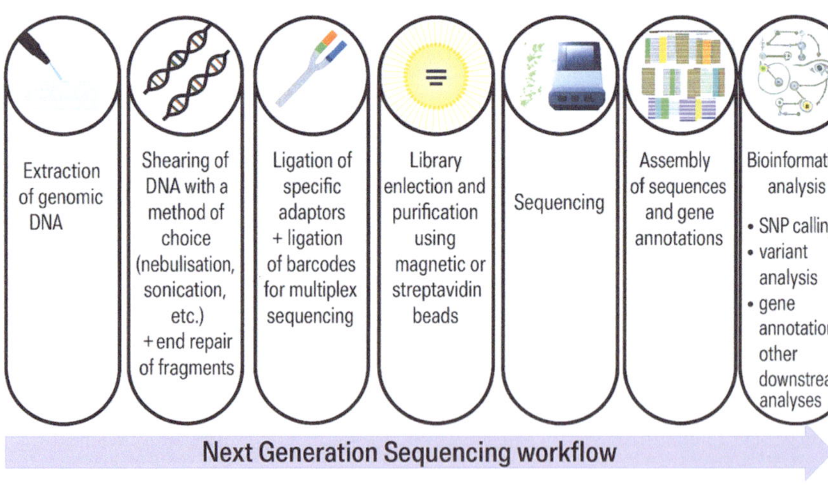

Fig. 6.1 Overview of the main steps involved in next-generation sequencing technology

Library Preparation Extracted DNA or cDNA is fragmented into smaller pieces, usually ranging from a few hundred base pairs to several kilobases, depending on the platform. These fragments are then ligated with specialized adapters that enable binding to the sequencing platform and allow for amplification. In some cases, barcoding (indexing) is introduced, which permits multiple samples to be sequenced in a single run (Wang 2021).

Template Amplification Many platforms, such as Illumina, rely on amplification of DNA fragments to generate a sufficient signal for sequencing. This is typically achieved through bridge amplification (on a flow cell) or emulsion PCR (in microdroplets). Other platforms, such as Oxford Nanopore or PacBio, bypass this step, sequencing single molecules directly (Casbon et al. 2011).

Sequencing Sequencing involves determining the nucleotide order of the DNA fragments. Different platforms employ different chemistries. Illumina uses sequencing-by-synthesis with fluorescently labeled nucleotides, Ion Torrent detects hydrogen ions released during nucleotide incorporation, and Oxford Nanopore reads the sequence by monitoring changes in ionic current as DNA passes through a nanopore. The outcome is a vast collection of short or long reads depending on the technology (Marudamuthu et al. 2023).

Primary Data Processing Raw signals from the sequencer are converted into nucleotide sequences (base calling). At this stage, quality scores are assigned to each base to indicate confidence in the read accuracy. Low-quality reads are filtered out, and adapter sequences are trimmed to ensure clean data for analysis (Cristea 2002).

Alignment and Assembly Reads are then aligned against a reference genome, or in cases where no reference is available, they are assembled de novo to reconstruct the genome or transcriptome. Specialized algorithms are used to handle repetitive regions, structural variants, and sequencing errors (Del Angel et al. 2018).

Variant Calling and Annotation Once aligned, the data can be mined for genetic variants such as single-nucleotide polymorphisms (SNPs), insertions/deletions (Indels), structural variations, or copy number changes. Annotation tools link these variants to genes, pathways, or functional categories, providing biological context (Kim 2015).

Downstream Analysis and Interpretation The final stage involves integrating sequencing results with biological questions. This could range from identifying candidate genes associated with specific traits in crops to uncovering regulatory networks or performing comparative genomics. In agricultural research, the interpretation often feeds into breeding programs, marker development, or functional genomics studies (Singh et al. 2025).

6.3 Evolution of Sequencing Technologies

The discovery of the DNA double helix in 1953 revealed that genetic material serves as the fundamental blueprint of life, sparking a long-standing need to decode DNA sequences (Watson and Crick 1953). However, it was not until nearly 25 years later that the first generation of sequencing techniques emerged (Sanger et al. 1977). Among these, Sanger sequencing became the most prominent, relying on a "sequencing by synthesis" principle, where the DNA sequence is read one nucleotide at a time based on the template strand. Over the following decade, improvements in the technique culminated in its automation by Applied Biosystems in 1987, which in turn enabled the launch of the Human Genome Project. This historic initiative required unprecedented global collaboration, took approximately 10 years to complete, and cost around $2.7 billion. By the project's completion in 2001, the limitations of first-generation Sanger sequencing had become apparent, as scaling it for routine whole-genome sequencing was impractical. These challenges laid the foundation for the development of more advanced approaches, collectively referred to as next-generation sequencing (Metzker 2010).

6.4 First Generation Sequencing

In the initial attempts at sequencing DNA and RNA, the molecules were chemically broken down or cleaved by enzymes to produce fragments that could be examined separately. In 1964, Robert Holley used *S. cerevisiae* ribonuclease to sequence Alanine tRNA, the first nucleic acid molecule (Holley et al. 1965). In a similar vein, Allan Maxam and Walter Gilbert created a chemical degradation method that made it possible to sequence the whole bacteriophage PhiX174 (Heather and Chain 2016). But Fredrick Sanger's invention of the chain termination-based sequencing technique was the true innovation (Sanger et al. 1977). This method produced sequence readings up to a few hundred nucleotides long using dideoxynucleotides, which stop DNA strands from lengthening their chains during replication. Sanger's technique was extensively used and transformed molecular biology by making it possible to sequence DNA and RNA (Schuster 2008) quickly. The Applied Biosystems ABI 370, the first automated sequencing equipment sold commercially, was introduced in the United States in 1987. This device greatly increased the speed and precision of DNA sequencing by automating the Sanger sequencing process using capillary electrophoresis and fluorescently labeled dideoxynucleotides. The ABI 370 swiftly established itself as the industry standard, and as technology advanced, higher throughput sequencers with longer read lengths were created. Even though more recent, higher throughput sequencing technologies have essentially replaced the first-generation technology, it nonetheless represents a significant historical turning point in the evolution of various sequencing methods. Numerous branches of biology and medicine have been transformed by the capacity to sequence DNA and RNA, which has also produced a wealth of new findings and advances in our knowledge of genetics and molecular biology (Satam et al. 2023).

6.5 Second Generation Sequencing

DNA sequencing has undergone a revolution since second-generation sequencing techniques allow for the contemporaneous sequencing of thousands to millions of DNA fragments. These techniques enable parallel sequencing, which distinguishes them from conventional Sanger sequencing (Satam et al. 2023).

Second-Generation Sequencing Technology Platforms Several significant platforms for second-generation sequencing have come up in the agriculture sector, and ways in which genomic data derived from these technologies are improving agricultural research (Marudamuthu et al. 2023).

Roche 454 Genome Sequencer FLX The 454 sequencing platform employs large-scale parallel pyrosequencing, in which the incorporation of a nucleotide by DNA polymerase releases pyrophosphate. This by-product drives a cascade of enzymatic reactions involving luciferase and ATP sulfurylase, ultimately producing light that signals nucleotide addition. For library preparation, genomic DNA is fragmented into pieces of approximately 800–1000 bp, followed by adaptor ligation and immobilization onto DNA-capture beads. Each bead carrying a single DNA fragment undergoes clonal amplification through emulsion PCR, and the resulting enriched beads are subsequently transferred to a PicoTiterPlate. Sequencing is then performed in the presence of DNA polymerase, luciferase, and ATP sulfurylase. Roche 454 Genome Sequencer (GS) FLX Titanium sequencing provides 100,000 reads or more with an average length between 330 and 500 bases per sample within 10 h. Subsequent enhancements to the GS FLX system increased read lengths to 700–1000 bp (Gupta and Gupta 2020).

Applications of such technology in different crop genomes could be performed. For instance, in barley (Hordeum vulgare), which is thought to have long stretches of repetitive DNA, sequences of 200–250 bp were assembled using 454 sequencing to yield assemblies of similar quality to HTGS, with good N50 statistics (Steuernagel et al. 2009). In addition, the transcriptome of the nonedible oilseed crop *Jatropha curcas* was generated using the GS FLX Titanium system (197.7 Mb from 383,918 reads, mean read length = 515 bp) (Natarajan and Parani 2011). The Potato Genome Sequence Collaborators also used a hybrid strategy of both 454 GS FLX and Illumina GA2 sequencing technologies in assembling the genome of Solanum tuberosum, the world's third most important food crop (Visser et al. 2009).

Ion Personal Genome Machine (PGM) Sequencer Ion Torrent sequencing, based on the PGM sequencer platform, is the detection of hydrogen ions that are produced when a polymerase incorporates a nucleotide into a freshly generated strand. The patented ion sensor detects the charge of the released ion, which alters the pH of the fluid. An ion torrent chip, featuring millions of ion-sensitive field-effect transistor (ISFET) sensors, enables the detection of sequencing processes in a massively parallel manner. According to Pereira et al. (2020), it is the first commercial sequencing process that does not include fluorescence or camera scanning.

Illumina Sequencing Analyzer The Illumina Genome Analyzer, originally developed by Solexa in 2006 and later acquired by Illumina in 2007 (Emiyu and Lelisa 2022), introduced a novel sequencing strategy known as reversible terminator-based sequencing. Unlike the classical Sanger method that relies on dideoxy terminators, this approach employs reversible dye terminators within a sequencing-by-synthesis framework. Illumina platforms utilize a glass flow cell coated with millions of immobilized oligonucleotides, to which DNA library fragments are covalently attached across eight separate lanes. Once hybridized, each fragment undergoes bridge amplification, generating millions to billions of identical clusters. Sequencing is then carried out through the incorporation of fluorescently labeled nucleotides, with each addition recorded by high-resolution imaging. The wavelength and intensity of the fluorescent signals permit precise base calling and generate massively parallel sequence information. A variety of Illumina machines are equipped to address different experimental volumes, ranging from high-throughput systems (HiSeq, NovaSeq, and NextSeq) to benchtop versions (MiSeq, MiniSeq, and iSeq) (Kulski 2016). Illumina is still the preferred platform for plant genome sequencing because it is cheap, efficient, and there are good assembly pipelines available. It has been effectively used to sequence several important crops, including barley (*Hordeum vulgare*), wheat (*Triticum aestivum*), rice (*Oryza sativa*), maize (*Zea mays*), and soybeans (*Glycine max*).

Applied Biosystems SOLiD Sequencer The Applied Biosystems SOLiD sequencer uses a sequencing by ligation approach, where DNA ligase stepwise delivers the order of incorporated nucleotides. In this process, DNA fragments are appended to adaptor sequences that anneal to complementary oligonucleotide-coated magnetic beads. The bead–DNA hybrid is amplified by emulsion PCR, resulting in thousands of copies of the fragment. Sequencing commences given a primer that reveals a free 5′ phosphate group, which permits the ligation of fluorescently tagged probes to the adaptor-linked fragments. The fluorescence generated by the incorporation of a probe following each ligation event determines the identity of the probe incorporated. Repeated ligation, fluorescence detection, and probe cleavage determine the reads (Gupta and Gupta 2020).

The emergence of second-generation sequencing platforms such as SOLiD has significantly accelerated DNA sequencing, offering high throughput and scalability. These innovations have broadened applications in genomics and clinical diagnostics, supporting advances in whole-genome sequencing, transcriptome profiling, targeted sequencing, and ultimately contributing to research on genetic diversity, disease mechanisms, and personalized medicine (Wang 2021).

6.6 Third-Generation Sequencing

The most recent developments in DNA sequencing are represented by third-generation sequencing technologies, which offer novel techniques that overcome the drawbacks of earlier generations. Third-generation sequencing is often referred to as single-molecule sequencing because it does not involve the amplification phase during sequencing library construction, like second-generation sequencing. Compared to previous techniques, these technologies' long-read sequencing capabilities allow for the sequencing of substantially bigger DNA fragments (Bleidorn 2016).

Third-Generation Sequencing Technology Platforms Third-generation sequencing includes Pacific Biosciences (PacBio) sequencing, Oxford Nanopore sequencing, and Helicos single-molecule sequencing.

PacBio Sequencing Pacific Biosciences introduced Single-Molecule Real-Time (SMRT) sequencing, a platform that directly observes DNA synthesis in real time (Wang et al. 2023). In this system, a single DNA molecule is anchored to a DNA polymerase, and the incorporation of fluorescently labeled nucleotides is continuously monitored as the enzyme synthesizes a complementary strand. The technology relies on Zero-Mode Waveguides (ZMWs), nanometer-scale wells etched into a metal film on a glass substrate, which allow the observation of single-molecule reactions in a highly confined space. Each SMRT cell run typically takes approximately 4 h, producing average read lengths of around 10 kb, with ultra-long reads of over 50 kb also achievable (Bleidorn 2016).

The most widely used library format is the SMRTbell template, which is created by ligating hairpin adaptors to both ends of a double-stranded DNA fragment, forming a closed circular molecule. This structure enables the polymerase to repeatedly traverse the circularized DNA, thereby generating multiple passes of the same sequence and improving accuracy. Alternatively, linear templates can be used for generating longer reads, albeit with reduced base-calling precision. As the polymerase incorporates nucleotides, the fluorescent tags are cleaved off and recorded optically in real time, with each emission event corresponding to a base call. This circular consensus sequencing approach captures information from both strands, increasing coverage depth and accuracy (Bleidorn 2016).

A unique feature of PacBio sequencing is its ability to detect epigenetic modifications, as altered bases influence the kinetics of nucleotide incorporation (Flusberg et al. 2010). While the technology was initially limited by low throughput, higher error rates (~13%), and cost (Ermini and Driguez 2024), continuous improvements have greatly enhanced performance. The Sequel series, for example, increased the number of ZMWs per chip from ~150,000 to over one million, while advances in sequencing chemistry have reduced error rates to below 3%. These developments have expanded the platform's applications, which now include whole-genome sequencing, targeted resequencing, full-length transcriptome analysis, microbial population studies, and epigenomic profiling.

PacBio's long-read capability provides clear advantages over short-read sequencing, particularly in resolving repetitive regions, characterizing structural variants, and sequencing DNA with extreme GC content. The platform has been instrumental in generating some of the most complete human genome assemblies to date, covering 2.89 Gb (~96%) and offering unmatched resolution of structural diversity (Mason 2021). In 2018, Illumina announced its acquisition of Pacific Biosciences to integrate true long-read sequencing into its next-generation sequencing portfolio.

Helicos Single-Molecule Sequencing The Helicos True Single Molecule Sequencing (tSMS) platform is a sequencing-by-synthesis technology that operates at the single-molecule level. In this approach, individual DNA template molecules are immobilized on a specially designed surface. Sequencing begins with the incorporation of fluorescently labeled nucleotides (A, T, G, or C) into complementary strands by DNA polymerase in a sequence-specific manner. After each incorporation step, unbound nucleotides are washed away, and the incorporated bases are detected through high-resolution imaging. The fluorescent tags are then cleaved, leaving the integrated nucleotide in place, and the process is repeated for the remaining bases. This cyclic incorporation and detection strategy generates short reads ranging from 25 to 55 base pairs, with an average read length of approximately 35 base. Despite the relatively short reads, the system can process between 600 million and 1 billion DNA molecules simultaneously, producing 21–35 gigabases of sequencing data with an accuracy rate of up to 99.995% (Gupta and Gupta 2020).

Oxford Nanopore Sequencing Oxford Nanopore Technologies launched the MinION in 2014, introducing the first truly portable sequencing platform based on nanopore technology (Lu et al. 2016). This method utilizes biological nanopores, which are tiny pores formed by transmembrane proteins embedded in a membrane. Sequencing begins when a hairpin adaptor is ligated to one end of a double-stranded DNA fragment, after which a motor protein and helicase unwind the DNA and guide single strands through the nanopore. Each pore is connected to an electrode and sensor chip that measures the ionic current across the membrane. As nucleotides pass through the pore, they cause distinct disruptions in current, and these characteristic signals are decoded to determine the DNA sequence (Kchouk et al. 2017). Because each nucleotide produces a unique electrical signature, sequence information can be derived directly from current fluctuations.

Although nanopore sequencing has historically been associated with higher error rates (up to 15%), continuous improvements have enhanced accuracy. One strategy, known as 1D^2 sequencing, enables the simultaneous reading of both template and complementary strands, thereby improving consensus sequence quality (Lu et al. 2016). The portability of the MinION has been one of its greatest strengths, enabling field-based sequencing during the 2015 Ebola outbreak and even in microgravity environments aboard the International Space Station (Castro-Wallace et al. 2017). To support larger projects, Oxford Nanopore developed the GridION X5 and

PromethION systems, offering 5 and 42 flow cell configurations, respectively, to deliver higher throughput. More recently, the Flongle, a disposable flow cell for the MinION, was introduced, enabling smaller scale sequencing applications such as targeted gene analysis, microbial genome sequencing, and diagnostics.

Nanopore sequencing offers another significant benefit, as it can sequence native DNA and RNA molecules directly, eliminating the need for amplification or labeling, and produces ultra-long reads, exceeding 2 Mb in some cases (Payne et al. 2019). It is particularly beneficial for de novo genome assembly and addressing complex structural variants (SVs). Nanopore platforms also carry over nucleotide modifications, like DNA methylation, and can detect a variety of epigenetic and RNA modifications without a priori chemical treatments (Jonkhout et al. 2017; Garalde et al. 2018). Together, these attributes position nanopore sequencing as an extremely versatile, dynamic, and rapidly evolving technology that applies to all areas of genomics, diagnostics, and epigenetics.

Synthetic Long Reads (Illumina and 10X Genomics) Recently, to circumvent the expensive operation and decrease the accuracy of the basecalling error in third-generation sequencing, 10X Genomics developed the Chromium synthetic linked-read technology. This approach can be combined with short-read sequencing technology in "reconstructing" long-range genomic information. The process starts with large molecular weight (usually greater than 100 kb) DNA fragments that are divided across the gel beads in microdroplets. DNA is fragmented into short-read libraries for each droplet, and all resulting fragments are marked with a unique molecular barcode.

Following Illumina short-read sequencing, these barcodes serve as identifiers that allow individual reads to be traced back to their original long DNA molecule. This barcode-guided reconstruction generates "synthetic linked-reads," which combine the precision of short-read sequencing with long-range context. By preserving haplotype and structural information, this approach improves read alignment, facilitates variant phasing, and enhances the detection of structural variants. While the method retains the accuracy and high throughput of Illumina sequencing, its specialized library preparation protocols make it comparatively more expensive (Kumar et al. 2024).

6.7 Hybrid Sequencing: Combining Second- and Third-Generation Sequencing

The combination of second and third-generation sequencing technologies presents a feasible and robust solution to cope with the limitations imposed by the limited read lengths of the Illumina platform and the increased error rates in long-read technologies. This hybrid sequencing, or "combining," strategy is very useful for the construction of a de novo genome. Generally, long-read sequencing is employed to construct a robust assembly scaffold, especially over repetitive or structurally complex regions that are unresolved with short-read data. Then, Illumina short reads are mapped to this framework and polished at the base level to create a high-quality consensus sequence (Minei et al. 2018).

Long-read technology can also be used to call genetic traits that are otherwise not identified with short-read sequencing, and the hybrid approach allows us to significantly decrease the amount of long-read data at the same time, thus offering a cost-effective trade-off. For instance, Illumina reads can be effectively used to rectify base-calling mistakes in very long nanopore sequences (Heather and Chain 2016), such as those that cross the centromere of the human Y chromosome (Jain et al. 2018). Likewise, the UK Genome 10 K consortium's Vertebrate Genomes Project highlights the advantage of a hybrid approach by using PacBio long-reads, 10X Genomics linked-reads, Bionano Genomics optical maps, and Hi-C proximity ligation data in combination with Illumina sequencing to generate high-quality genome assemblies (Koepfli et al. 2015).

6.8 Next-Generation Sequencing: A Tool for Agriculture Innovation

Next-generation sequencing (NGS) has become a cornerstone technology in plant genomics, offering unprecedented opportunities to accelerate genetic improvement. It has facilitated the sequencing of numerous agriculturally significant plant genomes, generating extensive datasets that serve as valuable resources for crop enhancement. NGS applications extend across whole-genome sequencing, transcriptome profiling, and metagenomic studies, each contributing to the identification of functional genes and molecular markers associated with desirable agronomic traits. These insights are increasingly applied in marker-assisted and genomic selection programs, thereby supporting crop breeding, food security, and the conservation of genetic diversity (Fig. 6.2).

Whole-Genome Sequencing In agriculture, there is a whole new realm of genetic information that has been explored, which can lead to revitalization of food security, poverty reduction, and species conservation. Novel technologies such as next-generation sequencing (NGS) have been particularly revolutionary, enabling cost-effective and high-throughput sequencing of complex plant genomes. The genomes of many crop species are particularly challenging due to their large size, high pro-

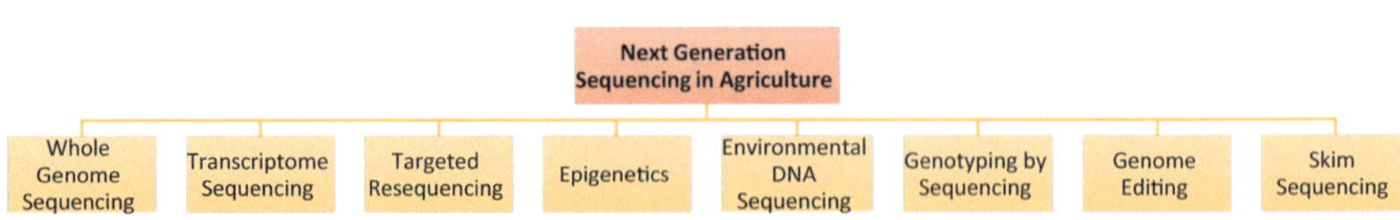

Fig. 6.2 Next-generation sequencing technologies in agriculture

portion of repetitive elements, presence of whole-genome duplication, and polyploidy, which makes gene identification and interpretation of gene function more problematic. However, NGS has made it possible for researchers to compile comprehensive catalogs of genes in several crop genomes, thereby aiding genetic advances and resource allocation.

Rice is the first crop species to have its entire genome sequenced, ushering in a new era of plant genomics. Goff et al. (2002) have sequenced two subspecies of *Oryza sativa* (japonica and indica) in combination with a hierarchical clone-by-clone and whole-genome shotgun strategy. Subsequently, the shotgun sequencing approach was also used for other economically important crops possessing bigger and more complex genomes than *O. sativa*, such as soybean (*Glycine max*), poplar (*Populus trichocarpa*), and grapevine (*Vitis vinifera*) (Tuskan et al. 2006; Velasco et al. 2007; Schmutz et al. 2010).

As NGS continues to advance, genome sequencing has been extended to various crops, including *Brassica oleracea*, *Theobroma cacao*, *Malus domestica*, *Cucumis melo*, *Cajanus cajan*, *Cicer arietinum*, potato, *Citrullus lanatus*, and *Citrus sinensis* (Michael and VanBuren 2015). The majority of these studies were carried out on short-read platforms like Illumina and 454 pyrosequencing, and contigs belonging to the short reads were assembled and mapped with reference genomes to detect genetic diversity and variation within the species. Although second-generation sequencing technology has reduced the cost and time associated with de novo genome construction, repetitive elements (such as TEs) are still the leading cause of gaps or fragmented genome assembly. To overcome such limitations, third-generation long-read sequencing technologies have proven to be highly useful, enabling the achievement of more continuous and nearly complete assemblies. At present, whole genomes of plants are mostly sequenced through whole-genome resequencing (WGR) for comparing multiple accessions with a reference or the de novo sequencing that constructs reference genomes without a prior reference.

Whole Genome Resequencing The whole genome resequencing (WGR) is an effective approach for the study of genetic diversity in crop species where a reference genome exists. Here, by anchoring the short reads on the reference, WGR allows the identification of all possible genomic variations, including copy number variations (CNVs), insertions/deletions (Indels), SNPs, and presence/absence variants (PAVs). Such markers serve as a valuable tool to study the genetic system of agronomic traits and to perform marker-assisted selection (MAS) for breeding.

The pace of evolution of next-generation sequencing (NGS) systems has led to the accumulation of high-quality reference genomes within a short period of time, which in turn facilitates WGR to serve as a powerful tool for analyzing genome-wide variation in many crops. This principle was first applied in the context of model staple crops, like *Oryza sativa* (Huang et al. 2009), *Zea mays* (Lai et al. 2010), and *Glycine max* (Lam et al. 2010), providing the basis for the conduct of large-scale comparative genomics studies. Subsequently, WGR has revealed a large number of agriculturally important variants, especially CNVs and PAVs, responsible for trait diversity and crop adaptation.

For example, Tong et al. built up a pan-genome variation map of cultivated lettuce by resequencing 445 *Lactuca* accessions from 47 countries. Their research pinpointed SNPs, Indels, and SVs, which provide insights into domestication and evolution of the genus (Wei et al. 2021). Resequencing of Cucurbita pepo also identified genes associated with early flowering, which is important for breeding new cultivars. Promoter-based indel markers have been used successfully in the differentiation of various cultivars (Abbas et al. 2022). In Sorghum, WGR allowed tagging SNPs and Indels associated with insect resistance traits, which were used for MAS to generate aphid-tolerant lines (Wei et al. 2021). Similarly, resequencing of the 588 *Brassica napus* accessions from 21 countries identified a large number of SNPs and Indel variations, some with significant associations for critical phenotypes related to yield and quality for improvement (Lu et al. 2019). In combination, these papers illustrate the fact that WGR has emerged as one of the cornerstones of modern crop genomics, providing high-resolution descriptions of genetic diversity that can chart out the strategies of breeding, underpin germplasm conservation, and deepen our comprehension of the domestication and evolutionary past.

De Novo Sequencing For crops that do not have a close relative reference genome available, whole-genome resequencing is no longer possible, and de novo sequences are used to investigate the genetic composition. This latter approach, however, suffers major limitations owing to the large sizes of many plant genomes and due to their (often) high content of redundant sequences. In de novo genome sequencing, contig quality and continuity will influence assembly quality as a whole. Short-read assembly strategies have demonstrated low efficacy for the generation of highly contiguous genomes, especially in the presence of repetitive DNA or complex polyploid genomes (Liao et al. 2019).

The arrival of long-read sequencing technologies has transformed de novo assembly, allowing for the conception of novel algorithms that can produce nearly complete, virtually scaffolded assemblies, even in genomes that are highly homozygous, polyploid, and of abundant repeat content (Gao et al. 2019). An example of the latter is *Oropetium thomaeum*, the first plant to have its genome sequenced using Pacific Biosciences' single-molecule real-time (SMRT) technology, achieving one of the most contiguous assemblies at the time, with approximately 30% of the centromeres remaining completely intact (VanBuren et al. 2015). Similarly, the ONT has been applied successfully for de novo sequencing in multiple crops such as *Solanum lycopersicum* (tomato), *Musa* (banana), Sorghum, and Brassica. More recently, an improved assembly for *Olea europaea* (olive) was developed using a hybrid ONT long-read sequencing and Hi-C scaffolding, which could serve as a high-quality reference genome for functional genomics, evolutionary biology, and molecular breeding studies in olive (Rao et al. 2021).

Targeted Resequencing is focused on sequencing specific genomic sites, often as a result of association mapping, exome sequencing, or a set of genes of interest.

Because it focuses on the well-defined parts of the genome that vary between individuals, this approach is more cost-effective than whole-genome sequencing, yet generates high-quality data across a large number of samples. It is particularly useful for the detection of important variants such assingle-nucleotide polymorphisms (SNPs) and copy number variations (CNVs) for selection and prediction of disease and parasite susceptibility (Bewicke-Copley et al. 2019). In light of its excellent coverage and specificity, targeted resequencing is particularly advantageous for identifying rare or elusive mutations that more extensive sequencing strategies could miss. These mutations may be beneficial alleles contributing to driving the selection in crop improvement, or causal variations linked to undesirable characteristics. For example, Pankin et al. (2018) performed targeted resequencing on 433 wild and cultivated *Hordeum vulgare* (barley) accessions, which resulted in the discovery of candidate domestication genes. Their study also shed new light on the phylogenetic relationships and the ancestral origins of cultivated barley haplotypes.

Transcriptome Sequencing Transcriptome sequencing (RNA-Seq) is the process of determining the nucleic acid sequence of all the RNA, both polyribosomal and non-polyribosomal, or just the polyribosomal mRNA, in a population of cells. This approach allows valuable information about gene expression patterns, functional genomics, and gene characterization in species whose background genomes are not yet completely sequenced (Afzal et al. 2020). According to the existence of a reference genome, transcriptome data show the possibility of mapping or de novo assembling. De novo transcriptome assembly and functional annotation are simpler than whole-genome assembly (Bryant et al. 2017). RNA-seq produces a transcriptome profile that dynamically responds to various organs, developmental stages, stress treatments, or external signals. However, it does not account for any structural variation in the genome (Schreiber et al. 2018). In crop science, transcriptome sequencing has been widely used for gene identification, regulatory network study, and examining the metabolic pathways leading to agronomic traits. For example, the transcriptomes of *Vitis amurensis*, *Aloe vera*, *Paeonia suffruticosa*, *Trillium govanianum*, *Polygonum minus*, *Cornus officinalis*, *Dracocephalum tanguticum*, and *Salvadora oleoides* have been sequenced, and the molecular regulation and functional mechanisms of these plants are now better understood (Fleury et al. 2012).

Epigenetics Factors such as drought, nutrient levels, and stress can prompt adaptive responses in plant and animal life that result in physiological effects that can go on to influence survival and reproduction directly. Epigenetic mechanisms, including DNA methylation, posttranslational histone modifications, and noncoding RNAs, frequently regulate these responses. Such processes modify chromatin, leading to gene-specific activation or silencing. The development of high-throughput NGS has allowed high-resolution investigation of such epigenetic modifications across the entire genome, leading to the emergence of epigenomics. Through the identification of these patterns of epigenetic variation, we may be closer to understanding how the dynamic epigenome contributes to phenotypic diversity, even in organisms with a relatively small vocabulary of inherited variation. In addition, the comparisons between epigenomic profiles have also provided evidence of epigenetic markers associated with specific genotypes or growing conditions, which has been named plant epigenetics (Kapazoglou et al. 2018).

For example, Forestan et al. (2018) profiled chromatin states with genomic and transcriptomic sequencings in *Zea mays* to decipher how maize senses, adjusts, and rebounds following drought stress. In *Triticum aestivum*, small RNAs (sRNAs, e.g., microRNAs (miRNAs), heterochromatic small interfering RNAs, and short RNAs with regulatory roles) are implicated in drought stress (Budak et al. 2015). Consistently, epigenomic analysis in *Hordeum vulgare* has uncovered that drought and salt stress also lead to differential methylation patterns in DNA, and the effects are more dramatic in leaves than roots (Debode et al. 2019). Other plants such as *Oryza sativa*, *Solanum lycopersicum*, *Brassica napus*, and various legumes have also been analyzed for the contribution of epigenetics to stress adaptation. For such insights, two primary NGS-based technologies that are commonly used are whole-genome bisulfite sequencing (WBGS) for the single-base resolution of DNA methylation and chromatin immunoprecipitation (ChIP-seq) for the physical mapping of histone modifications and the associated chromatin states (Marudamuthu et al. 2023).

Whole-Genome Bisulfite Sequencing The technology provides an evaluation of the genome-wide distribution of DNA methylation. Here, genomic DNA is subjected to sodium bisulfite treatment, which converts unmethylated cytosines to uracil (other than methylated cytosines). Following PCR amplification, library construction, and sequencing, the treated and untreated sequences are compared to identify the methylation status. This approach has been successfully extended to studies of genome-wide landscapes of DNA methylation in crops, including *Zea mays* and *Hordeum vulgare* (Li et al. 2017).

Chromatin Immunoprecipitation Thanks to chip-based technologies (see NEB-ChIP and ChIP-seq), we can now map out where a transcription factor binds and which histones are modified across entire genomes. The widely used technique for mapping chromatin components in epigenomes is chromatin immunoprecipitation (ChIP) followed by sequencing (ChIP-seq). Using this approach, scientists can map protein–DNA interactions and histone modifications at high resolution. For example, Zhao et al. (2020) conducted genome-scale profiling of five histone modifications, using an improved ChIP (eChIP) method to isolate RNA polymerase II, for 20 elite representative rice lines to produce systematic epigenomic landscapes.

Environmental DNA Sequencing Environmental DNA (eDNA) biology has gained popularity for studying the microbiome of soils, aquatic environments, and other environmental samples. With this method, scientists can observe ecosystem evolution, measure biodiversity, and assess the quality of water and the soil without disturbing the natural processes. Trace DNA traces of trace DNA are continually released by organisms into the environment, from which, by analyzing environmental DNA, one can identify the species living in a particular place. Next-generation sequencing (NGS) allows thousands of species to be profiled in parallel from a single sample and provides the sensitivity required to detect low-abundance DNA fragments (Veilleux et al. 2021).

The general scheme of the eDNA sequencing workflow normally involves PCR amplification with either species-specific primers for edition-specific studies or universal primers for community-level studies. The amplified DNA is subsequently sequenced to learn about the species makeup of the sample. With the development of NGS, the performance of high-throughput biodiversity surveying is becoming more efficient and affordable.

A highly relevant tool is eDNA metabarcoding, which enables the sequencing of multiple taxa in highly compositional samples. Nonetheless, metabarcoding allows for more comprehensive and faster detection of biodiversity than that attainable through Single-Species methods. Nevertheless, some challenges remain. For example, the use of universal primers can cause primer preferential amplification, leading to taxa of different phyla being amplified to an uneven extent. Furthermore, the analysis of the enormous data sets generated by high-throughput sequencing is a crucial tool for the field of bioinformatics (Marudamuthu et al. 2023).

Genotyping by Sequencing Next-Generation Genotyping (NGG) or Genotyping-by-Sequencing (GBS) is an inexpensive and high-throughput genome-wide genotyping approach, which is increasingly used in plant and animal genetics. This strategy involves the digestion of the genomic DNA with restriction enzymes and the sequencing of these "reduced representation" libraries to determine a genome-wide set of single-nucleotide polymorphisms (SNPs). Since GBS produces a large number of random markers, it is capable of high-resolution genetic analysis and genotype calling (at only a small fraction of the cost of traditional methods). Its straightforward performance, fast process, high reproducibility, and capability of targeting genomic regions frequently poorly covered by capture-based methods render it especially attractive in plant crop improvement programs (Scheben et al. 2017).

Among the unique benefits of GBS are reduced sample handling, little to no PCR and purification steps, inexpensive barcodes, and the absence of size selection. The procedure was first established for *Zea mays* and has been adapted to many other species, including the most complex ones. Currently, GBS is being used widely for SNP discovery, population genetics analysis, and marker development in crop breeding. As an example, it has been used in leafy vegetables (*Lactuca sativa*), oilseed crops (*Brassica napus*), bioenergy crops (*Panicum virgatum*), and legumes (Glycine max). And indeed, the method has been applied mainly in cereal crops, including *Triticum aestivum*, *Oryza sativa*, *Sorghum bicolor*, *Hordeum vulgare*, and *Avena sativa*, and tuber and industrial crops such as *Manihot esculenta*, *Solanum tuberosum*, and *Gossypium* spp. (Marudamuthu et al. 2023).

Skim Sequencing Skim sequencing (also named low-coverage genome sequencing) can reduce the cost and labor required for high-density genotyping of multiple genomes. It is more powerful than genotyping-by-sequencing (GBS) and does not rely on genome reduction steps, yet it reduces the suffering from low depth of sequencing caused by those methods. This approach is commonly implemented in

RILs and doubled haploid populations and offers better coverage of the genome than GBS. Skim sequencing can also be useful for SNP discovery, recombination mapping, QTL mapping, and GWAS.

It requires a reference genome for the first pair. The sequence reads from the target population are aligned to the reference genome, respectively, and then the variant is called. Variants having more than 80% missing data are eliminated, whereas SNPs for the remaining variants are called. To reduce error, SNPs are filtered, while false positives are discarded, followed by genotype inference using imputation. This protocol improves the robustness and accuracy of the SNP genotyping. Read resequencing has been successfully applied in various other crops, such as *Oryza sativa*, *Gossypium* spp., *Nicotiana tabacum*, and *Cicer arietinum*, indicating its widespread use in crop genetics and breeding programs (Kumar et al. 2021).

6.9 NGS Application in Agriculture

Advances in next-generation sequencing (NGS) have revolutionized agricultural research by providing a quick and cost-effective way of obtaining comprehensive knowledge of plant genomes and transcriptomes. It allows for fine mapping of genomic variability, contributes to the accelerated breeding in the form of genome-wide association studies, marker-assisted selection, and genomic prediction. NGS also makes it possible to discover stress-responsive genes, which may help in the development of crops possessing increased tolerance to drought, salinity, and diseases. In addition to crop improvement, it finds application in germplasm characterization, evolutionary studies, and the construction of reference genomes for other plants. Moreover, NGS has facilitated the elucidation of plant-microbe interactions, which is paving the way for better disease control and for the development of sustainable agriculture. Altogether, these applications place NGS as a foundation of agricultural innovation, supporting food security and climate resilience.

Genome Selection/Trait Scanning Genomic selection (GS), a contemporary molecular-marker-based breeding method, employs genomic loci, molecular markers, and haplotypes to accelerate plant improvement. In contrast to the usual methods, GS tests thousands of genomic markers simultaneously in an attempt to predict complex traits. With the development of sequencing technologies, the generation of high-throughput and cost-effective genome-wide markers has conferred to GS as a valuable tool in agri-genomics, which can be applied in model and non-model crops, including those with large genomes or complex genomes. An advantage of this method is that selection can be performed without a complete reference genome. In practice, markers linked to desirable traits are deployed to screen large populations, and this approach is particularly useful for characteristics controlled by many genes and difficult to improve through traditional breeding standardization services (Marudamuthu et al. 2023).

Iterative progeny testing can be used to detect individuals that carry the most favorable genes and can be preferentially selected for further breeding. GS has been successfully implemented across various crops, including wheat, rice, maize, soybeans, and switchgrass (Bhat et al. 2016). More recently, its performance has been tested in common bean, where elite Andean breeding lines were evaluated in a set of environments, including drought-stress, irrigated, and low-P conditions to assess their agronomic potential (Keller et al. 2020). The evolution of crop pan-genomes has increased the potential of GS even more, increasing the diversity of genetic variability associated with phenotypic traits. Pan-genomes of tomato, barley, and pigeon pea have been previously utilized in GS studies to enhance the association between genomic data and characteristics of interest (Bayer et al. 2021). The construction of such pan-genomes is now recognized as an essential step for enhancing the accuracy and efficiency of genomic prediction models.

Genome-Wide Association Studies Genome-wide association studies (GWAS or GWAS mapping) are a well-developed approach to associate genetic variants and phenotypes. This method involves scanning the full genome of a large population using molecular markers to detect alleles related to specific traits. The associations generated are of significant relevance that can be directly utilized in crop enhancement plans. Notably, GWAS is indirectly linked to other characteristics that directly influence productivity and resilience of commercial crops, including genetic fingerprinting, as well as marker-assisted selection and estimation of net merit. SNP arrays enable the simultaneous analysis of a subset of millions of variants in a single DNA release, allowing detailed information about gene pool diversity. For the last 10 years, GWAS has been widely used in agronomy for understanding the genetics of traits of interest. It is the case for flowering time, disease resistance, panicle type, fertility restoration, and other important agronomic traits (Berhe et al. 2021).

This approach has been effectively used in model plants, including barley, wheat, rice, maize, soybean, sorghum, and cotton (Liu and Yan 2019). The inclusion of next-generation sequencing has arguably enhanced GWAS by allowing the discovery of hundreds of thousands of SNPs quickly and cost-effectively across the entire genome, and it has provided a foundation for high-density SNP genotyping platforms. Recent reports demonstrate the potential of GWAS for detecting genes associated with important agronomic traits. For instance, Yang et al. (2022) performed a GWAS on cucumber and uncovered genes for stem diameter, which is closely related to vigor and potential yield of plants. Discoveries such as these not only further deepen our insight into complex traits but also offer breeders tangible goals for the genetic improvement of crops (Yang et al. 2022).

Genotyping and Marker-Assisted Breeding With the development of genome resequencing, genetic diversity was identified using different accessions of the same species, thereby serving as a valuable resource for plant breeding. With diverse genotyping tools, much segregation data for mapping populations can be produced to construct high-density genetic maps. Sequenced genomes can also be a launching pad for the discovery of a myriad of molecular markers to help in crop improvement.

Of these, it is of significant importance to the identification of ecotypes with favorable characteristics, the development of sensitive and accurate markers (Wang et al. 2017).

Marker-assisted selection (MAS) has been an important tool for expediting the breeding of superior crop lines. It emphasizes the introduction of the genes that provide resistance to biotic and abiotic stresses, as well as increased yield and quality. In reality, attractive traits for this purpose can be introduced from the donor parents to a breeding line through markers closely associated with the QTL of interest. It not only allows accurate identification of trait-linked alleles but also improves selection efficiency in breeding programs for developing commercially valuable cultivars or elite germplasm. Apart from direct utility for breeding purposes, molecular markers have wider implications for plant science. Some of these markers, which are termed anchor markers, are highly conserved across different organisms and can be utilized in evolutionary studies and comparative genomics. The information provided by these markers will fill the gap between breeding programs and basic research, and, hence, the marker-assisted breeding is one of the most generic and robust approaches of modern crop improvement (Sharma et al. 2025).

Characterization of GMO Analysis The genetic compositions of genetically modified organisms (GMOs) must be determined to demonstrate successful foreign gene integration and its safe utilization. Molecular characterization usually aims to identify the genomic location, format, and copy number of the introduced gene in the plant genome. These aspects are relevant not only for the level of the desired trait and its expression, but also for the biosafety assessment and the selection of the transgenic events. Commonly, several molecular technologies, including Southern blotting, digital PCR, Quantitative Real-Time PCR, and PCR-based genome walking, are used for that purpose. Although these methods can be effective, they also have disadvantages in terms of efficiency, effort, or resolution.

The introduction of next-generation sequencing (NGS) technologies has revolutionized GMO profiling as it is now possible to provide a more global solution, faster, and with higher throughput. Thanks to its capability to map insertion sites, estimate copy numbers, and identify flanking sequences with high accuracy, NGS has a clear advantage over traditional methods. Furthermore, NGS-based pipelines can be easily automated and standardized, thereby leading to more consistent and reproducible results across laboratories (Wang et al. 2023). Various methods have been applied in this framework, ranging from whole genome sequencing to targeted resequencing, and hybrid methods that integrate NGS with site-discovery PCR, or DNA walking. Together, these developments have enabled NGS to become a workhorse for accurate and comprehensive profiling of transgenic crops (Debode et al. 2019).

References

Abbas HMK, Zhou YY, Huang HX, Qamar MT u, Wang MM, Lu S, Xu C, Liu Z, Wu TQ, Zhong YJ. QTL mapping, whole genome resequencing, and marker-assisted selection provide basics of early flowering in pumpkin. Plant Breed. 2022;141(2):266–76.

Afzal M, Salem S, Alghamdi HH, Migdadi MAK, Mirza SB, El-Harty E. Legume genomics and transcriptomics: from classic breeding to modern technologies. Saudi J Biol Sci. 2020;27(1):543–55.

Bayer PE, Petereit J, Danilevicz MF, Anderson R, Batley J, Edwards D. The application of pangenomics and machine learning in genomic selection in plants. Plant Genome. 2021;14(3):e20112.

Berhe M, Dossa K, You J, Mboup PA, Diallo IN, Diouf D, Zhang X, Wang L. Genome-wide association study and its applications in the non-model crop Sesamum indicum. BMC Plant Biol. 2021;21(1):283.

Bewicke-Copley F, Kumar EA, Palladino G, Korfi K, Wang J. Applications and analysis of targeted genomic sequencing in cancer studies. Comput Struct Biotechnol J. 2019;17:1348–59.

Bhat JA, Ali S, Romesh K, Salgotra ZA, Mir SD, Jadon V, Tyagi A, et al. Genomic selection in the era of next generation sequencing for complex traits in plant breeding. Front Genet. 2016;7:221.

Bleidorn C. Third generation sequencing: technology and its potential impact on evolutionary biodiversity research. Syst Biodivers. 2016;14(1):1–8.

Bryant DM, Johnson K, DiTommaso T, Tickle T, Couger MB, Payzin-Dogru D, Lee TJ, et al. A tissue-mapped axolotl de novo transcriptome enables identification of limb regeneration factors. Cell Rep. 2017;18(3):762–76.

Budak H, Hussain B, Khan Z, Ozturk NZ, Ullah N. From genetics to functional genomics: improvement in drought signaling and tolerance in wheat. Front Plant Sci. 2015;6:1012.

Casbon JA, Osborne RJ, Brenner S, Lichtenstein CP. A method for counting PCR template molecules with application to next-generation sequencing. Nucleic Acids Res. 2011;39(12):e81.

Castro-Wallace SL, Chiu CY, John KK, Stahl SE, Rubins KH, McIntyre ABR, Dworkin JP, et al. Nanopore DNA sequencing and genome assembly on the International Space Station. Sci Rep. 2017;7(1):18022.

Cristea PD. Conversion of nucleotides sequences into genomic signals. J Cell Mol Med. 2002;6(2):279–303.

Debode F, Hulin J, Charloteaux B, Coppieters W, Hanikenne M, Karim L, Berben G. Detection and identification of transgenic events by next generation sequencing combined with enrichment technologies. Sci Rep. 2019;9(1):15595.

Del Angel VD, Hjerde E, Sterck L, Capella-Gutierrez S, Notredame C, Vinnere Pettersson O, Amselem J, et al. Ten steps to get started in genome assembly and annotation. F1000Res. 2018;7:ELIXIR-148.

Emiyu K, Lelisa K. Review on illumina sequencing technology. Austin J Vet Sci Animal Husbandry. 2022;9(1):1088–91.

Ermini L, Driguez P. The application of long-read sequencing to cancer. Cancer. 2024;16(7):1275.

Fleury D, Baumann U, Langridge P. Plant genome sequencing: models for developing synteny maps and association mapping. In: Plant biotechnology and agriculture. Academic Press; 2012. p. 83–97.

Flusberg BA, Webster DR, Lee JH, Travers KJ, Olivares EC, Clark TA, Korlach J, Turner SW. Direct detection of DNA methylation during single-molecule, real-time sequencing. Nat Methods. 2010;7(6):461–5.

Forestan C, Farinati S, Rouster J, Lassagne H, Lauria M, Dal Ferro N, Varotto S. Control of maize vegetative and reproductive development, fertility, and rRNAs silencing by HISTONE DEACETYLASE 108. Genetics. 2018;208(4):1443–66.

Gao L, Gonda I, Sun H, Ma Q, Bao K, Tieman DM, Burzynski-Chang EA, et al. The tomato pan-genome uncovers new genes and a rare allele regulating fruit flavor. Nat Genet. 2019;51(6):1044–51.

Garalde DR, Snell EA, Jachimowicz D, Sipos B, Lloyd JH, Bruce M, Pantic N, et al. Highly parallel direct RNA sequencing on an array of nanopores. Nat Methods. 2018;15(3):201–6.
Goff SA, Ricke D, Lan T-H, Presting G, Wang R, Dunn M, Glazebrook J, et al. A draft sequence of the rice genome (Oryza sativa L. ssp. japonica). Science. 2002;296(5565):92–100.
Gupta AK, Gupta UD. Next generation sequencing and its applications. In: Animal biotechnology. Academic; 2020. p. 395–421.
Head SR, Komori HK, LaMere SA, Whisenant T, Van Nieuwerburgh F, Salomon DR, Ordoukhanian P. Library construction for next-generation sequencing: overviews and challenges. BioTechniques. 2014;56(2):61–77.
Heather JM, Chain B. The sequence of sequencers: the history of sequencing DNA. Genomics. 2016;107(1):1–8.
Hess JF, Hess ME, Zengerle R, Paust N, Boerries M, Hutzenlaub T. Automated library preparation for whole genome sequencing by centrifugal microfluidics. Anal Chim Acta. 2021;1182:338954.
Holley RW, Apgar J, Everett GA, Madison JT, Marquisee M, Merrill SH, Penswick JR, Zamir A. Structure of a ribonucleic acid. Science. 1965;147(3664):1462–5.
Huang X, Feng Q, Qian Q, Zhao Q, Wang L, Wang A, Guan J, et al. High-throughput genotyping by whole-genome resequencing. Genome Res. 2009;19(6):1068–76.
Jain M, Olsen HE, Turner DJ, Stoddart D, Bulazel KV, Paten B, Haussler D, Willard HF, Akeson M, Miga KH. Linear assembly of a human centromere on the Y chromosome. Nat Biotechnol. 2018;36(4):321–3.
Jonkhout N, Tran J, Smith MA, Schonrock N, Mattick JS, Novoa EM. The RNA modification landscape in human disease. RNA. 2017;23(12):1754–69.
Kapazoglou A, Ganopoulos I, Tani E, Tsaftaris A. Epigenetics, epigenomics and crop improvement. In: Adv Bot Res, vol. 86. Academic Press; 2018. p. 287–324.
Kchouk M, Gibrat J-F, Elloumi M. Generations of sequencing technologies: from first to next generation. Biol Med. 2017;9(3) https://doi.org/10.4172/0974-8369.1000395.
Keller B, Ariza-Suarez D, De la Hoz J, Aparicio JS, Portilla-Benavides AE, Buendia HF, Mayor VM, Studer B, Raatz B. Genomic prediction of agronomic traits in common bean (Phaseolus vulgaris L.) under environmental stress. Front Plant Sci. 2020;11:1001.
Kim D. High-throughput detection, annotation, and interpretation of genomic variants using next-generation sequencing data. PhD diss., Johns Hopkins University; 2015.
Koepfli K-P, Paten B, Genome 10K Community of Scientists, and Stephen J. O'Brien. The Genome 10K Project: a way forward. Annu Rev Anim Biosci. 2015;3(1):57–111.
Kulski J. Next-generation sequencing—an overview of the history, tools, and "Omic" applications. In: Next generation sequencing-advances, applications and challenges; 2016. p. 3–60.
Kumar P, Choudhary M, Jat BS, Kumar B, Singh V, Kumar V, Singla D, Rakshit S. Skim sequencing: an advanced NGS technology for crop improvement. J Genet. 2021;100(2):38.
Kumar KR, Cowley MJ, Davis RL. Next-generation sequencing and emerging technologies. Semin Thromb Hemost. 2024;50(07):1026–38. Thieme Medical Publishers.
Lai J, Li R, Xu X, Jin W, Xu M, Zhao H, Xiang Z, et al. Genome-wide patterns of genetic variation among elite maize inbred lines. Nat Genet. 2010;42(11):1027–30.
Lam H-M, Xu X, Liu X, Chen W, Yang G, Wong F-L, Li M-W, et al. Resequencing of 31 wild and cultivated soybean genomes identifies patterns of genetic diversity and selection. Nat Genet. 2010;42(12):1053–9.
Li Q, Peter J, Hermanson a NM, Springer. detection of DNA methylation by whole-genome bisulfite sequencing. In: Maize: methods and protocols. New York: Springer; 2017. p. 185–96.
Liao X, Li M, Zou Y, Wu F-X, Wang J. Current challenges and solutions of de novo assembly. Quant Biol. 2019;7(2):90–109.
Liu H-J, Yan J. Crop genome-wide association study: a harvest of biological relevance. Plant J. 2019;97(1):8–18.
Lu H, Giordano F, Ning Z. Oxford Nanopore MinION sequencing and genome assembly. Genomics Proteomics Bioinformatics. 2016;14(5):265–79.

Lu K, Wei L, Li X, Wang Y, Wu J, Liu M, Zhang C, et al. Whole-genome resequencing reveals Brassica napus origin and genetic loci involved in its improvement. Nat Commun. 2019;10(1):1154.

Marudamuthu B, Sharma T, Purru S, Soam SK, Rao CS. Next-generation sequencing technology: a boon to agriculture. Genet Resour Crop Evol. 2023;70(2):353–72.

Mason T. PacBio: long read sequencing capability for whole genome sequencing. Master's thesis. Harvard University; 2021.

Metzker ML. Sequencing technologies—the next generation. Nat Rev Genet. 2010;11(1):31–46.

Michael TP, VanBuren R. Progress, challenges and the future of crop genomes. Curr Opin Plant Biol. 2015;24:71–81.

Minei R, Hoshina R, Ogura A. De novo assembly of middle-sized genome using MinION and Illumina sequencers. BMC Genomics. 2018;19(1):700.

Natarajan P, Parani M. De novo assembly and transcriptome analysis of five major tissues of Jatropha curcas L. using GS FLX titanium platform of 454 pyrosequencing. BMC Genomics. 2011;12(1):191.

Pankin A, Altmüller J, Becker C, von Korff M. Targeted resequencing reveals genomic signatures of barley domestication. New Phytol. 2018;218(3):1247–59.

Payne A, Holmes N, Rakyan V, Loose M. BulkVis: a graphical viewer for Oxford nanopore bulk FAST5 files. Bioinformatics. 2019;35(13):2193–8.

Pereira R, Oliveira J, Sousa M. Bioinformatics and computational tools for next-generation sequencing analysis in clinical genetics. J Clin Med. 2020;9(1):132.

Rao G, Zhang J, Liu X, Lin C, Xin H, Xue L, Wang C. De novo assembly of a new Olea europaea genome accession using nanopore sequencing. Hortic Res. 2021;8:64.

Sanger F, Nicklen S, Coulson AR. DNA sequencing with chain-terminating inhibitors. Proc Natl Acad Sci. 1977;74(12):5463–7.

Satam H, Joshi K, Mangrolia U, Waghoo S, Zaidi G, Rawool S, Thakare RP, et al. Next-generation sequencing technology: current trends and advancements. Biology. 2023;12(7):997.

Scheben A, Batley J, Edwards D. Genotyping-by-sequencing approaches to characterize crop genomes: choosing the right tool for the right application. Plant Biotechnol J. 2017;15(2):149–61.

Schmutz J, Cannon SB, Schlueter J, Ma J, Mitros T, Nelson W, Hyten DL, et al. Genome sequence of the palaeopolyploid soybean. Nature. 2010;463(7278):178–83.

Schreiber M, Stein N, Mascher M. Genomic approaches for studying crop evolution. Genome Biol. 2018;19(1):140.

Schuster SC. Next-generation sequencing transforms today's biology. Nat Methods. 2008;5(1):16–8.

Sharma S, Sharma J, Singh I, Bindra S. Role of marker-assisted breeding in crop improvement. In: Biotechnology innovations and sustainability for zero hunger. CRC Press; 2025. p. 326–43.

Shook B, Braff L, Nelson K, Aguilera K. Molecular biology and genetics. In: Explorations: an open invitation to biological anthropology. 2nd ed; 2023.

Singh D, Mathur S, Prasad M, Ranjan R. Next-generation sequencing in medicinal plants: recent progress, opportunities, and challenges. J Plant Growth Regul. 2025;44(4):1448–64.

Steuernagel B, Taudien S, Gundlach H, Seidel M, Ariyadasa R, Schulte D, Petzold A, et al. De novo 454 sequencing of barcoded BAC pools for comprehensive gene survey and genome analysis in the complex genome of barley. BMC Genomics. 2009;10(1):547.

Tuskan GA, Difazio S, Jansson S, Bohlmann J, Grigoriev I, Hellsten U, Putnam N, et al. The genome of black cottonwood, Populus trichocarpa (Torr. & Gray). Science. 2006;313(5793):1596–604.

VanBuren R, Bryant D, Edger PP, Tang H, Burgess D, Challabathula D, Spittle K, et al. Single-molecule sequencing of the desiccation-tolerant grass Oropetium thomaeum. Nature. 2015;527(7579):508–11.

Veilleux HD, Misutka MD, Glover CN. Environmental DNA and environmental RNA: current and prospective applications for biological monitoring. Sci Total Environ. 2021;782:146891.

Velasco R, Zharkikh A, Troggio M, Cartwright DA, Cestaro A, Pruss D, Pindo M, et al. A high quality draft consensus sequence of the genome of a heterozygous grapevine variety. PLoS One. 2007;2(12):e1326.

Visser RGF, Bachem CWB, de Boer JM, Bryan GJ, Chakrabati SK, Feingold S, Gromadka R, et al. Sequencing the potato genome: outline and first results to come from the elucidation of the sequence of the world's third most important food crop. Am J Potato Res. 2009;86(6):417–29.

Wang M. Next-generation sequencing (NGS). In: Clinical molecular diagnostics. Singapore: Springer; 2021. p. 305–27.

Wang W, Cao XH, Miclăuş M, Xu J, Xiong W. The promise of agriculture genomics. Int J Genomics. 2017;2017:9743749.

Wang Y, Ma X, Yang L, Ye H, Jia R. Direct Pacbio sequencing methods and applications for different types of DNA sequences. bioRxiv. 2023;2023:2012–23. https://doi.org/10.1101/2023.12.12.571020.

Watson JD, Crick FHC. The structure of DNA. In: Cold Spring Harbor symposia on quantitative biology, vol. 18. Cold Spring Harbor Laboratory Press; 1953. p. 123–31.

Wei T, Van Treuren R, Liu X, Zhang Z, Chen J, Liu Y, Dong S, et al. Whole-genome resequencing of 445 Lactuca accessions reveals the domestication history of cultivated lettuce. Nat Genet. 2021;53(5):752–60.

Yang Y, Dong S, Miao H, Liu X, Dai Z, Li X, Gu X, Zhang S. Genome-wide association studies reveal candidate genes related to stem diameter in cucumber (Cucumis sativus l.). Genes. 2022;13(6):1095.

Zhao L, Xie L, Zhang Q, Ouyang W, Deng L, Guan P, Ma M, et al. Integrative analysis of reference epigenomes in 20 rice varieties. Nat Commun. 2020;11(1):2658.

Chapter 7
Molecular Data Analysis

Sadaf Safder and Saqib Ali

7.1 Introduction

Molecular plant research follows the path from genotype to phenotype. It asks how information in DNA becomes structure, activity, and behavior in a living plant. The work spans genetics and the central dogma, so DNA, RNA, and proteins are all in scope. It also tracks changes over time, including development, evolution, and responses to the environment. Molecular markers remain core tools for measuring diversity within and between species. Platform choice depends on utilitarian concerns: low cost per assay, low hardware requirements, high throughput, user-friendliness, and the ease of workflow development and automation. Contemporary sequencing, particularly next-generation sequencing, combined with companion molecular assays, enables thorough analysis of plant genomes, gene expression programs, and population structure. These data create a reference for comparative research and clarify evolutionary relationships. Complex data files are now more accurate because of the combination of molecular techniques and artificial intelligence (AI) models for data analysis. Large-scale omics data analysis is further facilitated by cloud computing and big data, and 3D genomics visualization approaches specifically visualize spatial gene regulation and models. This chapter delves into transformational computational tools such as machine learning (ML) and deep

S. Safder (✉)
Center for Advanced Studies in Agriculture and Food Security, University of Agriculture, Faisalabad, Pakistan
e-mail: sadaf.safdar@uaf.edu.pk

S. Ali
Center for Advanced Studies in Agriculture and Food Security, University of Agriculture, Faisalabad, Pakistan

Department of Computer Science, University of Agriculture, Faisalabad, Pakistan
e-mail: saqib@uaf.edu.pk

S. Ijaz et al. (eds.), *Molecular Biology Methods*,
https://doi.org/10.1007/978-981-92-2642-9_7

learning for analyzing molecular plant data. This chapter lays out a roadmap for integrating AI techniques to explore insights from large molecular datasets and AI-based anomics. It begins with machine learning foundations. Deep neural networks convert sequences and structures into features. Network biology links these features into system maps. AI methods extend across omics. Genome, transcriptome, proteome, and metabolome align into one view. Cloud computing and big data carry the scale. Reproducible pipelines protect provenance and security. Multi-omics platforms fuse modalities and yield interpretable models. Three-dimensional genomics adds spatial context. Chromatin loops and compartments shape regulation. Visualization makes the structure testable. Quantum computing enters as a frontier. New algorithms promise faster search, simulation, and optimization.

7.2 Machine Learning (Deep Learning and Neural Networks)

Artificial intelligence is a field of computer science that builds systems that improve with data. Machine learning is the applied part of AI that learns patterns from examples rather than hard-coded rules. Two broad modes are used. In supervised learning, the training set includes labels or target values, and the model learns to map inputs to those targets. In unsupervised learning, the data lack labels, and the goal is to uncover structure, such as clusters or latent factors. Since most machine learning algorithms use supervised learning, complex computational techniques that can be divided into phases are needed. The following basic phases must always be completed for the development of supervised algorithms to be considered a successful process: (i) selecting data sources, which is crucial and necessitates more precise databases and the least amount of data redundancy; (ii) extracting features from the chosen data; (iii) assessing and choosing the primary attributes to be studied; (iv) selecting algorithms; (v) developing prediction/classification models; and (iv) evaluating prediction/classification performance (Silva et al. 2019).

Supervised, unsupervised, and semi-supervised machine learning techniques are available. While unsupervised methods are given unlabeled data for training and develop a prediction model by analyzing and structuring the input data, supervised algorithms create prediction models from previously labeled data in the training set. When the training set consists of both labeled and unlabeled data, semi-supervised algorithms combine the benefits of supervised and unsupervised methods.

The advancement of machine learning has led to the creation of numerous algorithms. Various algorithms work well for several types of data and activities. Using machine learning technologies requires selecting the appropriate algorithm or algorithms. Labeled output in the training data set is used in supervised machine learning, as previously mentioned. Its objective is to forecast unlabeled output by using novel inputs. Depending on whether discrete or continuous outputs are mapped, it can be further classified as either regression or classification. Support Vector

Table 7.1 The kinds of machine learning algorithms that are in use and their purpose

ML method	Functional class	Common methods	Role
Supervised	Categorization	Naive Bayes classifier, nearest neighbor, SVM	Assign observations to preset classes by learning decision boundaries from labeled instances
	Quantitative response modeling	Linear regression, support vector regression, ridge regression	Estimate how input variables affect continuous targets to model numerical results
Unsupervised	Grouping	K-means clustering, mean-shift clustering, DBSCAN	Identify natural clusters within data by grouping samples with similar structures without using labels
	Feature space compression	PCA, Kernel PCA	Data having several dimensions can be transformed into a space with fewer, more significant dimensions
	Anomaly and deviation detection	Index-based algorithm, nested-loop algorithm, cell-based algorithm	Find unexpected observations that deviate from more extensive sets of observations

Machine (SVM), neural networks, random forest, and penalized logistic regression are examples of classification techniques (Sun et al. 2018). Table 7.1 describes the supervised and unsupervised machine learning methods along with their functional class, commonly used methods, and role in molecular data analysis.

7.2.1 *CNN for Motif Discovery and Regulatory Signal*

Convolutional neural networks (CNNs) naturally adapt to biological sequences, as local patterns drive many molecular events. One-hot encodings or k-mer embeddings map bases or residues to numeric arrays, then small convolutional filters sweep across positions and act as detectors for short motifs. Early layers capture simple signals, such as GC-rich tracts or canonical splice motifs. Deeper layers combine these signals into higher order features that resemble cooperative binding or composite elements. Pooling aggregates evidence across variable positions and yields robust scores even when the motif position shifts. This shift tolerance is important for enhancer scanning and genome-wide prediction, where the same pattern may appear in many locations.

CNNs have become standard for transcription factor binding prediction, chromatin accessibility, and histone mark profiling. A single model can output multiple assays simultaneously and share features across tasks, improving sample efficiency and reducing overfitting when labels are scarce. Strand augmentation with reverse complements stabilizes learning and enforces the symmetry expected in double-stranded DNA. Dilated convolutions extend the receptive field without incurring a high memory cost, thereby allowing the network to connect motif clusters separated by hundreds of bases. Residual connections stabilize gradients during training and enable deeper stacks to learn nuanced regulatory grammars.

Interpretability methods help link predictions to biological insight rather than treating the network as a black box. Saliency maps, in silico mutagenesis, and integrated gradient scores mark positions that drive the output and often recover known motifs while revealing noncanonical variants. These tools guide experimental validation and refine hypotheses about cis-regulatory logic in specific cells or conditions. When multi-omic tracks are available, CNNs integrate sequence data with accessibility or methylation data and improve context-aware calls for active elements. The same architectures extend to RNA by scanning for signals that shape splicing or RNA-binding protein occupancy. In proteins, one-dimensional convolutions over residue sequences detect local structural propensities and secondary-structure patterns, which are then fed to downstream modules for contact or distance prediction. Despite strong performance, limits do appear when the task depends on long-range communication across kilobase spans or on complex three-dimensional constraints. Purely local filters struggle to capture distal enhancer–promoter interactions or to coordinate residues that are far apart in sequence yet close in space. Hybrid designs mitigate this by using wider dilations, attention blocks, or graph layers, but the gap points to architectures that model global context more directly (Sajitha et al. 2024).

7.2.2 Recurrent Neural Networks for Sequential Context and Structural Signals

Recurrent neural networks (RNNs) address temporal and ordered dependencies by maintaining state as the model reads a sequence. Long short-term memory and gated recurrent units handle vanishing gradients and can retain information over longer spans than vanilla RNNs. Bidirectional variants pass through the sequence in both directions, so the prediction at each position can use upstream and downstream context. This bidirectional view is well-suited for tasks in which the signal is symmetric, or the flanking sequence modulates the core motif.

In genomic analysis, RNNs support splice site recognition, promoter classification, polyadenylation site detection, and the identification of RNA regulatory elements. These tasks benefit from context because the same short word can indicate different functions depending on neighbors and distance to landmarks. RNNs summarize context into a compact state vector, providing a learned alternative to rule-based windows. For RNA, modules that estimate base-pairing tendencies capture sequence features that correlate with secondary structure and improve binding-site prediction for many RNA-binding proteins.

Protein modeling also benefits from recurrent encoders that convert residue sequences into informative embeddings. These embeddings capture local chemistry, hydrophobic patterns, and evolutionary signals when multiple sequence alignments are provided as input streams. The embeddings feed contact map predictors that infer residue–residue proximity and guide downstream structure assemblies. Some systems stack recurrent blocks with convolutional heads so that long-term memory handles global constraints while convolutions focus on short motifs such as turns and helices. Others add conditional random fields on top of RNN outputs to produce smooth secondary-structure labels along the chain.

RNNs are compact and often easier to train than very deep convolutional stacks on modest hardware. They stream long sequences without storing full attention matrices and can process reads or contigs as they arrive, which suits workflows that need online inference. Yet sequence lengths in genomics can reach hundreds of thousands of bases, and memory still becomes a bottleneck for strict recurrence. Furthermore, very long-range coupling across regulatory domains remains challenging when the model must rely on a finite hidden state. These factors opened the way for attention-based models that compute global relationships in parallel (Silva et al. 2019).

7.2.3 *Transformer Models for Long-Range Dependence, Structure, and Regulatory Logic*

Transformers replace recurrence with self-attention so each position can attend to every other position in one pass. This mechanism captures long-range interactions that shape regulation and folding. Positional encodings provide a map of order, while multi-head attention learns different relational views such as motif co-occurrence, spacing rules, or residue compatibility. Parallel computation improves throughput and stabilizes optimization on large datasets that mix many cell types or species.

Pretraining on unlabeled sequences has become a central strategy. Masked language modeling on DNA, RNA, or protein corpora teaches the network to recover masked tokens from context and thus to learn the statistics of the biological code. The pretrained model is then fine-tuned for specific tasks like variant effect prediction, enhancer activity scoring, splice alteration screening, or protein property estimation. This two-stage process shortens convergence time and often boosts accuracy in settings with limited labels. Tokenization choices influence success. Nucleotides can remain as single tokens or be grouped into k-mers to enrich the alphabet. Proteins naturally use a 20-token vocabulary, and some pipelines augment with special tokens for rare or modified residues. Long sequences require efficient attention. Sparse or sliding-window patterns, low-rank kernels, or recurrent memory segments reduce quadratic cost while preserving the ability to link distant segments.

Protein structure prediction illustrates the strength of attention in building consistent global geometry. Pairwise attention layers learn residue–residue relations and predict distances or orientations that define a contact graph. An end-to-end pipeline can then fold the chain by optimizing atomic coordinates under those learned constraints. The same approach supports complex modeling when coevolutionary signals are present across homologous sequences or when multiple chains interact. Fine-grained heads estimate disorder, solvent accessibility, and side-chain packing, which improves docking and design tasks.

In regulatory genomics, transformers scan long loci and connect motifs that operate in concert across enhancers, promoters, and insulators. Attention heads often specialize: some focus on core promoter neighborhoods, others track motif spacing in distal enhancers, while another set links enhancer signals to promoter outcomes across tens of kilobases. When combined with chromatin accessibility, methylation, or histone marks, the model grounds sequence signals in cell context and yields precise maps of active elements. Variant scoring falls within the same framework by running the model on reference and alternative alleles and comparing their predicted activities. It supports fine-mapping of causal variants in noncoding disease loci, where the effect is subtle and position-dependent (Ghosh et al. 2025).

Interpretability remains essential. Attention maps provide a first read of which positions interact, yet raw attention weights can be misleading. Gradient-based attribution and refitting tests add confidence and highlight condition-specific drivers. Controlled perturbations, such as motif knockouts or synthetic motif insertions, reveal grammar rules that the network has internalized. These experiments often recover known logic, like spacing preferences in composite motifs, and expose novel combinations that merit wet-lab follow-up.

Training choices must balance accuracy, stability, and computational limits. Curriculum strategies start with short segments and extend in length as learning progresses. Label noise is common in functional genomics and benefits from robust losses and calibration. Class imbalance occurs when positive regulatory sites are rare in a sea of negatives; hard-negative mining and focal losses help recover rare signals. Data augmentation improves invariance. Reverse complements enforce strand symmetry. Random shifts prevent overfitting to absolute positions. For proteins, masked spans and dropout over residues reduce shortcut learning and improve generalization to distant families. Across tasks, evaluation should reflect biology rather than only raw accuracy. The area under the precision–recall curve is informative when positives are rare. Calibration curves matter when predictions guide experimental design. For structure, distance errors, and TM-like scores, measure geometry consistency and report them together with confidence estimates. Reproducibility requires fixed splits by chromosome or by protein family to avoid leakage across training and test sets. External validation on independent cohorts or assays gives a realistic sense of deployment performance. The workflow for deep learning across sequence and structure is summarized in Fig. 7.1.

Inputs include DNA, RNA, and protein sequences, with optional multiple sequence alignments, predicted or known structures, and cell-context tracks such as accessibility or histone marks. Sequences are encoded with one-hot or k-mer tokens. Learned embeddings add positional information for order and distance. The neural core combines three families. CNNs act as motif detectors, capturing short-range signals that mark binding sites or structural propensities. RNN variants such as

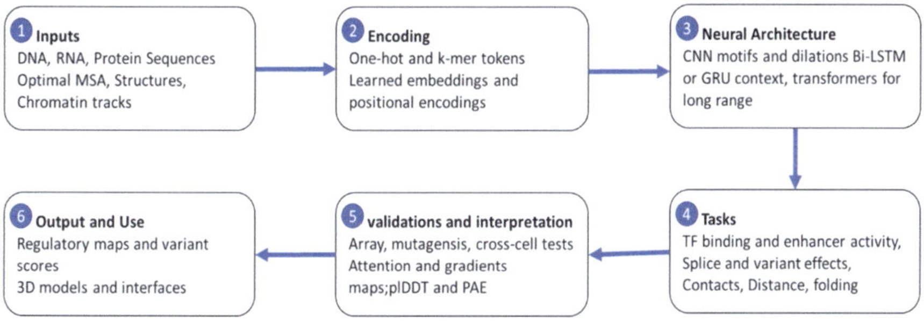

Fig. 7.1 Deep learning workflow for sequence and structural analysis

BiLSTM or GRU summarize sequential context, allowing flanking regions to modulate the local signal. Transformers model long-range dependencies through self-attention and link distant elements across kilobases or distant residues along a chain (Moeckel et al. 2024).

Task heads specialize in the shared representation. Regulatory heads predict transcription factor binding, enhancer activity, splice changes, and variant effects. Structural heads estimate residue contacts, distances, orientations, and fold quality. Validation uses orthogonal assays, mutagenesis, cross-cell or cross-species tests, and structure confidence metrics such as pLDDT or predicted alignment error. Outputs include genome-wide regulatory maps, fine-grained variant scores, folded models, and interface annotations. The figure highlights a single pass from raw sequence to actionable outputs and shows how the three architectures complement each other across motif, context, and long-range logic.

7.3 Network Biology

Biology is full of parts that work together. A protein binds another protein, and a signal moves through a pathway. A transcription factor binds to DNA and alters the level of gene expression. A metabolite feeds an enzyme, and a reaction runs faster. When these many small events are drawn one by one, the picture looks crowded. When the same events are drawn as a network, the picture becomes clear. Nodes stand for molecules or genes. Edges stand for the interactions that link them. The drawing is simple, yet it reveals how local events add up to system behavior.

Network views showed that most proteins in a cell are not isolated. Early maps of yeast showed that thousands of interactions form a single giant connected component. That simple fact changed how function was assigned. If an unknown protein sits next to many proteins of a known complex, the unknown one often shares the same job. The same logic holds for pathways and diseases. If many disease genes fall in the same region of a network, genes nearby may also be involved. Therefore, network biology became a core lens for genomics and proteomics. It is a way to handle complexity without losing detail.

High-throughput technologies made this possible. Two-hybrid screens and affinity purification gave the first large protein–protein interaction maps. Microarrays, then RNA-seq, profiled genome-wide expression. Next-generation sequencing mapped regulatory elements and chromatin landscapes. Each technology produced links among entities, thereby creating a network. Over time, these networks were curated in public resources that can be searched, downloaded, and joined. Today, a study can span several layers simultaneously and ask how regulation, signaling, and metabolism connect within a single model (Ehlers et al. 2025).

7.3.1 Structure of Biological Network

A network is a graph. The choice of nodes and edges depends on the question. Protein–protein interaction (PPI) maps treat proteins as nodes and their physical contacts as edges. Edges are usually undirected and can carry weights that reflect the strength or confidence of the evidence. Gene regulatory networks (GRNs) use transcription factors and their target genes as nodes, linked by directed edges that indicate activation or repression. Time-course data or perturbation screens add direction and strength. Signaling networks connect receptors, kinases, and substrates. Edges carry the type of action, such as phosphorylation or inhibition, and follow direction from upstream to downstream (Omony 2014).

Metabolic networks place metabolites or enzymes as nodes. Edges follow the stoichiometry of reactions. Bipartite and hypergraph models are used when reactions involve several inputs and outputs. They keep roles clear by separating molecule types or by linking many participants in a single edge. Co-expression networks link genes that rise and fall together across samples. Edges encode correlation, and the resulting clusters often mark modules with shared function.

These choices lead to different math but the same core ideas. Degree counts the number of partners a node has. Centrality ranks nodes that sit on many shortest paths and thus control flow. The clustering coefficient measures how tightly neighbors connect. Communities or modules group nodes that talk to each other more than they speak to the rest of the graph. Network motifs, such as feed-forward loops, appear more often than expected by chance and usually point to control logic. Many molecular networks show right-skewed degree distributions with a few hubs and many low-degree nodes. Such a structure makes networks efficient and robust yet sensitive when hubs fail.

Real systems change over time and across contexts. A regulatory edge may be strong in one cell type and weak in another. A signaling edge may appear only after a ligand binds. For this reason, modern studies often use context-specific networks. The backbone comes from curated interactions. These edge strengths are derived directly from available data, for example, from expression, chromatin marks, or from phospho-proteomics (Schirner et al. 2023), which generates an active map that follows the biology graph.

7.3.2 From Data to Discovery: Resources and Analysis Flow

A practical workflow starts with data collection and curation. Public resources gather interactions from experiments and from text mining with expert review. For PPIs, widely used collections provide large cross-species maps and often include evidence codes and confidence scores. For regulatory interactions, dedicated databases offer transcription factors, binding sites, and target genes for model organisms and humans. Pathway resources such as KEGG and Reactome curate signaling and metabolic routes with precise reaction semantics. Each source has strengths and blind spots, so merging them improves coverage. Versioning and provenance should be recorded so analyses remain reproducible (Omony 2014).

Next comes network assembly. Decide what counts as a node and an edge for the study. Choose whether edges are directed or undirected and whether they carry weights. Apply quality filters that keep high-confidence interactions. Integrate omics signals to adjust edge weights. For example, upregulated genes within a module may strengthen connections in a co-expression map. Chromatin accessibility can influence edge weights in a regulatory graph. Phospho-site evidence can lift edges in a kinase network (Liu et al. 2020).

With the graph in place, several analysis layers reveal structure and function.

- Topological profiling describes the graph with degree distributions, clustering, assortativity, path lengths, and centrality. Hub and bottleneck nodes often mark essential genes or drug targets.
- Module and community detection partition the graph into clusters. Modules frequently align with protein complexes or pathway segments. Enrichment tests then connect modules to known annotations and phenotypes.
- Network motifs and paths expose control logic. Feed-forward loops buffer noise and speed responses. Feedback loops stabilize or oscillate. Shortest-path and random-walk analyses reveal how signals may travel from a receptor to a transcriptional response.
- Diffusion and propagation map how perturbations spread. These methods prioritize disease genes by propagating evidence through the network to identify nodes that collect signals from known seeds.
- Reconstruction of GRNs infers edges from data. Approaches range from information-theoretic measures to sparsity-based regression to dynamic models for time-course data. Single-cell profiles add resolution by capturing state transitions and lineage.
- Network-based GWAS and variant interpretation project association signals onto the graph. Signals that cluster within a subnetwork point to pathways rather than single genes, helping move from a list of loci to a testable mechanism.

The final step is visualization and communication. Force-directed layouts give an intuitive map with hubs at the center and modules in colored islands. Edge thickness can represent weight. Arrows show direction. Tooling is now rich, allowing interactive filtering and figure export. Good visuals do not replace analysis, but they do expose patterns and outliers that numbers alone may hide.

7.3.3 Protein–Protein Interaction (PPI) Networks

The PPI network, also called the interactome, maps proteins as nodes and their physical contacts as edges. These links come from experiments such as yeast two-hybrid assays, affinity purification followed by mass spectrometry, or cross-linking approaches. Real interactomes are large: in budding yeast, the map spans thousands of proteins with tens of thousands of interactions, and in humans, it spans ~20,000 proteins with hundreds of thousands of edges. When plotted without filtering, even a small slice can resemble a "hairball," which is expected for dense systems and is why analysts use confidence weighting, clustering, or module layouts to make structure visible (Ma and Liao 2020).

GRN captures control of gene expression. Nodes are genes, including transcription factors (TFs) and their targets. Edges encode regulatory influence, activation, or repression linking a TF to a downstream gene. PPIs are often measured directly. GRNs are usually inferred from evidence like gene expression profiles, open chromatin maps, TF binding assays, and targeted perturbations. Regulation is context-specific. Edge strength and even edge presence can shift with cell type, tissue, developmental stage, environment, or disease. Well-curated GRN resources exist for major organisms, and study-specific GRNs can be reconstructed for the condition of interest (Zhivkoplias et al. 2022).

Most PPI maps are modeled as undirected graphs, because physical binding is mutual. GRNs are naturally directed graphs, with edges pointing from regulator to target; edges may carry a sign to distinguish activation from inhibition and a weight to reflect confidence or effect size. This formalism keeps the biology intact while enabling standard analysis. Mapping PPIs and GRNs shows how proteins assemble into complexes, how signals move toward the transcriptome, and where the key control points sit along a pathway. These maps then drive practice: rank likely disease genes, pick intervention targets, design clean perturbations, and build module-level biomarkers that hold up across cohorts.

7.3.4 Network Inference

Interest in biological network inference stretches back decades, alongside early developments in graph theory. What began as abstract work on nodes and edges is now central to reverse-engineering cellular wiring from genomic and proteomic readouts. The goal is simple to state yet hard to achieve: given data collected in a defined context, recover the structure of the network and the strength of its interactions. It is inferred that a genetic regulatory network links transcription factors to their targets. Some authors use "genetic network" more broadly to refer to any set of genes that regulate one another's activity. Formal definitions vary with edge semantics. Directed edges capture causal influence, as in TF-gene or receptor-kinase interactions. Undirected edges capture symmetric relations, as in protein–protein contacts. Signaling pathways are directed by nature. Protein–protein interaction maps are typically undirected.

Biological network methods constitute a large catalogue of approaches. At a high level, they fall into two classes. Discrete-state models represent each node with a small set of states and update nodes with logical rules (AND, OR, NOT, XOR). These models are transparent and useful for qualitative dynamics and control. Continuous-state models treat mRNA and protein levels as time-dependent variables and fit parameters in differential, regression, Bayesian, or neural formulations. These models capture graded effects and timing.

None of these methods is perfect. Each recovery recovers only part of the "true" wiring, and performance depends on the data type, sample size, noise level, and the calibration method. Choosing a process depends on the question, the assays available, and the practical constraints on computation and validation. Role of in-silico studies: Historically, limited experimental data led to heavy reliance on synthetic benchmarks and computational simulations. This practice remains valuable because models can be evaluated under controlled perturbations and their predictions compared against known or held-out reference data. Community efforts that compare methods on common tasks have accelerated progress and clarified strengths and weaknesses under different conditions. Multiple biological levels: Network inference can be applied across multiple biological layers, including gene regulatory networks, metabolic networks, and protein-protein interaction networks. Integrating these layers into a unified model remains challenging but is increasingly feasible through multi-omics data integration and advanced computational approaches. Gene-level information often anchors projections into biochemical networks, which then connect transcripts to reactions and complexes (Zhivkoplias et al. 2022).

Good network inference begins with careful experimental planning and solid data. Optimal Experimental Design (OED) is the quantitative plan for an experiment. It specifies which assays to run, when to sample, and how many specimens per condition, so the dataset carries maximum information. The aim is simple: enough technical and biological replicates to stabilize estimates and ensure reliable downstream inference. These choices protect the downstream steps of network reconstruction and parameter estimation from noise and bias. Figure 7.2 depicts a two-tier workflow for network inference. The upper panel covers planning and benchmarking: start with experimental design and data collection, incorporate prior knowledge, choose a GNR evaluation strategy, infer a candidate network, and, when a reference network exists, compare the inferred structure to the known one. The lower panel shows the perturbation-driven loop that delivers mechanistic insight: design informative excitations optimal experimental design (OED)/sequential experimental design (SED), simulate in silico data to test identifiability, run the planned experiments, and then assess model fit while estimating edge strengths, signs, and kinetics. Biological interpretation closes the loop. Insights from the model point to what to change and, when gaps remain, to a redesigned experiment. Taken together, the panels trace the path from study design to a validated network and show how simulation, prior knowledge, and targeted perturbations steady both edge discovery and parameter fitting.

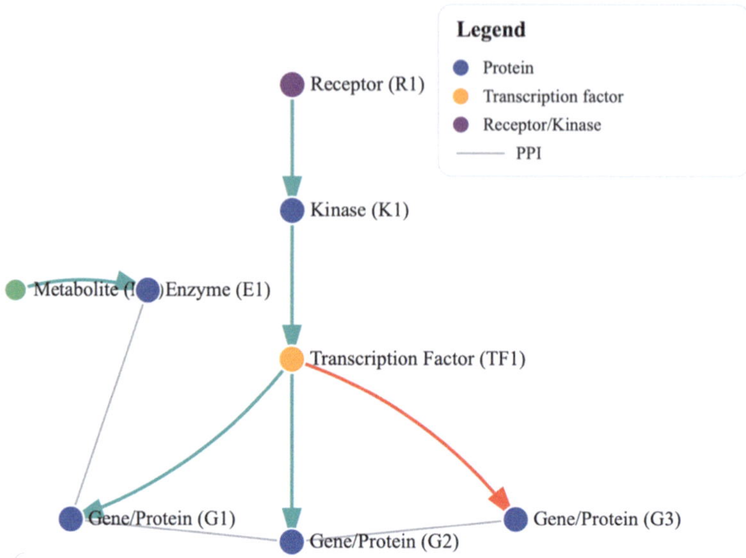

Fig. 7.2 Structure of network biology

7.4 Advances in AI Anomics

Large-scale measurement and analysis of an entire molecular layer in a biological system is known as Omics. It has recast approaches to biological questions by combining top-down surveys with bottom-up mechanistic explorations, providing an integrated understanding of system behavior. Complex diseases, such as cancer, are studied as dynamic, spatial, and temporal processes rather than as binary healthy versus malignant comparisons, with interconnected changes across the genome, transcriptome, proteome, and metabolome (Toussaint et al. 2023).

Early high-throughput tools like DNA microarrays set the stage, and the field has since moved along the central dogma. Genomics captures stable variation. Transcriptomics tracks expression over time and alternative splicing. Proteomics measures abundance, complex assembly, and posttranslational changes. Metabolomics reads out the pathway state. The toolkit now extends to epi-omics, including the epigenome, epitranscriptome, epiproteome, and a range of interactomes linking DNA, RNA, proteins, and metabolites. Domain views, such as the metabolome and immunome, add phenotype context. Multi-omics links molecular signatures to traits, while single-cell and spatial assays reveal heterogeneity that bulk assays miss.

Instead of advanced technology, cellular decision-making remains a complex process, driving continual development of advanced analytical methods. A consolidated view is timely. The article surveys the four core layers of genomics, transcriptomics, proteomics, and metabolomics, along with their epilayers and pairwise interactomics. It highlights emerging directions such as redoxomics, which connects cellular redox state to regulation and pathway flux. The aim is a concise map of methods and opportunities that shortens the path from measurement to mechanism (Ahmed et al. 2024).

7.4.1 Technology-Based Omics

Sequencing and mass spectrometry are the two primary technologies driving omics research. Sequencing enables the direct reading of DNA and RNA, forming the basis of genomics and transcriptomics. It also supports epigenomic and nucleic-acid interaction studies such as ChIP-seq, ATAC-seq, Hi-C, and CLIP. The output is base-level calls and digital counts. Mass spectrometry measures molecules by mass-to-charge. It drives proteomics, including PTMs such as phosphorylation and glycosylation, as well as metabolomics, lipidomics, and protein-centered interactomes such as AP-MS and cross-linking MS. Here, the readout is chemical identity, stoichiometry, and modification state.

Used together, sequencing covers DNA/RNA layers while MS resolves proteins and small molecules. Their complementarity enables a system-wide view of biology with both structural detail and functional readouts. Technology-based omics arise from central dogma tools and include: (i) the four core layers genomics, transcriptomics, proteomics, and metabolomics; (ii) epi-omics, epigenomics, epitranscriptomics, and epiproteomics; and (iii) interactomics, which capture pairwise links among layers (DNA–RNA, RNA–RNA, DNA–protein, RNA–protein, protein–protein, protein–metabolite). In the matrix, each interactome is formed by the two pink-labeled layers on its row and column (Dai and Shen 2022).

7.4.2 Advances in AI-Omics: Genomics, Transcriptomics, Proteomics, and Metabolomics

Artificial intelligence now sits at the center of omics analysis. Models take in raw reads, spectra, and tables, and extract patterns that link molecular shifts to phenotypes under stress. In plants, this pays off fast: gene discovery speeds up, regulatory maps get clearer, and biomarkers for drought, heat, salinity, pathogens, and nutrient limits hold up better. The point is not to replace classic analysis but to add pattern recognition and representation learning that scale with today's data.

High-throughput assays now produce massive, tangled datasets in which features are high-dimensional, noisy, and correlated. Classical statistics struggles when variables outnumber samples or when effects are nonlinear and context-dependent. Machine learning, especially deep learning, matches the scale and complexity of modern omics. It learns compact representations and links genotype, environment, and phenotype with stronger prediction. When labels are scarce, unsupervised and semi-supervised models still pull out structure. For integration, tree ensembles, kernel methods, matrix factorization, Bayesian graphs, network models, and deep nets fuse multi-omics to sharpen genomic prediction, prioritize traits, and probe immunity and pathogen effectors. Pattern recognition is a core advantage. Subtle signals surface as stable biomarkers and modules rather than isolated genes. Context matters.

In RNA-seq, ML improves sensitivity for differential expression, recovers co-expression programs, and supports cross-condition comparisons. Non-linear models can be hard to interpret, so explainable AI and gene set level readouts keep results actionable. Blending ML with prior biology motifs, pathways, and regulatory maps yields models that are both accurate and credible. Plant-omics demands careful feature selection because thousands of variables move together. ML ranks and compresses features, exposes submodules tied to stress traits, and builds compact predictors that generalize. Scaling multi-omics is still hard. Even so, current workflows align the genome, epigenome, transcriptome, proteome, and metabolome so signals echo across layers rather than stand alone. When that agreement is reached, new biomarkers emerge, and classifiers and forecasts hold up in the field.

ChIP-seq is sparse in plants, so regulatory networks rely on expression and accessible chromatin. Correlation and mutual information reveal association but miss direction and timing. Probabilistic graphs and dynamic models add causality at the cost of richer time-resolved data. Recent ML tools push further: integrative frameworks reconstruct temporal GRNs from transcriptomic, proteomic, and epigenomic data; single-cell methods infer cell-specific networks quickly and at scale. The net effect is clear: ML turns plant-omics into a predictive, integrative, and exploratory science that finds mechanisms, not just correlation (Du et al. 2024).

(i) *Genomic*

Genomic studies generate long lists of variants and regions. AI narrows these lists to the few that matter for function. Supervised learners and deep models fuse sequence context, evolutionary conservation, chromatin state, and gene proximity to rank candidates and to propose a mechanism rather than a mere association. When genome-wide association studies identify regions linked to drought tolerance or oil content, machine learning integrates sequence features, local chromatin context, conservation, and gene proximity to ensure that a candidate gene is not just near a single nucleotide polymorphism (SNP) but also mechanistically plausible.

In practice, plant genomics benefits from three recurring AI tasks. First, convolutional and transformer models read DNA windows to estimate how a base change alters transcription factor binding, chromatin accessibility, or splicing. It moves noncoding hits toward causal hypotheses and points to the likely direction of effect. Second, gradient-boosted trees and graph models combine LD structure, regulatory annotations, and network proximity to known stress genes. The output is a calibrated score that is easier to act on in breeding or editing programs. Third, AI guides CRISPR by scoring guide RNAs for on-target efficiency and off-target risk, accounting for local sequence, epigenetic marks, and genome uniqueness (Rodríguez-Durán et al. 2025).

(ii) *Epigenomics*

Epigenomics extends this view. Models trained on ATAC-seq or histone profiles predict cell-type-specific accessibility and enhancer activity from DNA alone. It is useful when experiments are costly or tissues are scarce. In developmental and stress time courses, temporal models encode lag and direction, linking changes in chromatin to later expression shifts. Together, these tools reframe genomics as a predict-then-validate cycle: propose a mechanism computationally, then test the top few edits or lines in the greenhouse. Two cautions keep models honest. Evaluate on held-out chromosomes or cultivars to prevent leakage through shared sequence, and report calibrated probabilities so decisions reflect uncertainty rather than raw scores (Sidira et al. 2024).

(iii) *Transcriptome*

Transcriptome data capture response over time and across tissues. AI makes this dynamic view actionable. Deep denoising models stabilize RNA-seq counts and remove batch effects without erasing true biological signals. Clustering with graph-aware or contrastive objectives finds co-expression modules that align with pathways and cell states. Recurrent and attention-based models track time-ordered responses, linking early sensors to late effectors in drought or heat stress. Interpretation improves when models expose what they learned. Attribution maps highlight motifs and regulators that drive a gene's expression change. Semi-supervised training allows unlabeled samples to shape embeddings, which is helpful when labeled data is uneven or noisy. Comparative analysis across species is supported by shared latent spaces that align orthologs. It exposes conserved defense programs while respecting species-specific branches.

Single-cell assays add resolution. AI handles dropout, recovers gene programs, and orders cells along pseudotime, making transient states visible. Joint models fuse scRNA-seq with scATAC-seq, matching expression to accessibility and revealing which transcription factors likely flip modules on or off. For bulk data, deconvolution estimates cell-type proportions and isolates the regulatory signal of each compartment within a mixed tissue. Two practical outcomes follow. First, regulatory network inference becomes more reliable when models combine expression with motifs and accessibility, producing directed hypotheses that can be tested with CRISPRi/a or reporter assays. Second, module-level biomarkers perform better across cohorts than single genes; lightweight classifiers trained on module scores transfer from growth chamber to field with fewer surprises.

(iv) *Proteomics*

Proteomics explains what the genome builds and how stress rewires those products. AI helps at every step. In identification, deep spectral predictors generate in silico MS/MS spectra for any peptide, improving peptide–spectrum matching and rescuing identifications in data-independent acquisition. In quantification, models learn instrument-specific patterns to correct for interferences and stabilize reporter ion ratios, thereby increasing confidence in differential abundance calls.

Posttranslational modifications are central to plant stress signaling. Classifiers trained on sequence context and structure features predict phosphorylation, ubiquitination, and glycosylation sites, then estimate which kinases or ligases are responsible. When paired with time-resolved phosphoproteomics, attention models map the temporal order of events along a pathway, linking upstream kinases and downstream targets with plausible timing (Sidira et al. 2024).

Structure adds another dimension. Protein language models and structure predictors give fast estimates of fold, disorder, and interfaces. These cues help interpret variants that may shift stability or disrupt complex assembly. In pathogen interactions, interface predictors point to likely contact surfaces between plant proteins and effectors, guiding follow-up experiments. Stress biology rarely respects a single omic layer, so proteomics works best when integrated. Joint models connect transcript changes to protein turnover and PTM waves, flagging cases in which translation or degradation, rather than transcription, explains a phenotype. It also prevents misleading conclusions from expression alone and focuses validation on the right mechanism.

(v) *Metabolomics*

Metabolites are close to the phenotype. They move quickly, they reflect flux, and they act as signals. AI brings order to complex chromatograms and spectra. Peak picking and deconvolution models separate co-eluting compounds. Classifiers assign molecular fingerprints from MS/MS. Retention-time and collision-energy predictors add orthogonal checks, so compound identification is not a guess. When reference spectra are missing, graph-based and generative models propose structures that fit both the fragments and the pathway's chemistry.

In plant defense, these tools reveal biomarkers and routes. Targeted analyses detect phenylpropanoids, alkaloids, or polyamines that rise under attack or drought. Untargeted studies guided by AI map networks of related features, so unknowns cluster with known defense chemicals. Linking metabolomics with transcriptomics and proteomics closes the loop: enzymes that change abundance under stress align with metabolite shifts, and the combined pattern explains tolerance or susceptibility within a cultivar. Field translation is a recurring theme. Lightweight models trained on sensor-friendly fingerprints and simple chromatographic methods allow screening at scale. Instead of deep structural proof for every molecule, the aim is a robust signature that distinguishes resilient from fragile lines, followed by a short list of validated markers that breeders can track.

Models help only when the pipeline is sound. Start with a clean design and adequate replication. Keep splits honest: by chromosome for sequence tasks, by family for protein tasks, and by time or field season for cohorts. Use external or time-split validation whenever possible. Report uncertainty and calibration rather than single numbers. Avoid shortcuts that leak information between the training and test sets by normalizing or selecting features on the full dataset.

Interpretability is not decoration. Saliency maps, motif recovery, counterfactual edits, and perturbation tests all raise trust and point to a mechanism. When a score drives an edit or a cross, write down the assumption and the expected direction of effect. If results hinge on a single feature or batch, reexamine the model.

Ethics and stewardship apply to plant data that record provenance and consent with care, especially when materials come from landraces or wild relatives. There is a need to share models, code, and run manifests to reproduce results and build the models. Moreover, release reusable embeddings and spectral libraries that reduce duplicate effort across labs and maintain the authenticity of benchmarks.

7.5 Cloud Computing and Big Data Analytics

High-throughput assays now let teams probe complex disease and physiology at scale. Still, the omics they produce are huge, high-dimensional, and in mixed formats, which strains storage, computing, and sharing. Cloud is the practical fix. Elastic resources keep data close to the CPUs, cut entry costs, and run typical jobs in molecular modeling and omics from RNA-seq to proteomics and metabolomics without standing up a local cluster. The field has moved from reductionist, single-target studies to data-driven, multilayered views in which DNA, RNA, proteins, and metabolites are analyzed together. Exploding sequence repositories and falling sequencing costs pushed that shift (Koppad et al. 2021).

The work still has technical and statistical hurdles: scale, noisy signals, spurious links, and measurement error. Modern pipelines break the job into three stages: acquisition and preprocessing, analysis, and interpretation, while big data backbones handle the load. Distributed storage, such as HDFS, and in-memory engines like Spark MLlib scale from a single node to whole clusters and public clouds. Production use also needs governance you can trust: security, privacy, access control, and reliable, reproducible workflow orchestration.

Major initiatives demonstrate the model. Multi-omics resources, such as TCGA, provide programmatic access and mirror datasets on public or private clouds, enabling users to run standard workflows close to the data, reduce transfer time, and improve reproducibility. National platforms, including the NCI Cloud Pilots (e.g., FireCloud), package reference tools, curated images, and audit trails, reducing the need for bespoke high-performance computing.

7.5.1 *Analysis of Biomedical and Multi-omics Data*

Biomedical data are expanding at an extraordinary pace, yielding molecular signatures that move care toward personalized diagnosis and treatment. Researchers now combine genomics, transcriptomics, proteomics, metabolomics, epigenomics, imaging, and rich clinical records that range from structured tables to free text. The result is large, uneven datasets that promise insight yet strain storage, compute, and analysis. One whole genome can yield hundreds of gigabytes of raw reads. Add phenotypes and longitudinal clinics, and the load climbs. What helps is solid infrastructure, secure repositories, fast networks, workflow engines, and query layers that keep data near the compute. Clinical impact remains uneven because tools must meet standards for reliability, privacy, and reproducibility before they change decisions at the bedside.

The next step is multi-omics integration. When the genome, transcriptome, proteome, and metabolome are connected, a variant or regulatory event can propagate through proteins to metabolites, producing clinical outcomes. The Integrated models convert that variant path into testable hypotheses, and support tools for risk, prognosis, and therapy choice. To run this at scale, projects need harmonized schemas, clear metadata, and analytics that grow with the data while handling noise and batch effects without washing out real biology (Liu et al. 2020).

7.5.2 Biomedical and Multi-omics Data Source

Building a production-grade environment for large-scale genomics in health care means stitching together several technology layers and using them well. AI does the modeling, and the triage that parallel stacks carry the weight. In an experimental setting, it means the Hadoop ecosystem, HDFS for distributed storage, MapReduce for batch processing, and data-mining and machine-learning libraries push petabytes without increasing cost or latency. Cloud adds elasticity and cuts setup time. Teams spin up managed compute and storage, park pipelines on platforms like Azure, and run custom tools next to the data instead of tending hardware.

None of this is turnkey. Moving genomic and clinical records into public or community clouds raises hard questions on privacy, security, and governance, and community mirrors can suffer from uneven curation, access control, and long-term upkeep. Most multi-omics frameworks still rely on classical statistics for retrieval, integration, and inference at today's scales. Those methods bend, so the stack must evolve toward truly scalable analytics.

Even core steps, such as read alignment, mapping, and de novo assembly, remain time-consuming at the cohort scale. MPI-based assemblers such as Ray, ABySS, and SWAP-Assembler helped distribute work early on, yet they can hit scalability, accuracy, or throughput ceilings as datasets grow and chemistries evolve. MapReduce-style DNA pipelines (for example, Halvade) ease scaling but may trade off efficiency or precision depending on configuration and workload. Similar constraints appear in single-cell RNA-seq, where millions of cells multiply the computational and memory demands across alignment, UMI handling, normalization, and clustering (Koppad et al. 2021).

Parallel platforms ease the load by using data locality and concurrency. Instead of dragging terabytes to a single node, jobs run on the nodes where the HDFS blocks live, and hundreds of workers handle shards in parallel, cutting wall-clock time and network traffic. Add an in-memory engine like Spark, and the I/O drops again, so iterative work model fitting, graph analytics, and feature engineering move faster. With the passage of time, the big-data technology stack has become increasingly complex, comprising Hadoop for distributed storage, MapReduce or Spark for large-scale data processing, Pig or Hive for querying, R and Python for statistical analysis and visualization, scalable machine learning frameworks, and commercial tools such as Microsoft SQL Server, Tableau, and Oracle Database where appropriate. Table 7.2 describes the big data tools, with their advantages and disadvantages, for large-scale omics dataset analysis.

Table 7.2 Big data tools for analysis of omics data

Application	Tool	Description	Advantages	Disadvantages
Whole-genome resequencing	Crossbow (Siretskiy et al. 2015)	End-to-end pipeline combining Bowtie alignment with SNP calling	Low cost, automated, fast, and memory-lean	Built for a single cluster; post-alignment steps can underuse threads
Parallel programming model	Dryad (Isard et al. 2007)	DAG-based parallel framework (runs on Hadoop/YARN) for NGS jobs	Simple scaling across big clusters	Only supports DAG workflows; new models can be hard to express
Short-read mapping	DistMap (Pandey and Schlötterer 2013)	Hadoop workflow for mapping short reads at scale	Fast parallel runs; accurate via graph-based routines	Two-stage I/O can demand very large disk space
Proteomics search	Hydra (Chapman et al. 2010)	MapReduce search engine for high-throughput MS/MS data	Leverages Hadoop; lowers infra overhead for big proteomics	Search rate growth can outpace cluster scaling
NGS analysis toolkit	GATK (Brouard et al. 2019)	Large-scale variant analysis (now with Spark options)	Mature data engine, auto-parallelism, efficient CPU/RAM use	Less flexible for custom data-access patterns
Sequence file I/O	Hadoop-BAM (Niemenmaa et al. 2012)	Library to read/write BAM/CRAM in Hadoop	Picard API support; easy MapReduce over BAMs	CLI-centric; steep learning curve for non-Hadoop users
Variant queries and workflows	SeqWare (O'Connor et al. 2010)	HBase-backed variant store with workflow tracking	Reproducible pipelines; link results to samples/studies	Overkill for small cohorts; few turnkey workflows included
Phylogenetics (distance)	MrsRF (Matthews and Williams 2010)	MapReduce computation of Robinson–Foulds distances on many trees	Cuts all-pairs distance output size; good for huge tree sets	Ignores communication overhead in the performance model

(Continued)

Table 7.2 (Continued)

Application	Tool	Description	Advantages	Disadvantages
Metagenomics/ phylo clustering	Nephele (Weber et al. 2018)	Cloud suite for amplicon and WGS microbiome analysis	Reproducible, cost-aware; tracks inputs and VM images	Limited fine-grained parameter control and output flexibility
GPU-accelerated search	GPU-BLAST (Vouzis and Sahinidis 2011)	GPU/CPU parallel BLAST implementation	Several-fold speedups on long/ short queries	High power draw; careful load balancing needed on big clusters
GPU short-read aligner	SOAP3 (Liu et al. 2012)	GPU-based short-read alignment	2–10× faster than many mappers; good sensitivity with few mismatches	Best for small mismatches; limited indel/gap handling
Hadoop BLAST at scale	Biodoop (Leo et al. 2009)	Hadoop framework to spin up virtual clusters for sequence alignment	Scalable, easier maintenance, good throughput	Start-up overhead; downstream BLAST post-processing needs work
MapReduce sequencing	BioPig (Nordberg et al. 2013)	Pig/MapReduce framework for sequence analytics	Scales as datasets grow; simpler scripting	MapReduce job start-up is slow for interactive work
Pig-based sequencing	SeqPig (Schumacher et al. 2014)	Pig scripts for parallel NGS processing on Hadoop	Concise scripts; easy parallelization	Interactive/iterative jobs are hard under MapReduce
Workflow engine	Nextflow (Di Tommaso et al. 2017)	Portable pipelines with containers; HPC/Cloud ready	Reproducible, fast prototyping; Docker/ Singularity; scales well	No CWL support; no built-in GUI; no native Spark task launcher
Workflow engine	Snakemake (Mölder et al. 2021)	Reproducible workflows from laptop to cloud/ cluster	Same workflow scales across backends with no code changes	No automatic CWL-to-Snakemake conversion (yet)
Cloud RNA-seq at scale	Falco (de Sena Brandine and Smith 2019)	Hadoop/Spark framework for parallel RNA-seq alignment, counts, QC	Uses spot instances; ~65% cost savings reported	Splitting very large files can be a bottleneck

7.5.3 *Cloud Computing Technologies for Omics Analysis*

Cloud, per NIST, means quick network access to a shared pool of compute servers, storage, networks, apps, and services that can be started or stopped with little effort. Five traits define it in practice: pooled resources, on-demand capacity, broad network reach, rapid elasticity, and metered use so costs track consumption. On that base sit three service layers. IaaS describes the raw computing, storage, and networking. PaaS provides prebuilt runtimes and tools so that anyone can build and run code without managing servers. SaaS delivers finished applications in the browser. Deployment varies by need: private for one institution, public from a provider, hybrid that mixes both, and community clouds shared by groups with similar requirements.

Bioinformatics uses the same stack with a domain twist. Labs often treat raw and curated datasets as DaaS, so analyses run where the data live. They pick SaaS for ready-made portals, PaaS for managed workflow engines and notebooks, and IaaS when complete control of clusters or GPUs is required. The draw is clear: elastic scale, built-in redundancy and reliability, metered cost control, and the flexibility to match compute to each analysis without overbuilding local infrastructure.

7.5.4 Challenges of Big Data Analysis

Big data work starts long before modeling. Instruments like sequencers and mass spectrometers generate torrents of raw files, much of which is redundant once high-quality reads or confidently identified spectra are in hand. The first challenge is smart reduction: filtering, compression, and subsampling that shrink data by orders of magnitude without throwing away signal. Getting this right depends on rich, automatic metadata: what was measured, how it was processed, the software and versions used, and key parameters. Therefore, downstream tools can reliably reconstruct context and interpret compressed formats. Because raw outputs rarely arrive analysis-ready, teams also need robust extraction and restructuring steps that turn heterogeneous logs and binaries into tidy, queryable tables.

Analysis itself is hard, not only because of size, but also because omics projects blend many data types and file formats. Scalable frameworks must align algorithms with available infrastructure and automate routine steps, while remaining flexible when projects deviate from the norm. Assumptions often fail: data are not perfectly clean; many popular omics algorithms do not integrate seamlessly with Hadoop or Spark; and rewriting code for cloud execution takes time. Distributed systems have their own wrinkles, such as replication overhead, I/O hotspots, and the need for data locality. If a NoSQL solution supports the stack, ingestion, and schema design, it becomes a first-class task rather than an afterthought.

For end users, interpretation is the make-or-break step. Results need provenance, readable reports, and clear visualizations, or they will not be trusted or reused. Well-documented workflows help others tune parameters, trace sources of noise, and reproduce figures; feedback channels let developers harden tools with real-world cases. Treating analyses as products with logs, versioning, and dashboards turns opaque pipelines into reusable assets that integrate smoothly with in-house datasets.

Finally, the ecosystem thrives on sharing. Public repositories such as GEO for transcriptomics and PRIDE/ProteomeXchange, PeptideAtlas, and MassIVE for proteomics have changed the default from "measure first" to "mine what exists." Open datasets enable reanalysis, method benchmarking, and deeper annotation, speeding discovery and raising quality across the field. The combination of careful data reduction, honest engineering for scale, clear interpretation, and open sharing is what turns raw big data into durable biological insight.

7.6 AI-Based Multi-omics Integration Platforms

All emerging technologies are being developed to address the need for large-scale omics data, encompassing metabolomics, genomics, and epigenomics. It is important to integrate these segregated datasets to improve treatments and gain deeper insights into complex biological systems. This critical analysis highlights the advantages and disadvantages of the multi-omics data integration platforms currently in use. The study examines how the latest digital platforms, including GraphOmics, OmicsAnalyst, and others, facilitate the seamless integration and analysis of omics data in genomics applications (Papadaki et al. 2024).

Platform for Multi-omics Analysis

Platforms specialize in integrating spatial multi-omics data in handling large datasets. The platforms are especially helpful for research projects that involve tissue-specific molecular processes, such as neurobiology or cancer research, where understanding the organization of molecular features within tissues is crucial.

(i) M-Omics: M-Omics combines spatial transcriptomics with antibody-based proteomics to uncover genetic heterogeneity across tissue regions. It is designed to process high-throughput spatial multi-omics data. For example, it has been used to map gene expression in tumor microenvironments in glioblastoma research. However, smaller labs with limited bioinformatics experience may find it difficult to use due to its command-line nature. To manage massive datasets, SM-Omics uses automated pipelines instead of a web-based interface. Gene expression and protein abundance within tissue samples can be mapped by combining spatial transcriptomics and antibody-based proteomics. It provides information on the molecular heterogeneity of complex biological systems, such as brain tissue or malignancies. By demonstrating how molecular characteristics differ across locations, this spatial integration provides a deeper understanding of biological processes specific to particular tissues.

(ii) SM-Omics is built for throughput. It chains multistep analyses with Snakemake or Nextflow, so big datasets run end-to-end with little manual handling. That automation scales well on powerful lab servers and suits large, spatially or geographically broad multi-omics projects. The trade-off is usability. It is command-line first with no intuitive web front end, so users need solid bioinformatics skills. Runs can be heavy on CPU, RAM, and storage, making them out of reach for smaller groups without dedicated infrastructure.

(iii) OmicsSuite takes a GUI-first approach, unlike SM-Omics, which uses a command-line interface. It still handles high-throughput integration and works across transcriptomics, proteomics, metabolomics, genomics, and single-cell data. The desktop app is aimed at experimental biologists who do not code every day. It bundles pipelines for data integration and statistics, and includes rich visualization tools such as heatmaps, PCA, t-SNE, and UMAP. The suite ships with ~175 sub-applications spanning many omics areas, which makes it flexible for broad studies. There is no web interface, but the GUI and the breadth of tools lower the barrier compared with command-line stacks. Installation is local, and real projects benefit from a capable workstation, so users without strong hardware or setup skills may find that a hurdle.

(iv) Single-Platform Multi-Omic Profiling: Proteomics and metabolomics can now be integrated on a single mass spectrometry platform thanks to the Single-Platform Multi-Omic Profiling system, another method in high-throughput multi-omics. Separate procedures were traditionally required for proteomics and metabolomics, which often led to greater variability and higher sample requirements. Single-Platform Multi-Omic Profiling integrates both analyses into a single workflow, reducing variability and improving analytical consistency. High-resolution mass spectrometry has been widely used to identify biomarkers in rare diseases. However, because this approach primarily relies on mass spectrometry, its integration with other omics layers, such as transcriptomics and epigenomics, remains limited.

(v) High-resolution nano-LC–MS sits at the center. It has the sensitivity to quantify proteins and metabolites in a single run. With a single setup, labs can track protein abundance and metabolite levels simultaneously and read pathway crosstalk almost in real time. It is practical in cancer and in rare disorders where the signal is scarce and multilayer measurements are essential. The platform couples the instrument with an end-to-end processing stack that includes normalization, retention-time alignment, and multi-omics fusion, so large batches remain consistent. Joint readouts of proteomic and metabolomic profiles connect shifts in protein abundance to changes in pathway flux and reveal links a single-omic assay would miss, giving a more precise mechanism and a shorter path from data to a testable experiment.

(vi) CHOmics: CHOmics addresses the specific challenges in optimizing Chinese Hamster Ovary (CHO) cells for the production of biopharmaceuticals, shifting from a broad multi-omics focus to a more specialized application. CHOmics is an enterprise-focused multi-omics platform for Chinese hamster ovary (CHO) cell lines, which are used as workhorses in therapeutic protein production. It condenses genomics, transcriptomics, proteomics, and metabolomics into a single view, enabling teams to monitor how molecular layers communicate and change throughout the culture cycle. It has a simple goal: raise yield and improve efficiency while maintaining product quality. The platform spans pathway analysis, metabolic optimization, and gene regulatory network exploration. Users can trace how a variant or expression shift moves through proteins to metabolites and then to bioprocess traits such as titer or growth. In practice, this turns scattered assays into a coherent map of CHO cell behavior and points to edits in media, process, or genome that push the line toward better performance. Although CHOmics has many advantages for the biotechnology sector, its use in more comprehensive multi-omics research remains constrained. Researchers can also become proficient in CHOmics relatively quickly thanks to its comprehensive training, preconfigured pipelines, and interactive visualizations. These resources ensure accessibility for multidisciplinary teams with varying levels of computational knowledge, especially when paired with web-based deployment and community interaction.

(vii) Majorbio Cloud: Large-scale multi-omics research can be supported by Majorbio Cloud, an integrated cloud platform providing centralized collections of bioinformatics tools and workflows for integrating data from metagenomics, transcriptomics, proteomics, metabolomics, and genomes. Preconfigured data processing, analysis, and visualization workflows are available in Majorbio Cloud to meet high-throughput demand, and many types of omics are supported, with accessibility and broad analytical capabilities prioritized. Since it is built at its core as a cloud-based system, researchers can analyze and work with large collections of data without dedicated local hardware, to the advantage of even modest institutions. With assistance from Majorbio Cloud's more than 300 unique analysis toolsets, analysts can perform sophisticated analyses in a single environment, like gene interaction mapping, pathway enrichment, differential expression, and clustering. Due to those features, it is well-suited to research areas such as cancer, where interactions across multiple levels of biology need to be considered. One of the key functionalities of Majorbio Cloud is its user-friendly web portal, which enables exploratory data analysis through customizable, interactive visualizations, including condition-correlation matrices, heatmaps, and PCA plots. Researchers can better understand disease causes and tissue-specific molecular activity by using this interactive environment to help them discover molecular linkages and pathway dynamics. Majorbio Cloud also places a strong emphasis on collaborative workflows, which allow researchers to share data and findings among teams and support multi-omics multidisciplinary research. Table 7.3 provides various computational platforms developed for multi-omics data analysis, highlighting their environments, analytical tasks, and important objectives.

7.7 3D Genomics Visualization

Data visualization sits at the heart of exploration and interpretation. Seeing the spread of signals and the links across layers helps make sense of pathways and structure in multi-omics data. The first two releases of IntelliGenes used 2D plots because they are easy to read and accessible to everyone. Each figure serves a purpose: showing model performance, how values are distributed, or how features relate. In those versions, the priority was a flexible AI/ML analysis flow that users could tune to their data. The next phase shifts the focus. Better annotations, clearer

Table 7.3 Various platforms for analysis of omics data

Platform	Environment	Tasks performed	Objectives
Biomix (Iperi et al. 2025)	R, R-Shiny	High-throughput omics data analysis, Single-omics processing, multi-omics integration (transcriptomics, metabolomics, methylomics), cohort comparison, subgroup analysis	Use the underlying latent-factor model (MOFA) for integration, pathway, enrichment, and visualization analyses, and literature-guided annotation makes it easier to interpret latent factors
EMitool (Xu et al. 2025)	R/Python	Multiomics integration for clinically based diseases	Provide superior subtyping accuracy, contribute scores for different omics data types
Contrastive Learning and Self Attention (CLCLSA) (Zhao et al. 2024)	Python/Deep Learning Framework	Integration of multiomics data for classification	Deal with missing multi-omics data as supervision, dynamically weights informative features across omics
SynOmics (Alif and Zhang 2025)	Python/Graph Neural Network	Multiomics integration	Adopt the parallel learning method to perform feature-level interactions at each layer of the model

summaries, and richer displays will carry more of the story, making AI/ML results easier to read, compare, and present.

2D views add real value to multi-omics work. They help with interpretation and keep figures easy to share. Yet they miss fine structure and often struggle to separate two patterns that matter equally. Moving from 2D to 3D captures more of the variation in the original space and more faithfully represents the latent shape. Other fields already use 3D to good effect for medical imaging for cancer detection, protein folding studies, and chemical space maps, but multi-omics has not seen the same push.

3D IntelliGenes fills that gap with a two-stage workflow designed for structure first, detail second: (1) Clustering and (2) Feature Plots. Each stage lives in its own tab and launches from the Visualization module. We run 3D IntelliGenes in a separate window to keep a responsive, interactive viewport and to make the split from 2D explicit. The split also preserves compatibility with the original GUI, so users can move between 2D and 3D without changing their routine.

(i) *Clustering Visualization Method*

Visualizing high-dimensional data in 3D reveals a structure that 2D obscures. Clusters will typically come apart more cleanly with an additional dimension, so overlapping disease states in a flat view can be separated in space, revealing features important for diagnosis or prognosis. Clustering also links patients by shared patterns and opens a path to downstream work on correlated pathways and shared mechanisms.

3D IntelliGenes builds on this idea by embedding a 3D dataset and providing an interactive workspace. Users can choose linear or nonlinear embeddings, inspect predictions from individual AI/ML classifiers, and drill into patient-specific profiles. The module renders four core views: Observed Data, AI/ML Predictions, Type I/II Errors, and Clusters. Each view can stand alone for detail or sit alongside the others for direct comparison (Ehlers et al. 2025).

(ii) *3D Embedding*

3D views start with embedding the data in a three-dimensional space. This step is dimensionality reduction. To understand the broad layout of a multi-omics cohort, linear models like principal component analysis (PCA) are a good first pass because they project the data into a subspace while minimizing the loss of variance, thereby preserving as much of the global structure as possible. Linear maps however struggle when the real geometry is curved or tangled and important relationships are nonlinear. Nonlinear embeddings tell a different story. Uniform Manifold Approximation and Projection (UMAP) preserves local neighborhoods, making it useful for measuring patient-to-patient similarity. It should not be used to claim global relationships across the whole cohort. Pairwise Controlled Manifold Approximation and Projection (PaCMAP) aims for a middle ground. It keeps local neighborhoods while also stabilizing long-distance relationships, and it still handles nonlinear shapes. Each method is a lens that emphasizes different aspects of the same data. 3D IntelliGenes lets you render any CIGT-formatted dataset using PCA, UMAP, or PaCMAP, and pick the view that best fits the question. For the chosen embedding, the app reports a preservation score indicating how well the 3D map preserves the original high-dimensional structure. PCA also reports the fraction of variance explained by the projection.

(iii) *Example Tools for 3D Visualization of Genomics Data*

VIS+AI brings visual analysis and AI together into a single loop. It pairs human-centered visual knowledge extraction with AI-centered data mining, so patterns are found and shown in ways people can read and act on. The framework is built for two-way communication: models surface structure through visualization; humans steer, correct, and refine through the same views. In practice, human judgment and machine inference coordinate during analysis, with visualization as the shared workspace (Wang et al. 2023).

GMOL is a genome viewer built for true multi-scale exploration. It lets researchers move across six levels of detail, from whole-genome context down to local structure, then back out again. At any scale, a user can select a unit, zoom in or out to inspect its shape, and pull the corresponding DNA sequence. The scene is interactive: rotate, pan, and measure features across the full model to test simple geometric hypotheses. Outputs are easy to reuse. GMOL can save publication images and export machine-readable structure files in PDB format from the same multi-scale model. In practice, this makes it simpler to study how 3D genome organization relates to gene activity and DNA function (Nowotny et al. 2016).

GenomeFlow is a graphical workspace that covers the full 3D genome pipeline end to end. It maps Hi-C reads to the 1D reference, builds and normalizes 2D contact maps, then reconstructs chromosomes and whole genomes in 3D and shows them in real time. Users can inspect the structure and tie it to function by painting gene annotations, expression, and methylation directly on the 3D model. Projects stay in one place: models can be saved, reopened, and annotated in the same interface, so building, viewing, and downstream analysis do not scatter across tools (Trieu et al. 2019).

CSynth is an interactive 3D genome browser and a real-time restraint-based modeller for any 3C-type dataset. It lets users adjust modeling parameters on the fly and watch the structure respond, which makes "what-if" experiments fast and transparent. Models from different tissues or cell states can be loaded side by side, and outputs from other 3D pipelines can be compared in the same view. A built-in VR mode allows researchers to step into complex chromatin structures, rotate and probe them at full scale, and it has proven useful for teaching and public demos as well (Todd et al. 2021).

7.8 Quantum Computing in Bioinformatics

Bioinformatics turned biology and medicine into data-driven disciplines by bringing computation to sequence, structure, and phenotype. A new layer is now in play: quantum bioinformatics. It applies quantum mechanics, quantum algorithms, and early quantum hardware to problems that strain classical methods, electronic structure in molecules, binding energetics for drug design, and the conformational search behind protein folding (Pallavi and Prasanna Kumar 2025).

There is a clear taxonomy that follows three axes: operation, platform, and data. In group 1, these operations fall into two groups: (i) calculations and data-mining tasks such as inference, clustering, and search; and (ii) design, modeling, or optimi zation where the goal is to build or tune tools or candidates. Additionally, platforms are split into classical and quantum machines, with hybrids in practice. Data is split into classical biological data sequences, expression matrices, spectra, and rarer quantum biological data, where the measurements encode quantum states. Across this space, quantum algorithms can reduce search, sample complex distributions, or model electronic structure with higher fidelity. That can speed up or improve core computational biology tasks. Recent work highlights concrete targets, including mapping multi-scale biological systems and genetic analyses onto quantum architectures so key subproblems run more efficiently than on conventional stacks. The principle is pragmatic: use quantum where it helps, keep the rest on stable classical pipelines (Mokhtari et al. 2024).

7.9 Application of Quantum Computing

(i) *Faster genomics and proteomics*

- Sequence alignment. Quantum search, like Grover, can copy and match from O(N) to O($\sqrt{N}$), which points to ultrafast genome-wide scans.
- Protein structure. Quantum annealing and variational quantum eigensolvers model interaction energies directly. They complement tools like AlphaFold by refining hard cases where electronic effects matter.

(ii) *Quantum ML for omics*

- Feature selection. Hybrid quantum–classical networks sift high-dimensional genomic panels and keep the variables that carry the signal.
- Classification. Quantum SVMs separate expression profiles and disease subtypes with quantum kernels that capture nonlinear structure.

(iii) *Optimization in network biology*

- Many tasks are combinatorial. Quantum methods target them head-on.

GRN inference. Search over sparse directed graphs.
Metabolic pathway optimization. Pick flux patterns under constraints.
Drug–target prediction. Score large ligand–protein pairs efficiently.
Algorithms in scope include quantum annealing, QAOA, and quantum walks.

(iv) *Quantum-enhanced multi-omics integration*

- Dimensionality reduction. QPCA compresses integrated omics while keeping variance that matters.
- Privacy-preserving learning. Quantum-secure federated schemes enable cross-site training without exposing raw patient or cultivar data.

7.10 Conclusion

The field has shifted from single assays and hand-tuned models to integrated analysis at scale. Machine learning sits at the center because it learns useful representations from sequences, structures, and spectra, and links genotype, environment, and phenotype with stronger prediction. Network biology adds a mechanism. PPIs show how complexes form. GRNs show how signals reach the transcriptome. Together, they expose control points that translate into perturbations, targets, and module-level biomarkers.

Scale changed the engineering. High-throughput assays produce heterogeneous data that demand data locality, concurrency, and reproducible workflows. Distributed storage and in-memory engines reduce I/O and wall time. Workflow systems minimize manual handling and enable audits. The same stack supports visual work. 2D plots remain the fastest way to check distributions and residuals. 3D views help when clusters overlap in 2D or when global and local structure must be read together. Good figures shorten the path from model output to biological interpretation.

Two cautions keep models honest. First, context matters. Edges in a regulatory graph strengthen or fade across tissues, time, and condition. Use data to weight edges and prefer context-specific networks over static wiring diagrams. Second, batch and bias matter. Harmonize inputs, record provenance, and report preservation scores for embeddings so readers understand what the map keeps and what it drops.

The agenda focuses on analyzing the genome, epigenome, transcriptome, proteome, and metabolome layers of stored data. Pretrained models, transfer learning code, run manifests, and reusable embeddings are commonly utilized so that others can replicate and extend. Treat datasets as DaaS when possible so analysis runs near the data. Use AI where it adds power to representation learning, fusion across omics, and calibrated prediction, and use explanation tools to trace the features and pathways behind each decision. Keep ethics in view: consent, lineage, and fair access, especially for data from landraces and wild relatives.

What comes next will widen the lens. Single-cell and spatial assays will turn static networks into time- and space-aware maps. Multi-omics platforms will move from side-by-side to truly fused analysis with uncertainty tracked end to end. 3D genome tools will tie structure to function with more substantial evidence. Quantum methods may seed new solvers for molecular modeling where classical search stalls. The through line does not change: precise methods, careful engineering, and plain, readable visuals that turn extensive, layered data into stable hypotheses and actionable biology.

References

Ahmed Z, Wan S, Zhang F, Zhong W. Artificial intelligence for omics data analysis. BMC Methods. 2024;1(1):4. https://doi.org/10.1186/s44330-024-00004-5.

Alif MN, Zhang W. SynOmics: integrating multi-omics data through feature interaction networks. Brief Bioinform. 2025;26(6) https://doi.org/10.1093/bib/bbaf595.

Brouard J-S, Schenkel F, Marete A, Bissonnette N. The GATK joint genotyping workflow is appropriate for calling variants in RNA-seq experiments. J Anim Sci Biotechnol. 2019;10(1):44. https://doi.org/10.1186/s40104-019-0359-0.

Chapman JA, Kirkness EF, Simakov O, Hampson SE, Mitros T, Weinmaier T, Rattei T, Balasubramanian PG, Borman J, Busam D, Disbennett K, Pfannkoch C, Sumin N, Sutton GG, Viswanathan LD, Walenz B, Goodstein DM, Hellsten U, Kawashima T, et al. The dynamic genome of Hydra. Nature. 2010;464(7288):592–6. https://doi.org/10.1038/nature08830.

Dai X, Shen L. Advances and trends in omics technology development. Front Med. 2022;9:911861. https://doi.org/10.3389/fmed.2022.911861.

de Sena Brandine G, Smith AD. Falco: high-speed FastQC emulation for quality control of sequencing data. F1000Res. 2019;8:1874. https://doi.org/10.12688/f1000research.21142.1.

Di Tommaso P, Chatzou M, Floden EW, Barja PP, Palumbo E, Notredame C. Nextflow enables reproducible computational workflows. Nat Biotechnol. 2017;35(4):316–9. https://doi.org/10.1038/nbt.3820.

Du P, Fan R, Zhang N, Wu C, Zhang Y. Advances in integrated multi-omics analysis for drug-target identification. Biomolecules. 2024;14(6):692. https://doi.org/10.3390/biom14060692.

Ehlers H, Brich N, Krone M, Nöllenburg M, Yu J, Natsukawa H, Yuan X, Wu H-Y. An introduction to and survey of biological network visualization. Comput Graph. 2025;126:104115. https://doi.org/10.1016/j.cag.2024.104115.

Ghosh N, Santoni D, Saha I, Felici G. A review on the applications of Transformer-based language models for nucleotide sequence analysis. Comput Struct Biotechnol J. 2025;27:1244–54. https://doi.org/10.1016/j.csbj.2025.03.024.

Iperi C, Fernández-Ochoa Á, Barturen G, Pers J-O, Foulquier N, Bettacchioli E, Alarcón-Riquelme M, Cornec D, Bordron A, Jamin C. BiomiX, a user-friendly bioinformatic tool for democratized analysis and integration of multiomics data. BMC Bioinf. 2025;26(1):8. https://doi.org/10.1186/s12859-024-06022-y.

Isard M, Budiu M, Yu Y, Birrell A, Fetterly D. Dryad: distributed data-parallel programs from sequential building blocks. In: Proceedings of the 2nd ACM SIGOPS/EuroSys European conference on computer systems 2007; 2007. p. 59–72. https://doi.org/10.1145/1272996.1273005.

Koppad S, B A, Gkoutos GV, Acharjee A. Cloud computing enabled big multi-omics data analytics. Bioinf Biol Insights. 2021;15(2):117793222110359. https://doi.org/10.1177/11779322211035921.

Leo S, Santoni F, Zanetti G. Biodoop: bioinformatics on Hadoop. In: International conference on parallel processing workshops, vol. 2009; 2009. p. 415–22. https://doi.org/10.1109/ICPPW.2009.37.

Liu CM, Wong T, Wu E, Luo R, Yiu SM, Li Y, Wang B, Yu C, Chu X, Zhao K, Li R, Lam TW. SOAP3: ultra-fast GPU-based parallel alignment tool for short reads. Bioinformatics. 2012;28(6):878–9. https://doi.org/10.1093/bioinformatics/bts061.

Liu C, Ma Y, Zhao J, Nussinov R, Zhang Y-C, Cheng F, Zhang Z-K. Computational network biology: data, models, and applications. Phys Rep. 2020;846:1–66. https://doi.org/10.1016/j.physrep.2019.12.004.

Ma C-Y, Liao C-S. A review of protein–protein interaction network alignment: from pathway comparison to global alignment. Comput Struct Biotechnol J. 2020;18:2647–56. https://doi.org/10.1016/j.csbj.2020.09.011.

Matthews SJ, Williams TL. MrsRF: an efficient MapReduce algorithm for analyzing large collections of evolutionary trees. BMC Bioinformatics. 2010;11(S1):S15. https://doi.org/10.1186/1471-2105-11-S1-S15.

Moeckel C, Mareboina M, Konnaris MA, Chan CSY, Mouratidis I, Montgomery A, Chantzi N, Pavlopoulos GA, Georgakopoulos-Soares I. A survey of k-mer methods and applications in bioinformatics. Comput Struct Biotechnol J. 2024;23:2289–303. https://doi.org/10.1016/j.csbj.2024.05.025.

Mokhtari M, Khoshbakht S, Ziyaei K, Akbari ME, Moravveji SS. New classifications for quantum bioinformatics: Q-bioinformatics, QCt-bioinformatics, QCg-bioinformatics, and QCr-bioinformatics. Brief Bioinform. 2024;25(2) https://doi.org/10.1093/bib/bbae074.

Mölder F, Jablonski KP, Letcher B, Hall MB, Tomkins-Tinch CH, Sochat V, Forster J, Lee S, Twardziok SO, Kanitz A, Wilm A, Holtgrewe M, Rahmann S, Nahnsen S, Köster J. Sustainable data analysis with Snakemake. F1000Res. 2021;10:33. https://doi.org/10.12688/f1000research.29032.2.

Niemenmaa M, Kallio A, Schumacher A, Klemelä P, Korpelainen E, Heljanko K. Hadoop-BAM: directly manipulating next generation sequencing data in the cloud. Bioinformatics. 2012;28(6):876–7. https://doi.org/10.1093/bioinformatics/bts054.

Nordberg H, Bhatia K, Wang K, Wang Z. BioPig: a Hadoop-based analytic toolkit for large-scale sequence data. Bioinformatics. 2013;29(23):3014–9. https://doi.org/10.1093/bioinformatics/btt528.

Nowotny J, Wells A, Oluwadare O, Xu L, Cao R, Trieu T, He C, Cheng J. GMOL: an interactive tool for 3D genome structure visualization. Sci Rep. 2016;6:1–8. https://doi.org/10.1038/srep20802.

O'Connor BD, Merriman B, Nelson SF. SeqWare query engine: storing and searching sequence data in the cloud. BMC Bioinformatics. 2010;11(S12):S2. https://doi.org/10.1186/1471-2105-11-S12-S2.

Omony J. Biological network inference: A review of methods and assessment of tools and techniques. Ann Res Rev Biol. 2014;4(4):577–601. https://doi.org/10.9734/ARRB/2014/5718.

Pallavi G, Prasanna Kumar R. Quantum natural language processing and its applications in bioinformatics: a comprehensive review of methodologies, concepts, and future directions. Front Comput Sci. 2025;7 https://doi.org/10.3389/fcomp.2025.1464122.

Pandey RV, Schlötterer C. DistMap: A toolkit for distributed short read mapping on a Hadoop cluster. PLoS One. 2013;8(8):e72614. https://doi.org/10.1371/journal.pone.0072614.

Papadaki E, Kakkos I, Vlamos P, Petropoulou O, Miloulis ST, Palamas S, Vrahatis AG. Recent web platforms for multi-omics integration unlocking biological complexity. Appl Sci. 2024;15(1):329. https://doi.org/10.3390/app15010329.

Rodríguez-Durán A, Andrade-Silva V, Numan M, Waldman J, Ali A, Logullo C, da Silva Vaz Junior I, Parizi LF. Multi-omics technologies applied to improve tick research. Microorganisms. 2025;13(4):795. https://doi.org/10.3390/microorganisms13040795.

Sajitha P, Andrushia AD, Anand N, Naser MZ. A review on machine learning and deep learning image-based plant disease classification for industrial farming systems. J Ind Inf Integr. 2024;38:100572. https://doi.org/10.1016/j.jii.2024.100572.

Schirner M, Deco G, Ritter P. Learning how network structure shapes decision-making for bio-inspired computing. Nat Commun. 2023;14(1):2963. https://doi.org/10.1038/s41467-023-38626-y.

Schumacher A, Pireddu L, Niemenmaa M, Kallio A, Korpelainen E, Zanetti G, Heljanko K. SeqPig: simple and scalable scripting for large sequencing data sets in Hadoop. Bioinformatics. 2014;30(1):119–20. https://doi.org/10.1093/bioinformatics/btt601.

Sidira M, Agriopoulou S, Smaoui S, Varzakas T. Omics-integrated approach (metabolomics, proteomics and lipidomics) to assess the quality control of aquatic and seafood products. Appl Sci. 2024;14(22):10755. https://doi.org/10.3390/app142210755.

Silva JCF, Teixeira RM, Silva FF, Brommonschenkel SH, Fontes EPB. Machine learning approaches and their current application in plant molecular biology: a systematic review. Plant Sci. 2019;284:37–47. https://doi.org/10.1016/j.plantsci.2019.03.020.

Siretskiy A, Sundqvist T, Voznesenskiy M, Spjuth O. A quantitative assessment of the Hadoop framework for analyzing massively parallel DNA sequencing data. GigaScience. 2015;4(1):26. https://doi.org/10.1186/s13742-015-0058-5.

Sun S, Wang C, Ding H, Zou Q. Machine learning and its applications in plant molecular studies. Brief Funct Genomics. 2018;19(1):40–8. https://doi.org/10.1093/bfgp/elz036.

Todd S, Todd P, McGowan SJ, Hughes JR, Kakui Y, Leymarie FF, Latham W, Taylor S. CSynth: an interactive modelling and visualization tool for 3D chromatin structure. Bioinformatics. 2021;37(7):951–5. https://doi.org/10.1093/bioinformatics/btaa757.

Toussaint PA, Leiser F, Thiebes S, Schlesner M, Brors B, Sunyaev A. Explainable artificial intelligence for omics data: a systematic mapping study. Brief Bioinform. 2023;25(1):1–16. https://doi.org/10.1093/bib/bbad453.

Trieu T, Oluwadare O, Wopata J, Cheng J. GenomeFlow: a comprehensive graphical tool for modeling and analyzing 3D genome structure. Bioinformatics. 2019;35(8):1416–8. https://doi.org/10.1093/bioinformatics/bty802.

Vouzis PD, Sahinidis NV. GPU-BLAST: using graphics processors to accelerate protein sequence alignment. Bioinformatics. 2011;27(2):182–8. https://doi.org/10.1093/bioinformatics/btq644.

Wang X, Wu Z, Huang W, Wei Y, Huang Z, Xu M, Chen W. VIS+AI: integrating visualization with artificial intelligence for efficient data analysis. Front Comp Sci. 2023;17(6):176709. https://doi.org/10.1007/s11704-023-2691-y.

Weber N, Liou D, Dommer J, MacMenamin P, Quiñones M, Misner I, Oler AJ, Wan J, Kim L, Coakley McCarthy M, Ezeji S, Noble K, Hurt DE. Nephele: a cloud platform for simplified, standardized and reproducible microbiome data analysis. Bioinformatics. 2018;34(8):1411–3. https://doi.org/10.1093/bioinformatics/btx617.

Xu Y, Wu J, Chen C, Ouyang J, Li D, Shi T. EMitool: explainable multi-omics integration for disease subtyping. Int J Mol Sci. 2025;26(9):4268. https://doi.org/10.3390/ijms26094268.

Zhao C, Liu A, Zhang X, Cao X, Ding Z, Sha Q, Shen H, Deng H-W, Zhou W. CLCLSA: cross-omics linked embedding with contrastive learning and self attention for integration with incomplete multi-omics data. Comput Biol Med. 2024;170:108058. https://doi.org/10.1016/j.compbiomed.2024.108058.

Zhivkoplias EK, Vavulov O, Hillerton T, Sonnhammer ELL. Generation of realistic gene regulatory networks by enriching for feed-forward loops. Front Genet. 2022;13:1–11. https://doi.org/10.3389/fgene.2022.815692.

Chapter 8
Techniques in Molecular Diagnosis

Imran ul Haq, Nabeeha Aslam Khan, Siddra Ijaz, and Hayssam Mohamed Ali

8.1 Introduction

The molecular diagnostics revolution has transformed the treatment of modern diseases, biological research, agriculture, and disease surveillance, enabling the rapid and highly specific detection of genetic and molecular markers. Molecular approaches are an alternative method to diagnosis, which is frequently dependent on culture, microscopy, or visible disease symptoms, and they detect disease-associated DNA, RNA, mutations, proteins, and other biomarkers directly from biological samples. The development of nucleic acid amplification and biomarker analysis has also enhanced the diagnostic sensitivity, speed, accuracy, and opportunity for point-of-care testing.

This chapter explains key emerging techniques in molecular diagnosis, including Loop-Mediated Isothermal Amplification (LAMP), circulating tumor DNA (ctDNA) analysis, exosome-based diagnosis, Digital and Droplet PCR, and Recombinase Polymerase Amplification (RPA). They have growing applications in infectious-disease detection, cancer diagnosis and monitoring, genetic analysis,

I. u. Haq
Department of Plant Pathology, University of Agriculture, Faisalabad, Pakistan
e-mail: imran_1614@uaf.edu.pk

N. A. Khan
Plant Pathology Section, PPRI, Ayub Agricultural Research Institute, Faisalabad, Pakistan

S. Ijaz (✉)
Centre of Agricultural Biochemistry and Biotechnology (CABB), University of Agriculture, Faisalabad, Pakistan

H. M. Ali
Department of Botany and Microbiology, College of Science, King Saud University, Riyadh, Saudi Arabia
e-mail: hayhassan@ksu.edu.sa

S. Ijaz et al. (eds.), *Molecular Biology Methods*,
https://doi.org/10.1007/978-981-92-2642-9_8

plant pathology, food safety, and precision agriculture. Their basic principles, workflows, key applications, benefits, drawbacks, and potential future are presented, emphasizing the ongoing shift toward field-based, sensitive, minimally invasive, and rapid molecular diagnostic systems.

8.2 Isothermal Amplification (LAMP)

The need for rapid and cost-effective diagnostic techniques has led to the development of isothermal amplification techniques. Loop-Mediated Isothermal Amplification (LAMP) is one of the most popular and widely used methods of this type. Since its discovery in 2000, the application of LAMP and the resulting capability, which eliminates the need for costly and sophisticated thermal cyclers, of creating and amplifying DNA or RNA at a specific temperature, has been acclaimed as one of the significant technological breakthroughs in molecular biology. Unlike a more traditional PCR (Polymerase Chain Reaction), in which one must complete a series of temperature cycles via denaturing, annealing, and extension, LAMP makes it easier by amplifying genetic material under isothermal conditions, typically between 60 °C and 65 °C (Hassan et al. 2022). This sequencing of nucleic acids at low but constant temperatures makes LAMP particularly appealing as a technique for point-of-care diagnostics, remote laboratories, and field settings, especially in areas with limited or infeasible sophisticated laboratory infrastructure (Ding et al. 2019).

8.2.1 Principle

LAMP has several steps that are significant and make it different from the standard PCR. The former uses a pool of four to six specifically designed primers for the various regions of a DNA sequence. These primers are F3, B3, and FIP, BIP and (optionally) loop or LF, LB to accelerate the amplification (Wang et al. 2015). It begins by binding the primers to the target DNA and then proceeds with the formation of more DNA fragments through a DNA polymerase enzyme that possesses the ability to dislodge this strand of DNA. As the polymerase copies the strand, the DNA will take the shape of a loop, which accelerates the rate of reaction because it is possible to move the strands continuously. It is also essential when maintaining the amplification cycle, which aids in the accumulation of DNA product within a short period of time (Ijaz et al. 2022; Hardinge and Murray 2019). LAMP produces many amplicons, and the different detection methods can test the reaction. Colorimetry is one of the more common methods by which the magnesium pyrophosphate (a product of the reaction) builds up, resulting in a solution that turns opaque and provides a visual signal. Alternatively, real-time amplification can be traced using fluorescent dyes and real-time monitoring (Lee et al. 2009).

8.2.2 Applications

Many applications have been developed using LAMP, and one of the most notable uses of LAMP has been in diagnosing infectious diseases. The technique is quick, easy, and sensitive, making it ideal for detecting pathogens in medical care. For example, LAMP has been successfully used to identify bacteria like *Mycobacterium tuberculosis*, the cause of tuberculosis, and *Plasmodium falciparum*, which causes malaria. It has also helped detect RNA viruses such as the Zika virus and SARS-CoV-2 (Le and Vu 2017). Outside infectious disease diagnostics, LAMP is being developed for noninfectious applications in agriculture, where it is used to detect plant pathogens. On-site diagnostics are better for managing and preventing diseases because they allow testing without sending samples to a central lab. LAMP has also been used in food safety to identify foodborne pathogens like *Salmonella* and *Escherichia coli*, helping ensure safety in remote locations. Additionally, in genetic testing, LAMP is being studied for early detection of cancer-related genetic mutations and monitoring disease progression (Le and Vu 2017).

8.2.3 Advantages

Basic: LAMP is basic, and simplicity is one of the most significant advantages of LAMP. Compared to the standard PCR, which requires thermal cyclers to experience temperature cycling, LAMP requires nothing more complicated than a simple source of heat; hence, it can be simplified to be feasible in low-resource settings (Varona and Anderson 2021).

Speed: LAMP can generate results in 30 min or less, which is about 1000 times less than other conventional PCRs that may take many hours (Kreutz et al. 2019).

Sensitivity: LAMP is a sensitive test that can detect even traces of nucleic acid. It is sensitive enough to detect a disease that is at an early stage or in a circumstance where the target DNA is in small concentration (Novi et al. 2025).

Cheap: LAMP is cheaper than PCR. It does not need a thermal cycler that costs a lot of money and can be applied to considerably expanded resources and field-based diagnostics (Curtis et al. 2009). LAMP is a diverse tool, and its usefulness may not be overrated in most other fields of study and clinical practice (Curtis et al. 2009).

8.2.4 Challenges

Despite all those advantages, there are certain issues with LAMP. One of such problems is the possible improper amplification. LAMP requires a number of primers, and unless the primers are carefully selected, there is a risk of binding to an incorrect sequence, and this can cause false positives. To prevent this, the primer should be designed well, and the reaction conditions should also be optimized. The other challenge is the method of detection. Even though the presence of a substance could be detected visually using colorimetry, more sensitive and quantitative methods, such as fluorescence-based detection, are necessary. Moreover, the technology of detecting LAMP products is also not as developed as the one used in quantitative PCR, and it may limit the type of data that can be obtained with the help of a LAMP test (Yang et al. 2024).

8.3 Circulating Tumor DNA (ctDNA)

Introduction Circulating tumor DNA (ctDNA) is a fragment of DNA that is released into the blood by tumor cells through different pathways, including cell death (apoptosis) and cell necrosis. It is also important to consider that a traditional tissue biopsy is an invasive surgery. Consequently, ctDNA can be detected noninvasively via a simple blood test, making it a potential biomarker for cancer detection, monitoring, and targeted therapy. The fact that ctDNA is the foundation of the concept of liquid biopsy has transformed the landscape of cancer diagnostics by suggesting an alternative to the classical forms of biopsy, as well as enabling the real-time molecular mapping of tumors with in vivo use, without having to repeat the invasive nature of performing the biopsy (Mishra et al. 2024). Over the past decade, ctDNA has gained increasing significance in clinical oncology, and its potential in early cancer diagnosis, minimal residual disease (MRD) detection, genetic profiling of cancers, and assessment of tumor responses to multiple treatments has been investigated in numerous studies. This is a less invasive procedure that is rapidly becoming a necessity of personalized cancer treatment (Alimirzaie et al. 2019).

Origin of ctDNA ctDNA is a product of the genetic material of tumor cells that are released into the blood through various cellular processes. They are referred to as apoptosis (programmed cell death), necrosis (unplanned cell death), and as an active secretion by tumor cells into the circulatory system. The ctDNA released by tumor cells is higher compared to that released by normal cells due to the higher rate of turnover. They are very short DNA fragments relative to the normal genomic DNA in a cell, ranging between 100 and 1000 base pairs in size. Those small fragments are more likely to be identified in blood than larger DNA fragments (Peng et al. 2017). DNA of ctDNA fragments reflects the genetic changes of the tumor, including mutations, deletions, amplifications, and rearrangements. Alexandrines involving all tumor subpopulations freed ctDNA, making it a very informative biomarker to comprehend the heterogeneity of the tumor and its reaction to treatment (Gao et al. 2017).

8.3.1 Heterogeneity in Tumor and ctDNA

One of the key advantages of ctDNA over tissue biopsies is the possibility of disclosing the genetic heterogeneity of the tumor. Tumors are often comprised of genetically different subclones, which do not respond to therapy in the same manner. Only a small, focal portion of the tumor is seen in tissue biopsies, and this may not identify the subclones that could exist elsewhere in the tumor or in the metastases. CtDNA, by contrast, is a release of the entire tumor, which presents a view of the whole genetic panorama of the tumor. In particular, this need is necessary in relation to the cancers that are highly heterogeneous, such as the nonsmall cell lung cancer (NSCLC), breast cancer, and colon cancer (Gao et al. 2017). Through ctDNA analysis, clinicians can uncover new mutations or genetic changes that may be sustaining the malignancy and its resistance, since any mutations or genetic changes occur concealed in the bulk of the initial tumor or in a minor sub-population of the cancer cells (Ohmura et al. 2024).

8.3.2 Clinical Applications

The potential clinical uses of ctDNA are numerous and continue to increase. ctDNA use in cancer medicine is also changing the game entirely, and some of its advantages outweigh the antiquity of the old techniques of diagnosis. The following are the most remarkable clinical applications of ctDNA.

8.3.3 Early Cancer Detection

Identifying cancer in its initial stages, when it can be treated most effectively, is one of the biggest challenges in oncology. The traditional cancer diagnosis methods, such as imaging and tissue biopsies, may not detect the cancer at a later stage, or not at all, which may have spread. However, many years ago, ctDNA could be detected in the blood, typically well before the onset of clinical symptoms or visible abnormalities in imaging, such as pancreatic cancer, ovarian cancer, and lung cancer (Bronkhorst et al. 2019). It has already been proven that ctDNA analysis can detect genetic mutations and alterations associated with these cancers in a nonsymptomatic individual, which could be a factor in saving lives and improving their outcomes (Fiala and Diamandis 2018).

Furthermore, with ctDNA, it is possible to screen high-risk populations, such as those who have a family history of cancer or whose genetic background is a known predisposition to cancer. The future promise of ctDNA to diagnose cancer earlier could 1 day lead to more cancers being treated at an earlier stage, increased survival rates, and improved long-term outcomes for the disease (Fiala and Diamandis 2020).

8.3.4 Genetic Mutation Detection and Genetic Tumor Profiling

ctDNA is an invaluable genetic tumor profiling tool. It enables clinicians to determine certain things such as the genes that can be mutated, amplified, deleted, and rearranged and spread the development of the cancer, e.g., in cancers, e.g., non-small cell lung cancer (NSCLC). ctDNA can be studied to reveal mutations in specific genes, such as EGFR, ALK, and KRAS, which are targeted by therapies. The detection of these mutations can contribute to the development of the most suitable choice of the treatment, which increases the risk of receiving a positive response (Restrepo et al. 2024). Besides the identification of amenable mutations, there was also profiling of genetic changes that made them resistant to treatment. It plays a paramount role chiefly in the pursuit of monitoring the treatment response and the identification of a resistance mechanism that could evolve during the treatment. To exemplify this, ctDNA can be examined in patients who are undergoing specific curative treatment or immunotherapy to develop the existence of mutations or alterations that could make the otherwise potentially ineffective treatment effective (Gray et al. 2015).

8.3.5 Monitoring Disease Progression and Treatment Response

One of the most important assets of ctDNA is the ability to observe the progression of the disease and the reaction to treatment in real time. As the tumors develop and are resistant to the development of the treatment, the amounts of ctDNA in the bloodstream can vary. By routinely measuring ctDNA, clinicians may observe the fluctuations in the tumor burden and the first indication of the relapse, along with responding to the variations in the treatment strategies (Gray et al. 2015). An example would be to apply ctDNA to patients undergoing chemotherapy or targeted therapy to determine whether the therapy is actually shrinking the tumor.

When the ctDNA concentrations are supremely lowered, it means that the treatment is working. On the other side, an increase in ctDNA levels may serve as a marker of tumor progression or resistance to therapy and the necessity to alter the course of treatment (Reece et al. 2019). ctDNA can also be used in the monitoring of minimal residual disease (MRD). Patients who are under primary treatment and still in remission may still have residual tumor cells circulating in their bodies. These types of cells can release ctDNA into the bloodstream, and the presence of it can indicate the existence of MRD even before the occurrence of clinical symptoms or any imaging. Early identification of MRD can prevent relapse and possibly give some guidance on further treatment.

8.3.6 *Minimal Residual Disease (MRD) Liquid Biopsy*

Minimal residual disease (MRD) refers to the small number of remaining malignant cells that can be present in the body following treatment, which can lead to relapse. ctDNA is a useful biomarker to determine MRD, with a low-level concentration of ctDNA in the body, demonstrating that there may be residual cancer cells in the organism. It has also been shown that ctDNA can detect MRD earlier than conventional imaging methods, which allows intervening earlier and improving long-term results (Semenkovich et al. 2023). ctDNA-based detection of MRD is specifically applicable to cancer that is related to hematologic issues, which include but are not limited to leukemia, lymphoma, etc., where a relapse may happen several months or even years after the therapy (Wen et al. 2023).

8.3.7 *Prognosis and Predicting Relapse*

ctDNA can also be used to predict the outcome the patient will receive and the likelihood of the disease happening again. By examining the genetic mutation and alteration that occurs in the ctDNA, clinicians are able to ascertain the aggressiveness of the tumor and the response of the tumor to multiple treatments. The concentration of CtDNA has been shown to be associated with the tumor stage, tumor grade, and prognosis, and can be used by clinicians to make more informed choices about how they treat the tumor (Li et al. 2020). What was more, this ctDNA might be applied in the relapse forecasting of the already treated patients who received remission. A low or undetectable ctDNA level would suggest that the patient is in remission. On the other hand, the elevation of ctDNA levels may signal the persistent presence of tumor cells or the emergence of a secondary disease, for which the symptoms necessitate continued treatment (Ijaz 2011; Li et al. 2020).

Technical CtDNA Analysis Issues. CtDNA analysis has numerous strong points, but there are quite a number of technical setbacks that must be overcome before it can become a possibility in clinical practice. Among the key issues are:

8.3.8 *Sensitivity and Detection Limits*

The sensitivity of ctDNA, especially in low-tumor-burden patients, or the ctDNA level in the blood, which remains very low in percentage, especially in patients with early-stage cancer, are among the major problems faced by ctDNA testing. They involve extremely sensitive methods like digital PCR or next-generation sequencing (NGS) and can detect a small amount of genetic material (Bronkhorst et al. 2019). It is necessary to make ctDNA detection more sensitive so that ctDNA can be easily observed in all patients regardless of their grade of disease stage or amount of tumor DNA in blood. Some scientists are optimizing detection methods and increasing the sensitivity of ctDNA analyses to expand their application area (Myint et al. 2018).

8.3.9 *Background DNA Contamination*

The second drawback of ctDNA analysis is that normal cell background DNA is present in the blood and that ctDNA may be mixed with the normal cells' DNA that is nonmalignant. Researchers are working on a solution to the problem, not only to produce ctDNA that is more highly enriched with other types of DNA, but also to improve the quality of ctDNA in analysis. The particular methods used are targeted sequencing and special cancer cell-specific biomarkers (Mouliere et al. 2018). Although ctDNA analysis has considerable potential, its clinical application remains limited because of the high cost of the technology. Next-generation sequencing (NGS) techniques used to analyse ctDNA, potentially costly and involving required equipment and expertise. As the technology becomes more advanced and standardized, the cost of ctDNA testing is likely to drop, which will then allow the technology to be offered to a greater number of patients and healthcare providers (Lin et al. 2021).

8.4 Exosome-Based Diagnosis

Introduction Exosome-based diagnostics is a novel technology in molecular diagnostics, a noninvasive technology in the diagnosis of many diseases, such as cancers, heart diseases, neurodegenerative diseases, and infectious diseases. Exosomes are small vesicles that are secreted by virtually all types of cells, and that can be found in most body fluids, including blood, urine, saliva, and cerebrospinal fluid. These vesicles transport various molecular cargos such as proteins, lipids, RNA, and DNA, and reflects a signal of the physiological state of the cells the vesicles are generated (Mishra et al. 2024). Exosome-based diagnostics has received considerable interest, taking into consideration the fact that the method can potentially enable the detection of disease at an early stage, be minimally invasive, and capture disease-related biomarkers. Exosomes have the promise of becoming providers of biomarkers of many pathologies, such as cancer, Alzheimer's disease, and cardiovascular diseases, and provide an alternative, potentially less-invasive, less time-consuming, and less-expensive bio-substitute to the most common biopsy methods (Cheung et al. 2018).

8.4.1 *Biology of Exosomes*

Exosomes are 30–150 nm-sized extracellular vesicles. They are released by the cells into the extracellular space through a process called exocytosis, once the multivesicular bodies (MVBs) fuse with the plasma membrane. The molecular cargo (lipids, proteins, mRNA, microRNA, and DNA) found in exosomes is heterogeneous. Exosomes may also be employed to determine the molecular characteristics of the parent cell, an explanation why it is highly informative in the context of the disease. Exosomes serve the function of intercellular communication, and they are apparatus utilized to transport their molecular content to another cell. The potential applications of this intercellular signaling are very broad, which can influence immune responses, tumor progression, and tissue repair. The contents of exosomes can be changed with the physiological or pathological state of the cell that contains the exosome, and therefore, exosomes hold interest when looking at finding biomarkers and monitoring of disease.

8.4.2 Exosomes Biogenesis and Release

Exosomes are produced by the initiation of the process within the endosomal compartment of the cell, where the endosomal membrane is inwardly pulled toward the formation of early endosomes. Late endosomes or MVBs are formed out of these primitive endosomes. Smaller vesicles are formed in MVBs, which will be discharged to the extracellular space at some point when the MVB fuses with the plasma membrane. This release sets the vesicles free that are later referred to as exosomes (Skryabin et al. 2020). Biogenesis exosomes are choosy in packaging molecular cargo in the process of biogenesis. It is a selective packaging by which the sorting of the vesicles of the specific proteins, lipids, and nucleic acids is carried out. They typically include some proteins on the surface of exosomes, e.g., tetraspanins (CD9, CD63, and CD81), heat shock proteins (Hsp70 and Hsp90), and a number of others, found on the surface of membranes. Furthermore, exosomes can carry microRNAs (small, noncoding RNAs that regulate gene expression) and the latter can control gene expression in recipient cells (Zhang et al. 2019).

8.4.3 Exosome Composition

The exosomes carry a highly complex complement of biomolecules under close supervision by the parent cell. Exosomes can be altered to suit different cell types, physiological conditions, and disease conditions. Following are the most important signature parts of exosomes:

Proteins: The exosomal proteins are the products of the parent cell membrane and of the cytoplasm. These proteins include membrane receptors, enzymes, cytoskeleton, and signaling proteins. Some of the exosomal proteins are involved in cell adhesion, signal transduction, and immune modulation (Wang et al. 2015).

Lipids: Phospholipids, cholesterol, and sphingolipids compose the lipid bilayer of the exosomes and are involved in the maintenance of the vesicle and enable the fusion of membranes to accomplish the targeting of cells.

RNA: Exosomes also contain a variety of RNA species, including messenger RNA (mRNA), microRNA (miRNA), and long noncoding RNA (lncRNA). These RNA molecules can be transferred between cells and can regulate the gene expression of the target cells.

DNA: The exosomes in some cases may also carry DNA fragments, which may be genetic information about the parent cell or tumor DNA in the case of cancer. Exosomes have a very useful purpose in diagnosis because the cargo contained within their exosomes is an indicator of the molecular well-being of the parent cell. To illustrate, tumor-secreted exosomes can include cancer-specific mutations, and the exosome mRNA or miRNA can be used to predict the pathophysiology of the underlying disease (Cheng et al. 2022).

Exosome-Based Diagnostics: Diagnosis of Cancer One of the most viable exosome-based diagnostic applications is cancer. The exosomes of the cancer cells are released into the blood, which could carry with tumor-specific biomarkers, including proteins, nucleic acids, and other molecules that are related to the presence, development, and spread of the illness. As exosomes serve as signs of the molecular aspects of the tumor, they can be utilized as noninvasive cancer biomarkers that could be used to detect, monitor, and evaluate the treatment outcome (Andre et al. 2024).

8.4.4 Cancer Detection with Exosomes

The ability to detect cancer using exosomes is based on the reality that tumors possess biomarkers that distinguish tumors and are found in the blood. The exosomes released by the tumors are usually filled with modified proteins, RNA, and DNA that are reflective of the genetic mutations, which are linked with cancer (Dermody et al. 2022). They can be diagnostic of cancer easily and noninvasively because exosomes are found in many body fluids, such as blood, urine, and saliva. Another alternative to tissue biopsies is liquid biopsy because this is done by analyzing exosomes in the plasma or in the serum (Wandrey et al. 2023). A second use of exosomes is the detection of minimal residual disease (MRD), tiny clusters of cancer cells that remain invisible to the body post-treatment but can be identified by employing exosomes. MRD is also required to assess treatment response and the prognosis of relapse. Tests based on exosomes may contribute to the discovery of detect the genetic changes that may point to the existence of a remaining tumor cell to distort the initial-line clinical routine and to modify care options.

8.4.5 Exosomal RNA in Cancer Diagnostics

RNA in exosomes can also be used to detect genetic abnormalities, regulatory gene expression patterns, noncoding RNAs, including microRNAs, associated with cancer (Zhang et al. 2019). Exosomal miRNAs show promise as biomarkers for different cancers. MiR-21 is a frequently overexpressed example of miRNA that is detected in exosomes in breast cancer, liver cancer, and lung cancer patients. Monitoring miRNA levels in exosomes has enabled the detection of cancer-specific miRNAs, which can be used as diagnostic and predictive biomarkers for cancer (Hu et al. 2020). Messenger RNA (mRNA) may also be transmitted through exosomes and can be exploited to uncover the expression profiles and mutations of tumor cells. Quantitative PCR, RNA sequencing, or microarray analysis allows studying the composition of mRNAs in exosomes and identifying the upregulated/downregulated genes in cancer and their mutations that lead to the condition (Xia et al. 2019).

Neurological Diseases Exosome Diagnostics Exosome-based diagnostics are not cancer-specific, but encompass the spectrum of most neurological conditions, such as Alzheimer's disease, Parkinson's disease, and multiple sclerosis. It has also been demonstrated that exosomes can transport biomarkers of this disease, including neuroinflammatory, neurodegenerative, and cellular signaling proteins, lipids, and RNA (Abreu et al. 2018).

Alzheimer Disease Exosomes In Alzheimer's disease (AD), exosomes have been shown to transport amyloid-beta (Aβ) peptides, which are a major pathological hallmark of the disease, between cells. The formation of plaques that lead to neurodegeneration is related to the accumulation of A beta in the brain. The exosomes of neurons or glial cells in the brain may be investigated in light of A2's presence, which could become a potential biomarker for the early onset of Alzheimer's disease. The tau protein, which is also found in exosomes, is another significant biomarker of Alzheimer's disease. Tau proteins cause tangles in brain cells, an aspect that contributes to the development of AD. It has been determined that the amount of tau in exosomes indicates disease progression, which is why the exosome-based assays can potentially be used in monitoring the disease progression or even analyzing the treatment efficacy (Colvee-Martin et al., 2024).

Detection Techniques Methods of exosome analysis are not standardized as well. The methods of isolating and analyzing exosomes in various labs may be different and may produce variations in outcome. The standardization of protocols of exosome isolation, biomarker detection, and data analysis methods is important to implement exosome-based diagnostics in clinical practice.

8.4.6 Exosomes in Agriculture and Plant Pathology Diagnostics

Exosome-based diagnostics has become another direction of agricultural and plant pathological diagnostics, which offers new opportunities to monitor plant health and early detection of plant diseases. Similar to animal cells, plant cells also secrete extracellular vesicles, such as exosomes, which contain diverse biomolecules, such as proteins, lipids, RNA, and small metabolites. They are intercellular communication vesicles through which plants can transmit genetic and molecular messages regarding their current intercellular status, like responses to environmental stress and pathogen infection. They are diagnostic exosomes due to their makeup based on the physiological and pathological state of the plant, and can be applied to detect both abiotic (e.g., drought, salinity) as well as biotic (e.g., viral, bacterial, and fungal) stresses (Morcillo and Manzanera 2021). The other area where exosome-based diagnostics is particularly promising is in diagnosing plant diseases at an early stage in plant pathology. Traditionally, plant diseases are identified by visual inspection, although this may be performed at a later stage once the disease has caused noticeable damage, or using more complicated molecular diagnostic techniques, which may require a sophisticated laboratory. However, exosome-based diagnostics is a noninvasive, real-time pathogen detection technique that involves the prior detection of the pathogen before it yields any symptoms. This may include plants which are infected with additional viruses such as the tobacco mosaic virus (TMV) or cucumber mosaic virus (CMV). The released exosomes in plants in this case contain the viral RNA and particular proteins, which are associated with the virus. The isolation and analysis of these exosomes may facilitate the early detection of pathogens, enabling more targeted interventions, such as antiviral treatment or the development of pathogen-resistant plant varieties (Berdugo et al. 2014). There is also a potential to use exosome-based diagnostics in the process of detecting bacterial and fungal infections in plants. Plant exosomes of infected plants may have pathogen-specific proteins or metabolites, which are applicable in determining the type of pathogen infecting a plant. To illustrate, exosomes secreted by plants infected with *Xanthomonas* species, causing bacterial wilt, may contain proteins/signaling molecules that are relevant to the plant defense response and pathogen virulence factors. These signals may help detect the disease much sooner than a classic symptom-based diagnosis and therefore provide early remedy and better control of plant diseases (Leonard et al. 2017). The other promising exosome-based diagnostics application in agriculture is that exosomes can be applied to trace the stress response of vegetation to environmental conditions. Plants release exosomes in response to abiotic stresses such as drought, salinity, extreme temperatures, and nutrient deficiencies. The molecular cargo of these exosomes can provide insight into how plants adapt to environmental conditions and may help identify early biomarkers of stress associated with reduced crop yields. For example, exosomal RNA can reflect stress-induced changes in gene expression and cellular signalling. The proteins or metabolites carried by the exosomes may serve as indicators of changes in metabolic state or stress-related signal transmission. These findings may support the regulation of molecular responses or the selection and development of more resilient crop varieties to mitigate the adverse effects of environmental stress (Varadharajan et al. 2025). Exosome diagnostics also has the potential to increase precision agriculture. Exosome-based systems can be integrated into field diagnostic devices, like handheld equipment or cell phone applications, enabling quick, on-site testing for a wide range of plant diseases and plant stresses. Agriculturists can use the technologies to diagnose and detect diseases or instances of stress during the early period of the growing season before the problems go out of control and, as such, act as a preventative or remedial measure. This could cause additional use of resources such as fertilizers, pesticides, and water, and eventually more harvesting and a smaller ecological footprint (Fahey et al. 2020).

8.5 Droplet PCR and Digital PCR: Advanced Techniques for DNA Quantification and Mutation Detection

8.5.1 Introduction

PCR techniques have also evolved, and highly sensitive and reliable procedures like droplet PCR and digital PCR have been developed. These innovative techniques have altered the ability to quantify DNA as well as filter abnormal genetic mutations at a level of sensitivity and specificity that previously could not have been accomplished utilizing the earlier technique of PCR. Additional similarities of digital PCR and droplet PCR are that a sample is divided into smaller and independent reactions, and that DNA can be relatively quantified without a standard curve being needed, a key innovation in many applications, including disease diagnostics and genomic research (Tang et al. 2016).

8.5.2 Digital PCR (dPCR)

Principle of Digital PCR Digital PCR (dPCR) is an innovative method for measuring or analyzing DNA or RNA by fragmenting a sample into hundreds of separate reaction chambers, and a minor fraction of DNA molecules is bound to one reaction chamber. The basis of the technique is that of end-point PCR, i.e., each partition is either positive (amplified) or negative (unamplified). The counting of the positive reactions to determine the absolute abundance of the target DNA in the original sample is the post-amplification step. The sample is typically sorted using a microfluidic system that divides the sample into hundreds or thousands of small reaction cavities or droplets (Nie et al. 2019). Also, the digital PCR provides an absolute estimate of the number of copies of DNA instead of a relative estimate of

DNA concentration, as with traditional PCR, no calibration curves are needed, and thus digital PCR is well suited in detecting rare mutations, low abundance targets, or where bulk change in DNA concentration is required. This has especially come in handy in biology applications such as gene expression research, mutation detection, and studies of variation in copy numbers and tracking minimal residual disease in cancer patients (Jensen et al. 2020).

8.5.3 Digital PCR Workflow

The digital PCR begins with the step of dissociating the DNA sample into reaction chambers or droplets. Each chamber or droplet contains a small fragment of the target DNA, which is then amplified by PCR in all the partitions simultaneously. Once the amplification process is complete, each partition is analyzed to either have or not have the target DNA, typically by fluorescence or colorimetric means. A positive partition implies that the target DNA was located and it was effectively amplified, and a negative partition implies that it was not amplified. The positive reactions/total number of partitions is the absolute quantification of DNA. Digital PCR is a highly sensitive method that can be resorted to determine some rare DNA sequences or low-frequency mutations, even when it is possible to do so on plenty of wild-type DNA. It may be regarded as an invaluable clinical diagnostic tool, particularly in oncology, to identify small residual disease (MRD), identify rare mutations within a heterogeneous population, and measure treatment response (Parsons et al. 2020).

8.5.4 Applications of Digital PCR

Cancer Diagnosis and Monitoring dPCR provides great function in cancer biology and is utilized in cancer-related gene mutations, such as EGFR, KRAS, and BRAF. It notably works with the unusual mutations or the ones with low frequency that are difficult to identify with the conventional methods within the liquid biopsies, such as the blood, the urine, or the plasma. The specified capability permits turning dPCR into an important component of the monitoring of minimal residual disease (MRD) in cancer-afflicted patients to detect the emergence of a recurrence in time and monitor the quality of treatment (Parsons et al. 2020).

Gene Expression Analysis Absolute quantification of gene expression can be done through digital PCR, and in the expression analysis of specific genes, the researchers can analyze, without reference genes or calibration curves. The aspect may be used where precise quantification is needed, as in the case of biomarker discovery, disease diagnostics, and environmental monitoring (Manoj 2014).

Infectious Disease Diagnostics: Diagnostics of infectious diseases is another field of application of dPCR, where it can be used to detect pathogens at extremely low concentrations. An example could be to identify viral RNA in already infected individuals like HIV, hepatitis, or COVID-19, even in a low viral load (Tan et al. 2021).

Diagnosis of Genetic Diseases Digital PCR is used to diagnose genetic diseases with rare genetic defects or SNPs associated with genetic diseases. It has been used in screening of diseases in fetuses, such as CFTR (cystic fibrosis transmembrane conductance regulator) and DMD (dystrophin), which can be caused by abnormalities in such genes (Bennett et al. 2001).

8.5.5 *Advantages of Digital PCR*

Absolute Quantification dPCR can also provide absolute quantification of DNA or RNA as compared to the normal PCR, which only provides relative quantification of the DNA or RNA in question, without any standard curves. This makes the method highly accurate and reliable for detecting low-abundance targets.

Sensitivity dPCR can identify low-frequency alleles and/or rare mutations in the presence of a high background concentration of DNA, and is thus applicable in fields such as the detection of minimal residual disease or the identification of rare genetic variations.

High Precision Digital PCR is capable of measuring the concentration of DNA with a great deal of accuracy, and provides more accurate results than certain other tools, including quantitative PCR (qPCR), which can be affected by variations in the amplification efficiency.

No Calibration Curves dPCR eliminates the use of standard curves and simplifies the working process without introducing the errors that can be encountered when producing a standard curve (Cilloni et al. 2019).

8.5.6 *Challenges of Digital PCR*

Cost Digital PCR apparatus (including microfluidic methodologies and custom reagents) can also be expensive, and thus, the technique is less accessible in all settings.

Complexity It can be difficult to carry out technically, and the mechanical task of splitting the sample up into a large number of individual chambers can require both special equipment and skill.

Low Throughput Digital PCR is highly sensitive, but dPCR systems have lower throughput compared to conventional PCR or qPCR, and thus are not used in high-throughput applications (Nie et al. 2019).

8.5.7 Droplet PCR Principle

A type of digital PCR, droplet PCR, involves the sub-division of a PCR sample into thousands of small droplets, each containing a small quantity of the target DNA. The droplets are amplified with a PCR, and the droplets are examined to be either harboring the target DNA or not. A fluorescence-based assay is usually used in cases where the presence of the target is reflected in the support of a fluorescence signal. The number of positive droplets and the total number of the target DNA are established by a ratio of the number of positive droplets to the total number of droplets. The standout variation between the droplet PCR and the standard digital PCR is the partitioning mode. In droplet PCR, a droplet generator (emulsifier) is employed to prepare the sample in very small droplets; each droplet is considered to be a single PCR reaction. It is a partitioning method that allows the simultaneous exploration of hundreds or thousands of separate reactions in parallel, which provides a more reliable and repeatable count of the copy number in DNA (Tang et al. 2016).

8.5.8 Applications of Droplet PCR

Detection of Rare Mutations Droplet PCR can be highly helpful in detecting rare mutations or mutations with low frequency, e.g., in cancer genomes. It can detect these mutations even when large quantities of wild-type DNA are present, and can therefore find application in the detection of cancer, particularly on liquid biopsy samples.

Copy Number Variation (CNV) Analysis Droplet PCR can also be used in the analysis of copy number variations (CNVs) or changes in the number of copies of a given gene or genomic region. The approach would allow fine dosing of the genes, and this is critical in the study of diseases like cancer, where the CNVs could trigger tumorigenesis.

DNA Quantification Droplet PCR can provide absolute quantification of DNA needed in the use of DNA quantification in studying gene expression, infectious disease surveillance, and genetic engineering. Droplet PCR is a useful tool in clinical diagnostics and research due to its high sensitivity and accuracy of measurement of DNA (Manoj 2014).

Advantages of Droplet PCR Plant diseases are also a huge problem to crop productivity and food security, both in agriculture and the entire world, due to the wide array of pathogens that include fungi, bacteria, viruses, and nematodes. A lot might be lost through the harvesting of crops, poor yields, and food quality because these diseases affect the livelihood of the farmers and the food web in the world. The traditional methods of plant disease detection and control are typically founded on visualizations, diagnosis of symptoms, and laboratory-grounded practices that can be time-consuming, expensive, and require knowledge and skills to implement. The timely and correct diagnosis of plant problems prevents a large loss, and here it is possible to consider such innovative diagnostic procedures as exosome-based diagnostics, PCR, and next-generation sequencing. These technologies may be used to prevent disease early, more efficiently treat the disease, and even design disease-resistant crop types by detecting disease-specific biomarkers or genetic signatures in the plants before the disease starts to show itself. As an illustration, by applying exosomes and other biomarkers to plant tissue or plant fluids, the scientists can diagnose plant infections by pathogens such as *Xanthomonas* (the causative organism of bacterial wilt) or *Phytophthora* (the causative organism of blight) on the molecular level. In addition, the diagnostics can be included in the precision agriculture systems and become capable of reporting real-time data about the plant condition, allowing farmers to make evidence-based decisions about the usage of fertilizers, pesticides, and irrigation, and, hence, slowing down the use of chemicals, increasing the crop yield, and supporting the sustainable farming systems. Integrating high-level diagnostic technologies with agriculture will not only maximize the efficiency of disease control but also make agricultural systems across the world more resistant to the latest challenges of the changing climate, which will ensure the sustainability of farms and agricultural output at a high level (Wang and Guo 2023).

8.5.9 Challenges of Droplet PCR

Cost and Equipment Droplet PCR system requires specialized equipment, including droplet generators and microfluidic systems, which are costly and might not be available in some laboratories.

Needs Technicality to Install Droplet PCR is not simple to install, and its maintenance remains a demand of expertise that may hinder its popularization.

Optimization Droplet PCR involves the optimization of the emulsification process to ensure that the sample is evenly distributed into droplets, which can be a technically complex task (Attali et al. 2016).

8.6 Recombinase Polymerase Amplification (RPA): A Revolutionary Technique in Molecular Diagnostics

8.6.1 Introduction

Recombinase polymerase amplification (RPA) is an isothermal apparent time of nucleic acid which exhibits striking similarities to the classical PCR in rapidity, performance, and availability. RPA was invented in the mid-2000s to prepare large quantities of amplified DNA and RNA at a constant temperature, usually 37–42 °C, and not thermally cycled, and, so, should be an acceptable and simpler substitute to PCR. RPA can increase genetic material under steady-state temperatures and does not necessitate expensive thermocyclers (which are required in the PCR to regulate temperatures throughout the denaturation, annealing, and extension of strands). RPA involves a recombinase enzyme, the presence of single-stranded binding proteins (SSBs), and an enzyme polymerase of DNA strand displacement capability. This ability to scale up nucleic acids at a single temperature can significantly find application to laboratory and field-based diagnostics, particularly in resource-constrained environments (Mota et al. 2022). The benefits of RPA are its simplicity, speed, and strength, which have found uses in various diagnostic tasks in clinical diagnostics as well as in environmental monitoring and food safety. This ability to work with minimal equipment, high sensitivity, and speed of the RPA has been of great benefit to a wide range of applications in the field of infectious disease diagnostics, genetic testing, and plant pathology. The next chapter will examine the principles and applications, advantages, and drawbacks of RPA and refer to the application of the discussed technology in agriculture and plant pathology in terms of the potential of such a technology to revolutionize the process of plant disease detection and control (Boluk et al. 2020).

Mechanism of RPA RPA is an effective and precise method of amplification; however, it requires certain significant elements to amplify nucleic acid under isothermal conditions. These are primarily recombinase enzymes, single-stranded binding proteins (SSBs), and a DNA polymerase enzyme that replaces the strands and primers that traditionally pair with particular sequences of DNA or RNA. The RPA mechanism could be separated into several steps:

Recombinase Binding This is where the binding of the recombinase enzymes to the primers or the short fragments of the DNA that bind to specific regions on the DNA of interest begins. The recombinases carry out the process of identifying and conjugating with the DNA sequences involved, thus forming a primer–template complex (Higgins et al. 2018).

Strand Displacement The recombinase aids in displacing one of the strands of the DNA after the attachment of the recombinase to the DNA, with the assistance of the single-stranded binding proteins (SSBs). The SSBs stabilize the single-stranded DNA (ssDNA) structure formed during the strand replacement stage, in which the DNA must be in a single-strand format, which is required in the amplification process (Ferrari et al. 2009).

DNA Synthesis The DNA polymerase with the feature of strand displacement activity subsequently stretches the primer against the target single-stranded DNA template, synthesizing a complementary strand. This extension can be repeated several times and does not need to be subjected to temperature cycling, making RPA simpler and quicker than traditional PCR.

Amplification and Extension This displacement/synthesis cycle of strands of DNA can be completed both in the cyclic format, and the target DNA/RNA may be replicated exponentially. The RPA is performed at a temperature that can be held constant, meaning that it does not need very complicated equipment, e.g., a heat block or battery-heated heaters, and therefore it can be used in the field (Yan et al. 2014). This increases the target nucleic acid sequence and can be detected using various methods like colorimetric, fluorescence, or even lateral flow detection methods (Li et al. 2020).

Applications of RPA Such versatility is the reason why RPA is applicable to diverse areas and, currently, the tool may be used in clinical diagnostics, veterinary medicine, environmental monitoring, and food safety. However, one of the most promising areas to implement RPA can be plant pathology and agriculture, as the detection of diseases can be essential to control and avoid others.

8.6.2 *Infectious Disease Diagnostics*

RPA has been shown to be immensely applicable in the rapid detection of infectious diseases, especially those caused by bacteria, viruses, and parasites. The standard conventional methods of diagnosis tend to use culture-based tests or PCR, which require the use of special laboratory equipment and exhaustive processes. Rather, RPA allows speedy and reliable detection of pathogens, avoiding heating the sample in the temperature cycle or needing much sample preparation. This saves it particularly during field diagnosis, whereby some complicated laboratory equipment may be limited. Indicatively, RPA has been successfully utilized to identify various human pathogens including the Zika virus, the Ebola virus, and *Mycobacterium tuberculosis*. Veterinary medicine too has promise and was applied to aid the diagnosis of diseases in livestock and pets: in cattle, it diagnoses foot-and-mouth disease, and in pigs, African swine fever.

RPA has been an efficient tool in disease control and health monitoring across the globe because it has the ability to rapidly detect and identify pathogens in people and animals (Wang and Guo 2023).

8.6.3 Agriculture and Plant Pathology

RPA has also shown so much promise in the agricultural sector, which deals with plant diseases that are a significant threat to crop production and food security in all parts of the world. Traditional methods of diagnosing plant diseases are usually associated with protracted procedures, such as the culture of the pathogen or the emergence of observable symptoms, therefore, postponing the response, and eventual wastage of crops. RPA offers high turnaround time and is portable, so it is a highly sought-after alternative to on-site diagnostic and early diagnosis of disease, which is a major component of disease management (Lau et al. 2016). RPA has been successfully used in detecting multiple types of plant pathogens, including bacterial pathogens, fungal pathogens, and viral pathogens. RPA has been used to determine bacterial causal agent *Xanthomonas* spp., which causes bacterial blight in rice, *Phytophthora infestans*, which causes late blight in potatoes and tomatoes, and cucumber mosaic virus (CMV), which infects many crops, among others. A timely detection of these pathogens in the disease cycle enables farmers and other agricultural workers to undertake corrective measures, which involve the use of resistant varieties, chemical treatment, or quarantine of the disease and loss of crops to the bare minimum (Lau et al. 2016). Secondly, the capacity to detect plant diseases in real time and at the point of occurrence is a game-changer in precision agriculture. This is particularly important in the context of world food production, which is anticipating a growing need for sustainable agricultural practices and proper control measures against diseases. The traditional methods of disease detection normally include sample collection to centralized labs that can take a long duration to provide their findings, thus reducing the effectiveness of interventions. When farmers apply RPA, they can produce timely harvests in the field so that they can respond promptly and minimize the impact of diseases on the crop (Scortichini 2022).

8.6.4 Food Safety and Quality Control

RPA is also being used in detecting pathogens and quality control in the food industry. Rapid and accurate identification of pathogens such as *Salmonella*, *Escherichia coli*, and *Listeria monocytogenes* is relevant in the food safety sector so as to be able to assure food product safety and prevent food-borne diseases. RPA can be used to identify these pathogens in food samples within hours, which makes it a cheaper and faster alternative to traditional culture-based methods (Wang and Salazar 2015). Additionally, food adulteration and GMO traceability with the help of RPA can be applied. It is possible to use the potential of testing the occurrence of specific genetic markers in food products to ensure adherence to the food safety legislation and prevent food labeling fraud (Xu et al. 2014).

8.6.5 *Environmental Monitoring*

RPA can also be applied in environmental monitoring, where it can be applied to detect the presence of environmental pollutants, bacteria, viruses, or pollutants in air, water, or soil sample. This is more so when it comes to the health of an ecosystem and the security of water. As the approach that allows investigating the pathogen of contamination to the environment in the short term and in a sensitive way, RPA can be used during the early detection of the contamination incident and prevent the development of the threat to the health of the population (Del Río et al. 2017).

8.6.6 *Benefits of RPA*

Isothermal Characteristics This is one of the main benefits of RPA in comparison to traditional PCR because it can be done at a constant temperature, traditionally 37–42 °C. This removes the costly nature of thermal cyclers, making it therefore most applicable where field work is required and resource-constrained.

Speed RPA can amplify both DNA and RNA in less than an hour, compared to the hours it can take to do so by the traditional PCR method. It has been utilized well in applications that are time-dependent, like testing of infectious diseases, plant pathogens, and food safety, due to the large turnaround time (Euler et al. 2012).

Portability Simple, mobile equipment, including battery-operated heaters or even portable PCR systems, can be used to perform RPA, and can be easily transported to the field, into isolated locations, or even to third-world countries with less ready access to lab facilities.

Sensitivity RPA is very sensitive and can detect small quantities of the target DNA or RNA and does not require the presence of an inhibitor or convoluted sample matrix (Mesquita et al. 2024).

Usability RPA is a readily accessible design with simple applicability, with a small preparation of samples and no special devices needed. Most of the available commercial RPA protocols are primers with predefined designs and programs that are already optimized, based on which the study can be performed by a layperson (Higgins et al. 2018).

8.6.7 RPA Issues and Disadvantages

Despite all the benefits, RPA is not devoid of its own challenges. One of the largest weaknesses is the possibility of nonspecific amplification. Considering that RPA relies on a recombinase enzyme, which induces amplification, there exists the possibility that the primers will bind to other regions of nontarget, leading to false positives. This can however be overridden by tailoring primers and the optimization of reaction conditions. Another issue is that the quality of reagents and enzymes could be high, which could become expensive and limit the use of RPA in less-developed nations. In addition, RPA is delicate, and its specificity may be influenced by the complexity of a sample and the presence of other offensive agents, e.g., environmental or plant samples (Munawar 2022).

8.6.8 Applications in Agriculture and Plant Pathology

The way plant diseases are detected and managed in the agriculture and plant pathology sector can be changed by RPA. The traditional methods of plant disease detection are culture methods or visual methods of detection, which may be time-consuming, labor-intensive, and may require special laboratory facilities. RPA, on the other hand, offers a fast, simple, and field-focused model where the recognition of plant pathogens can be carried out in the field without the expensive use of equipment, and sample processing is also time and labor-intensive (Lau et al. 2016). It has also been used in the diagnosis of viral infections like cucumber mosaic virus (CMV) and tomato spotted wilt virus (TSWV), which are devastating to crop production globally. RPA allows the farmers to act on such kind of pathogens at an early stage of the disease, which introduces a specific cure or eliminates affected crops, thereby preventing the disease transmission and reducing the losses of the crop (Strayer-Scherer et al. 2019). Moreover, the immediate nature of RPA pathogen detection, at the point of care, is an invaluable resource to precision agriculture, whereby the diagnosis of disease at the right point and time is critical in order to optimize the crop management process. RPA is a very portable and usable technology and can be applied to remote or rural areas where there is little access to the traditional diagnostic laboratory. Embedded in the precision agriculture platform, thanks to RPA, farmers will simply be able to detect disease outbreaks, monitor pathogen infection, and contain it, which, in the end, will lead to more yield and sustainable agriculture (Ahmed et al. 2018).

8.7 Liquid Biopsy

Introduction Liquid biopsy may be a potential game-changer in the field of diagnostic medicine, providing a noninvasive, viable, and, possibly, life-saving procedure for diagnosing and monitoring a wide variety of diseases, especially cancer. Diagnostic techniques like tissue biopsies are invasive and, in the majority of cases, presuppose the use of surgery, which is dangerous and not always possible in cases of metastasis or inaccessible tumors (Yazdi et al. 2024). Liquid biopsy is a new, promising biotechnology that can allow diagnosing early-stage cancer, therapy responses, residual disease of minimal residual disease (MRD), and even the presence of genetic mutations that can be used to tailor individual treatment regimes. The effectiveness of such a method to deliver noninvasive real-time diagnostics will be able to revolutionize disease and cancer detection and treatment in particular (Ge et al. 2024). The liquid biopsy is the procedure of gathering body fluid and examining it to generate biomarkers that may later be utilized to diagnose, keep track, and arrive at therapeutic decisions regarding diseases. The most widespread components that are examined in liquid biopsies are:

Circulating Tumor DNA (ctDNA): ctDNA are fragments of DNA that become liberated into the bloodstream by the tumor cells. ctDNA is a genetic mutation and alteration that reflects the molecular profile of the tumor. Circulating tumor cells are viable cancer cells shed from a primary or metastatic tumor into the bloodstream. They can be isolated and analysed to investigate the characteristics, progression, and behaviour of the tumor (CTCs).

Exosomes: These are small vesicles that surround the proteins, lipids, and nucleic acids of the parent cell. Exosomes can deliver biomarkers related to cancer formation, metastasis, and resistance to treatment.

MicroRNA (miRNA): There are small noncodifying molecules of RNA that control genes. Other miRNAs are cancer- and metastasis-related and can be detected in blood samples (Ge et al. 2024). Liquid biopsy offers a potentially good alternative to standard tissue biopsy, that is, it is noninvasive, involves only a simple blood sample (or other bodily fluid) and may be repeated, allowing the patient to be monitored in real-time, which in the case of many diseases, such as cancer, it is desired that the patient reacts to treatment and progresses with the disease over time.

8.7.1 Advantages of Liquid Biopsy

Noninvasive One of the greatest advantages of liquid biopsy is its noninvasiveness. In comparison to conventional biopsies that require the implementation of the surgical procedure or needle aspiration, liquid biopsy only requires a blood sample or other types of bodily fluid, which is not as traumatic to the patient and excludes the risks of invasive procedures (Li et al. 2021).

Real-Time Monitoring Without the regular necessity of invasive methods, liquid biopsy can be applied to real-time disease progression, treatment response, and recurrence. The ability to implement changes in treatment plans in real time and improve the outcomes of patients is possible under such real-time monitoring.

Early Detection Liquid biopsy will be in a position to diagnose diseases like cancer early, before it progresses to more advanced stages. Liquid biopsy makes it possible to identify the cancer before its appearance or the emergence of the tumor is noticeable in the radiograph (Yaghoubi Naei et al. 2023).

Tumor Heterogeneity Watch Liquid biopsy has the application of identifying genetic mutations in tumor subpopulations, which can be utilized to comprehend tumor heterogeneity. This helps clinicians have a better understanding of the complexity of cancer and identify the most effective treatment modalities that act on a portion of the entire tumor and not on a particular part.

8.7.2 Limitations and Problems with Liquid Biopsy

Sensitivity and Detection Limits Liquid biopsy is highly sensitive, but it is not always able to detect very low concentrations of ctDNA or CTCs, especially at the earliest stages of the disease or in small tumors. Liquid biopsy techniques will have to continue to improve in sensitivity to reach low genetic material detection levels in order to detect it and track its path in early-stage detection (Yaghoubi Naei et al. 2023).

Tumor Shedding Variability Tumors do not uniformly shed out ctDNA or CTCs into the blood, and certain types of cancers shed minimal detectable material which leads to the inability to produce reliable findings. This variability can be a limitation to the usefulness of liquid biopsy in certain groups of patients or within certain types of cancer.

False Positives/Negatives Just as with all tests of diagnosis, liquid biopsy is not an exception. To give an example, ctDNA or CTCs evidence in the blood might not always imply the presence of cancer since other factors may include the presence of a nonmalignant tumor or inflammation.

Standardization and Validation Liquid biopsy is a relatively novel technology, and it will require standardization of testing procedures and validation of biomarkers in various cancers and patient populations to be used in all clinical settings (Veldore et al. 2018).

8.8 Metagenome Profiling: A Gateway to Understanding Complex Microbial Communities

8.8.1 Introduction

Metagenome profiling is a revolutionary approach and involves the analysis of the genetic material within a sample of the environment without the first microorganisms being lab cultured. The method provides a closer look at the variation and functional capability of the microbial communities that have adapted to a particular environment, whether it is soil, water, the human gut, or even those communities associated with plants. The total genome of all the microorganisms present in any given environment, such as bacteria, fungi, archaea, viruses, and protozoa, has been defined as the metagenome (Suenaga 2011). Formerly, microbiologists had been restricted to studying only microbes that had undergone cultivation, and accordingly, this restricted the knowledge about many of the unculturable or difficult to grow microbes. Metagenome profiling is one use of high-throughput sequencing technologies to characterize the number of microbial species and the functional genes in complex communities to allow the study of entire ecosystems without a prior understanding of their structure. Impacts of this comprehensive strategy have proved to be revolutionary with respect to most fields of science, particularly in microbial ecology, plant sciences, and human health (Suenaga 2011). Studying the genetic content of a microbial community is called metagenome profiling, and does not demand isolation and culture of individual organisms. It lets the researcher examine communities of microbes with environmental samples such as soil, water, air, or biological samples such as feces or plant tissues. It involves analyzing the DNA of the sample collected and sequencing the collected DNA, then computationally analyzing the obtained DNA to determine the type of species in it and its functionality (Behzad et al. 2015).

8.8.2 There Are Two Approaches to Metagenome Profiling

Shotgun Metagenomic Sequencing This technique incorporates whole genome sequencing of a sample that, on man-hands-alone, has been split into disjointed bits and sequenced randomly. This process gives the general view of the complete genomic structure of the microbial community and the identification of a tremendously diverse array of organisms, including bacteria and viruses. Furthermore, it can provide information on functional genes in the community and reveal the metabolic and ecological functions of the microbes (Donovan et al. 2018).

Amplicon-Based Sequencing It is based on the idea of analyzing a certain number of microorganisms with the help of examining a certain marker gene, which can be the gene encoding 16S ribosomal RNA (rRNA) of a bacterium or the internal transcribed spacer (ITS) of a fungus. It is more focused and often more affordable than shotgun sequencing, and it is not able to show a global community view of the functional diversity in the same way (Aßhauer et al. 2015). Bioinformatics analysis of the output sequencing data of the two methods is further done to recapitulate the sequences and provide the microbial taxonomic names and their potential functions using the available level of the sequences of individual genes (Ghaly et al. 2025).

Techniques of Metagenome Profiling The metagenome profiling processes might be considered to have three core steps, i.e., sample collection, DNA extraction, sequencing, and data analysis.

Metagenome Profiling Application Most areas of science use the metagenome profiling. Its use in both plant sciences and microbial sciences is discussed in the subsequent sections, with the focus on the fact that it has been applied to understand complex biological systems better and address global issues.

8.8.3 Metagenome Profiling Plant Sciences

The plant microbiome, or community of microorganisms that live in symbiosis with plants, is found to be one of the greatest determinants of both plant health and growth as well as resilience. The metagenomics-based profiling of the microbiomes attached to the plants has provided an inkling of the complex interactions of the relationship between the plant and the community of attached microorganisms, i.e., bacteria, fungi, and viruses. These interactions can impact plant resistance to plant diseases, nutrient cycling, and even plant growth (You et al. 2024).

Learning about Plant-Microbe Interactions: Metagenome profiling offers the researcher the opportunity to explore the interactions between the plant and the microbes that live within it. As an example to illustrate, plants form symbiotic associations with their beneficial microbes, e.g., rhizobacteria and mycorrhizal fungi, to gain access to soil-originated nutrients and to combat pathogens. By analyzing a microbial community within the environment under varying conditions, the microbial species that are most common and active in plant roots, leaves, or other tissues may be studied through metagenomic techniques. Based on this understanding, microbial inoculants or biofertilizers can be developed that can upsurge the development or tension tolerance of crops (Berg 2009).

Plant Disease Diagnosis and Management Metagenome profiling also may play an essential role in investigating plant diseases such as fungi, bacteria, and virus pathogens. Traditional methods of identifying plant pathogens include culture of the microorganisms, which is tedious, and there is the likelihood that unculturable microbes will not be identified. On the other hand, the presence of the pathogen DNA in a sample of plant material is immediately detected by means of the metagenomic analysis, which makes it possible to be able to find plant diseases in a time and be able to cope with the diseases more successfully. In such a way, by profiling the microbiota of tomato plants with *Phytophthora infestans* (causing late blight), researchers can characterize potential secondary microbial communities, which may facilitate the growth of the disease, and more focused interventions may be pursued (Bahramisharif and Rose 2018).

Crop Breeding Disease Resistance Knowledge of how the microbiome can contribute to disease resistance can lead to breeding crops with greater disease resistance. Through the study of microbiomes of naturally resistant plants to some pathogens, breeders will be introduced to the useful microorganisms present that can be used to protect plants. They can be introduced into crops or can transfer their usefulness to the breeding cycle of new cultivars (Du et al. 2024).

Environmental Stress Tolerance in Plants Metagenome profiling has a potential use in defining how microbial communities contribute to environmental stresses, including drought, high salinity, and extreme temperatures. The plant microbiome can be a central variable to ameliorate plant recovery to unstable conditions during stress by transmitting the signal of stress tolerance in terms of its ability to retain water, internalize nutrients, and synthesize stress-mediated metabolites. By profiling the microbiome of a plant grown in different environments, researchers could identify which type of microorganism or functional gene correlates with stress resistance and use the information to enhance stress resistance in plants through breeding or microbial interventions (Noman et al. 2021).

8.8.4 *Metagenome Profiling of the Science of Microbes*

Metagenome profiling has helped to uncover the diversity, functions, and interactions of microorganisms of natural and human-associated microbiomes in diverse environments, both natural environments like soil and water, and human-associated microbial communities. Microbial ecology has been using metagenomics to assist in better understanding the structure of microbial communities, their ecological functions, and their evolution.

Microbial Diversity and Maintenance Works Profiling of metagenomes has been applied to the analysis of microbial diversity in previously poorly known systems. One of them is the soil microbiota, which is extremely rich and contributes greatly to nutrients and carbon cycling and plant health (Dindhoria et al. 2024).

Microbial Communities of Extreme Environments One of the most astonishing findings of metagenomics is the revelation of microbial communities of extreme environments like deep-sea vents, acidic hot springs, and polar ice caps.

Special microorganisms are in such environments, and they have evolved their own form of operating the metabolism in these harsh environments. Scientists use profiling of metagenomes to identify new enzymes, metabolic pathways, and microbial species that can be used biotechnologically to produce enzymes, bioremediation, and bioenergy on an industrial-scale level (Behzad et al. 2015).

Human Microbiome The metagenomics of our microbial population has enabled us to know so much about the microbes that inhabit our body, more specifically, those that are found in the gut, skin, and respiratory tract. Human microbiome forms an essential part in keeping the whole healthy by digestion, production of vitamins, defense against pathogens, and immune system regulation. Metagenomics has contributed to the study of human microbiome structures and mechanisms, and this has opened up new frontiers in personalized medicine where microbiome structure can be linked to health conditions such as obesity, diabetes, and autoimmune diseases (Devaraj et al. 2013).

Bioremediation and Environmental Health Bioremediation is also another microbial process that can be studied using metagenomic profiling; it involves the degradation or breakdown of environmental contaminants like oil spills, heavy metals, and pesticides by microorganisms. By examining the functional abilities of microbes that inhabit polluted environments, scientists can determine which individual microorganism or genes contribute to the degradation of pollutants that will result in more effective bioremediation protocols (Elshafei and Mansour 2024).

8.8.5 Challenges

Although ground-breaking, metagenome profiling is associated with challenges:

Data Complicacy Data of metagenomics are generally very large and convoluted, comprising millions of sequencing reads that must be correctly assembled, annotated, and interpreted. The complexity of the analysis process is complicated by microbial species abundance and the overlap/incompleteness of sequences. Processing and analysis of metagenomic data require the most advanced bioinformatics framework and computational power (Dong et al. 2023).

Sampling Bias The metagenomic profiling is dependent on the quality and the representativeness of the sample. Ecological biases may arise due to the method used to extract the DNA or an uneven abundance of certain groups of microbes that are more readily quantifiable than others (Verma et al. 2017).

Standardization and Validation All laboratories have varying protocols and bioinformatics pipeline strategies for metagenomics, and the result should be reproducible and comparable to enable standardization. Moreover, metagenomic findings must often be verified by culture or other techniques to determine the presence and functions of the microbial species (Tourlousse et al. 2021).

8.9 Bioimaging: Unlocking the Mysteries of Biological Systems

8.9.1 Introduction

Bioimaging refers to a series of technologies and modalities, which are deployed to visualize the biological structures and processes between scales, i.e., between individual molecules and entire living organisms. It is a rapidly developing field that has transformed the manner in which scientists understand biological systems and has provided an improved view of the way cells function, how organs perform, and the movement of molecules in living organisms. Bioimaging applications in medicine, plant sciences, neuroscience, and molecular biology have provided insights into how diseases, treatment responses, and physiologic processes occur in real-time (Shapiro et al. 2021). Bioimaging has made a revolution in the sciences of plants, as it has made it possible to examine how plants grow, develop, and how they respond to stresses, and how they can interact with their surrounding environment on a new scale. Using bioimaging, scientists have managed to understand the complex plant cell structures, how water and nutrients enter and exit plants, and how plants respond to environmental conditions, pathogens, and diseases (Hu et al. 2020).

Concepts and Techniques of Bioimaging The use of different technologies to see biological samples, including the structure of cells and their internal components, or tissue and organ structure, is called bioimaging. The various imaging types suit various research purposes based on the space resolution, imaging depth, and requirements of the sample preparation. The main bioimaging methods applicable in plant sciences include the following (Tenorio Berrío and Dubois 2024).

Light Microscopy (LM) Light microscopy is one of the most typical imaging techniques applied in plant biology. It relies on visible light and optical lenses in making visible and magnifying structures in the biological samples. The technique is particularly employed to observe the morphology of plant cells, the structure of tissues, and the morphology of plant organs, e.g., roots and stems. The key advantage of light microscopy is that it is not an invasive method, such that living plant tissues can be examined without making any substantive sample preparation. Light microscopy also has limitations since it can only see the subcellular components, say organelles, due to resolution, which is typically in the range of 200 nanometers (Gallei et al. 2025).

Fluorescence Microscopy Fluorescence microscopy is a more specialized form of light microscopy, in which a fluorescent dye or protein is used to either stain or label selected molecules or structures of the cell. These fluorescent probes take up the light at a specific wavelength and release the light at a longer wavelength to allow a researcher to see the particular components of the cells, such as proteins, lipids, or DNA (Choi et al. 2021). Microscopic research in the plant sciences has come to a different facet with the aid of fluorescence. It enables the researchers to locate the position of some of the proteins within the plant cell, monitor the expression of a gene, and see how plant cells and pathogens interact. The technique can be invaluable in the research on how plants respond to environmental stimuli, nutrient intake, and how the plant metabolism process unfolds (Walter et al. 2004).

8.9.2 Confocal Microscopy

Another type of fluorescence microscopy is confocal microscopy, which is more resolute and has the ability to provide optical sections of thick samples. Confocal microscopy also eliminates the depth ambiguities in the traditional fluorescence microscopy that gathers light at all depths of the observed object, and at the same time, confocal microscopy uses a laser by scanning a single spot at a time and collecting light in a focal plane, reducing the blurring effect of out-of-focus light. It results in high-resolution images and three-dimensional images (Ter Veer et al. 2017). Confocal microscopy has been widely used in plant research to study the internal structure of plant tissues, including the cellular structure, vascular system, and the movement of fluids and solutes. It can be particularly applied to the study of plant roots and the root microbiome, and the study of interactions between plant cells and pathogens or symbiotic microorganisms (Rich et al. 2014).

Electron Microscopy (EM) Of all the imaging processes, electron microscopy is the most resolved since it is capable of visualizing structures on a nanometer scale. There are two broad categories of electron microscopy, namely, transmission electron microscopy (TEM) and scanning electron microscopy (SEM) (Strnad et al. 2015).

Transmission Electron Microscopy (TEM) TEM helps the researcher to view the structure of the interior of the cells on the ultrastructural level that comprises organelles such as chloroplasts, mitochondria, and endoplasmic reticulum. TEM has rendered itself essentially necessary in the investigation of the structuring of plant cells in minute detail, e.g., the cell wall, the cytoskeleton, and the vesicular transport systems. One can also analyze the interaction between plants and viruses or other cellular processes at the molecular level with its help (Graham and Orenstein 2007).

Scanning Electron Microscopy (SEM) SEM helps to provide a close-up image of the surface morphology of plant tissues. It can also be used to investigate plant organs like leaves, stems, flowers, and seeds, and plant surface structures such as trichomes, stomata, and cuticular wax. SEM is important in the environmental interaction of plants, including plant reaction to stresses in the environment and defense against pathogens (Stabentheiner et al. 2010).

Magnetic Resonance Imaging (MRI) Magnetic resonance imaging (MRI) is a noninvasive imaging technique and involves the application of strong magnetic fields and radio waves to produce undistorted images of organs and tissues. Although MRI has been utilized in medical imaging, it has turned out to be an invaluable tool in the plant sciences pertaining to the study of water flow and distribution in plant tissues (Moustaka and Moustakas 2023). MRI can visualize the plant organs 3D, i.e., roots, stems, and leaves, and track the distribution of water and nutrients in real time. It is an effective method, particularly in studies that can be on the water relations of plants, including the effect of drought stress on the well-being of plants and on salinity and the uptake of nutrients. It also figures prominently in the research of plant reactions to environmental changes, such as temperature and soil conditions (Bolaji Umar et al. 2022).

Positron Emission Tomography (PET) Positron emission tomography (PET) is an imaging technique in which radioactive tracers have been employed to determine the circulation of specific substances in plants. PET finds applications in plant sciences to either trace the movement of carbon and other metabolites in the process of photosynthesis, or where the location of nutrient and other molecules in plant tissues is to be determined. PET can help to provide meaningful data concerning plant metabolism and ways in which the organism adapts to the evolving environment (Joseph and Tawakol 2016).

Magneto-Optical Imaging (MOI) A more recent method is called magneto-optical imaging (MOI) and involves a mixture of magnetic field sensors and optical imaging to visualize the changes in plant tissues in dynamics. The technology may be particularly useful in studying the plant root formation and response to the environmental factors of gravity, light, and access to nutrients. MOI is able to provide noninvasive imaging of cell signaling and the development of roots in plants in real-time.

8.9.3 *Applications of Bioimaging in Plant Sciences*

The sciences of plants have found bioimaging methods so useful that it has become indispensable. The most significant uses of bioimaging have left an impressive mark on the understanding of plant biology in the following ways:

Growth and Development of Plants Without bioimaging, one cannot be able to understand the processes of plant growth and development, seed germination up to flowering. Scientists can study what controls plant growth using a visualization of cellular processes such as cell division, cell elongation, and cell differentiation. They can provide an example of examining the structure of the plant tissue in development, as confocal microscopy can be applied to the study of the structure of vascular tissues and cell layers in roots and stems (Huang and Xu 2015).

Carbon Fixation and Photosynthesis Photosynthesis is the fundamental process whereby plants absorb light energy from the sun, thereby converting it into chemical energy. Bioimaging, namely fluorescence microscopy, is commonly used to test the movement of the chloroplasts, the organelles that facilitate photosynthesis, in plant cells. The dynamics and energy exchange of chloroplasts assist the researchers to have a better idea of how plants can optimize photosynthesis in various environmental settings. The given data are crucial in not only increasing crop yield but also creating crops with a higher photosynthetic rate (Li et al. 2022).

Plant Stress Responses Drought, high salt content, excessive temperature, and attacks by pathogens are all stress-causing agents to which a plant is exposed at all times. In real-time, visualizing the plant response to such cellular and molecular-scale stresses can be done with bioimaging. Translocation of reactive oxygen species (ROS) and other stress species can be detected and traced in the cell, and the fluorescence and confocal immunofluorescence help to reveal how the plant cell responds to environmental stress factors through its defenses (Burlakoti et al. 2024).

Plant–Microbe Interactions Plants are highly interactive with both positive microorganisms (rhizobacteria and mycorrhizal fungi) and with the adverse. Studies of these interactions in real time by bioimaging techniques that provide the opportunity to understand the methods of protecting plants against parasites and the ways in which beneficial microorganisms can stimulate plant growth are incredibly useful. One example is the fluorescent tagging of single proteins/microbial species in plants, by which a researcher is able to monitor the movement of microbes in plant tissues and detect the establishment of symbiotic interactions, as well as how plants can modify their immune reaction to an infection (Shekhawat et al. 2022).

Growth of Roots and Their Interactions with Soil The absorption of nutrients and water is also crucial, and bioimaging has helped to understand the composition and structure of roots and their functions. MRI and confocal microscopy can assist researchers in tracking the development of roots and even root hairs, which are vital in the uptake of nutrients. Another example of a sample that can be investigated using bioimaging to learn more about the connection between microorganisms that colonize the rhizosphere and the relationships between the roots and plant development and root health. This kind of knowledge would have implications to enhance soil health and yield crops that are more efficient to absorb nutrient in the soil (Sun et al. 2021).

8.9.4 *Crop Improvement and Biotechnology*

Another field where bioimaging can be a useful tool is crop improvement and biotechnology. That will allow scientists to track the effects of genetic manipulations on the growth and development of plants and their reaction to stresses. Examining the case example of scientific analysis, the effects of genetic modifications made on plants to alter their physiology can be monitored in an attempt to make them more disease or nutrient uptake resistant. There is also the application of bioimaging in analyzing transgenes in genetically modified plants, and in this case, we may assume that the desired trait or traits are successfully being expressed in a plant (Yin and Qiu 2019).

Plant Disease Diagnostics Bioimaging can be used to its full extent within the area of plant diagnostics due to pathology, e.g., fungi, bacteria, and viruses. Fluorescence microscopy and confocal microscopy enable researchers to not only see how the pathogens attack the plant cells but also the key changes that occur in the cells during the process. It has led to the invention of rapid diagnostic tools in diagnosing plant diseases to help in early treatment and prevention of disease transmission (Kashyap et al. 2016).

8.9.5 *Challenges*

Despite the influential role played by bioimaging in the research of the plant, there are certain issues. In some research conditions, high-resolution imaging may be prohibitively expensive and require extra equipment and skills. In addition, the nature of plant tissue and the absence of imaging can make it difficult to obtain clear, detailed images of specific structures, especially small plant tissues like stems and roots.

8.10 CRISPR-Cas Based Diagnostics: SHERLOCK and DETECTR

8.10.1 *Introduction*

CRISPR-Cas technologies swept through the realm of gene editing, molecular diagnostics, and biotechnology. Originally, CRISPR-Cas systems were designed to have the capability to act as a system of gene-editing, but now they serve as diagnostic tools and provide an affordable, rapid, and accurate method of detecting an immense number of pathogenic organisms, genetic mutations, and other biological markers. CRISPR-based diagnostics, such as SHERLOCK (Specific High-Sensitivity

Enzyme Reporter UnLOCKing), were developed, and DETECTR (DNA Endonuclease Targeted CRISPR Trans Reporter) has stood out as a significant innovation in the scope of molecular diagnostics, with a point-of-care system that can potentially democratize the process of diagnostic testing in the world, even in a resource-restricted environment (Yoshimi et al. 2022). Such diagnostic techniques exploit the molecular specificity of CRISPR-Cas systems, which proves to relate to the alternative techniques of signal amplification, to determine the presence of some nucleic acids in a quick and efficient way. Both SHERLOCK and DETECTR can use the CRISPR-Cas recognition and cleavage power of target DNA or RNA sequences with easy and simple readouts, such as colorimetric or fluorescence, making them highly accessible and convenient to operate (Kellner et al. 2019).

CRISPR-Cas system. This is an endogenous immune defense mechanism of bacteria and archaea that can be used to combat viral infections by storing fragments of viral genome in the CRISPR array as the so-called spacers. When the same virus reinvades, the CRISPR system is also able to target the destruction of viral DNA by using these spacers to guide the Cas proteins to the viral DNA. The CRISPR-Cas9 system, the most renowned of all the numerous CRISPR systems, was adapted as an instrument in gene editing due to its ability to cause precise double-strand breaks at a particular site within the genome. Other CRISPR-Cas systems, including Cas12 (Cpf1) and Cas13 (C2c2), by their specific properties and abilities to identify and cut the nucleic acid, can be repurposed for use in diagnostics (Ghorbani et al. 2021). CRISPR systems, in the field of diagnostics, are expected to associate with a particular DNA or RNA sequence because they are based on the ability of the Cas family proteins to cleave. Once the target nucleic acid is identified, the CRISPR-Cas proteins can also cleave a reporter molecule that will generate a signal that can be detected to indicate that the target sequence indeed exists (Gootenberg et al. 2018).

8.10.2 SHERLOCK (Specific High-Sensitivity Enzymatic Reporter UnLOCKing)

SHERLOCK is a diagnostics technology relying on CRISPR developed at the Zhang Lab at the Massachusetts Institute of Technology (MIT) and the Broad Institute of MIT and Harvard. The Cas13 enzyme, which has been characterized by having the ability to cut and target RNA, is utilized in SHERLOCK. The most significant aspect of SHERLOCK is the ability to detect not only DNA but also RNA, which is why the platform is ideal to detect RNA viruses, including influenza, Zika, and SARS-CoV-2, among others (Gupta et al. 2021).

Mechanism of SHERLOCK The mechanism utilized by SHERLOCK in diagnosis is listed as follows:

Target Recognition: SHERLOCK uses CRISPR-Cas13 system, which is used to recognize specific RNA or DNA sequence. The guide RNA (gRNA) that is complementary to the target sequence hybridizes with the target sequence (Gootenberg et al. 2017).

Cas13 activation: When the RNA molecules around a Cas13 are identified they are activated, and the Cas13 cuts indiscriminately through them. This collateral cleavage capability is one of the major traits of Cas13 in comparison with other CRISPR systems like Cas9 that cleaves only the target DNA (Patchsung et al. 2020).

Signal Amplification: SHERLOCK is amplified using a procedure called isothermal amplification (typically through the assistance of LAMP-Loop-Mediated Isothermal Amplification). The amplification uses target nucleic acids multiplications that enhance the sensitivity of the detection (Nguyen et al. 2020).

Reporter System: Cleavage activity of Cas13 can be utilized to cleave a reporter molecule e.g. a fluorescent or colorimetric signal freeing a detectable signal. This signal can then be measured by simple methods, including a color change, fluorescence, and lateral flow, depending on the application (Patchsung et al. 2020).

8.10.3 Applications of SHERLOCK

Viral Detection SHERLOCK has been used to detect viral infections as well as specifically RNA viruses. One such use is the use of the platform to detect the Zika virus, dengue virus, and even the SARS -CoV-2 virus causing COVID-19. SHERLOCK is a swift and effective method to test infections as it detects the RNA of the virus, especially when a standard lab test may be too slow or too expensive to perform (Adedokun et al. 2024).

Point-of-Care Testing SHERLOCK may offer a chance at a possibility of point-of-care diagnostics. Resource-constrained settings (e.g., remote areas or low-resource hospitals) can use the platform on the basis of its ability to work with minimal equipment and the presence of an easy-to-read out (i.e., the colorimetric or fluorescent signal), where access to expensive lab facilities is limited (Fonjungo et al. 2015). In DNA detection, DNA Detection SHERLOCK may also be utilized, and this has taken the SHERLOCK further in genetic testing, pathogen detection, and monitoring of minimal residual disease (MRD) in cancer patients (Bengtson ct al. 2022).

8.10.4 DETETR (DNA Endonuclease Targeted CRISPR Trans Reporter)

The other CRISPR-based diagnostic instrument is known as DETECTR, and its purpose is to detect DNA sequences using the assistance of the Cas12 enzyme, which has become popular in diagnostic applications due to its capacity to cut DNA sequences so efficiently. DETECTR is a diagnostic platform developed by scientists at Mammoth Biosciences, and it is a simple, fast, and inexpensive diagnostic platform that can be used in a wide range of settings to detect genetic material with a variety of applications, including in clinical diagnostics, food safety, and environmental monitoring (Kellner et al. 2019).

Mechanism of DETECTR The DETECTR platform relies on the Cas12 protein that has an unusual attribute: it becomes activated when the protein detects its own target DNA code, but this time it becomes activated in its nonspecific cleavage, that is, in cleaving the surrounding single-stranded DNA. This splitting at the collateral can be put to use to generate a signal.

Target DNA Recognition DETECTR is also based on SHERLOCK, except that the guide RNA (gRNA) directs the Cas12 enzyme to a specific DNA sequence. The gRNA is coded to complement a piece of a DNA sequence of interest and enables a highly specific and exact choice of the target (Chen et al. 2025).

Cas12 Activation and Cleavage When complexed with the target, Cas12 is activated to cleavage activity by a conformational change. Cas12 will subsequently randomly cut the single-stranded DNA surrounding it and that can be functionalized with a fluorescent or colorimetric reporter molecule.

Signal Generation When the reporter DNA is cleaved, a signal (generally fluorescence or a color change) that indicates the presence of the target sequence is released. This signal is easily measurable and observable, and, thus, there is a prospect of rapid diagnostics (Gong et al. 2021).

Applications of DETECTR Detection of Viruses and Bacteria DETECTR has been particularly useful in the detection of DNA-based pathogens. It has been used as an example to identify the DNA of *Streptococcus pyogenes*, responsible for strep throat, and other bacterial and viral infections. DETECTR will be able to offer a prompt solution to the issue of the presence of the pathogen in the clinical sample and guarantee prompt diagnosis and treatment to prevent the transmission of the disease.

Food and Environmental Safety DETECTR can also be applied to food testing, including rapid testing of bacteria like *Salmonella*, *Listeria*, and *E. coli* in food or water. It is also applicable in the monitoring of the environment, e.g., to find the presence of bacteria in food or water. It is an effective instrument of food safety as one can obtain the result within several hours, thereby turning into an efficient instrument of real-time monitoring and risk management.

Genetic Testing Genetic testing can also be performed through DETECTR since it can detect some specific DNA sequences. Combined with the affordability and rapid outcomes, the platform is a valuable component of the massive genetic screening initiatives (Shi et al. 2010). CRISPR-Cas Based Diagnostics have a number of advantages.

High Sensitivity and Specificity Both SHERLOCK and DETECTR are highly sensitive and can be applied to the detection of small amounts of DNA or RNA in complex samples. This makes them appropriate in both early-stage infection detection and the detection of rare genetic mutations.

Quick and Simple The CRISPR-based diagnostics products can be received within several hours, which is significantly quicker than the traditional PCR-based methods. The simplicity of the test, which involves very little sample preparation and little equipment, allows point-of-care testing with limited resources.

Low Cost The CRISPR-based diagnostics are cheap when compared to other molecular diagnostics methods, such as PCR and next-generation sequencing. Both SHERLOCK and DETECTR have fairly low-cost reagents and equipment, and therefore, can be run on a massive scale and in large quantities.

Versatility SHERLOCK and DETECTR can be used to talk with a broad range of DNA and RNA targets, and viral, bacterial, and genetic targets are examples of these. They can be utilized in numerous different applications, such as clinical diagnostics, environmental monitoring, and agriculture due to their versatility (Das et al. 2023).

8.10.5 Challenges and Limitations

Sensitivity to Contamination CRISPR-based systems, like any other diagnostic system, are susceptible to contamination that leads to a false positive result. Attention to the sampling and contamination should be very strict to achieve the right results.

Limitations of Targets In Limitation of CRISPR-based diagnostics is that it relies on the presence of particular target sequences. Even though guide RNA design is highly customizable, genetic differences between the pathogens or the patient's sample may affect the sensitivity of the test.

Regulatory Approval and Validation Although CRISPR-based diagnostics have proven to have a lot of potential, there remains the issue of regulatory approval and clinical validation. Such systems must be validated and tested to determine that they meet the requirements to make them massively applicable in clinical settings (Kaur et al. 2024).

Also, the diagnostics developed by CRISPR would play an essential role in reducing the total health concerns of the planet, including pandemic preparedness, personalized medicine, and precision agriculture. The platforms have the potential to revolutionize diagnostic testing since they are still evolving and pave the way toward a faster, more precise, and cost-effective healthcare.

References

Abreu CM, Relvas JB, Soares-Dos-Reis R, Mendes Pinto I, Guimarães J, Melo PN, Cruz AP, Sá MJ. Emerging biosensing technologies for neuroinflammatory and neurodegenerative disease diagnostics. Front Mol Neurosci. 2018;11:164. https://doi.org/10.3389/fnmol.2018.00164.

Adedokun G, Sidhu G, Alipanah M, Wang GP, Fan ZH. A handheld HIV detection platform using paper-based sample preparation and real-time isothermal amplification. Microsyst Nanoeng. 2024;10(1):181. https://doi.org/10.1038/s41378-024-00822-1.

Ahmed FA, Arif M, Alvarez AM, Larrea-Sarmiento A. Genome-informed diagnostics for specific and rapid detection of Pectobacterium species using recombinase polymerase amplification coupled with a lateral flow device. Sci Rep. 2018;8(1):15972. https://doi.org/10.1038/s41598-018-34275-0.

Alimirzaie S, Bagherzadeh M, Akbari MR. Liquid biopsy in breast cancer: a comprehensive review. Clin Genet. 2019;95(6):643–60. https://doi.org/10.1111/cge.13514.

Andre M, Caobi A, Miles JS, Vashist A, Ruiz MA, Raymond AD. Diagnostic potential of exosomal extracellular vesicles in oncology. BMC Cancer. 2024;24(1):322. https://doi.org/10.1186/s12885-024-11819-4.

Aßhauer KP, Wemheuer B, Daniel R, Meinicke P. Tax4Fun: predicting functional profiles from metagenomic 16S rRNA data. Bioinformatics. 2015;31(17):2882–4. https://doi.org/10.1093/bioinformatics/btv287.

Attali D, Haynes C, Bryan J, Bidshahri R. Ddpcr: an R package and web application for analysis of droplet digital PCR data. F1000Res. 2016;5:1411. https://doi.org/10.12688/f1000research.9022.1.

Bahramisharif A, Rose LE. Efficacy of biological agents and compost on growth and resistance of tomatoes to late blight. Planta. 2018;249(3):799–813. https://doi.org/10.1007/s00425-018-3035-2.

Behzad H, Mineta K, Gojobori T. Challenges and opportunities of airborne metagenomics. Genome Biol Evol. 2015;7(5):1216–26. https://doi.org/10.1093/gbe/evv064.

Bengtson M, Van Der Torre J, Dekker C, Schallig H, Franch O, Bharadwaj M, Meerdink V. CRISPR-dCas9 based DNA detection scheme for diagnostics in resource-limited settings. Nanoscale. 2022;14(5):1885–95. https://doi.org/10.1039/d1nr06557b.

Bennett RR, Kunkel LM, Darras BT, O'Brien KF, Den Dunnen J. Detection of mutations in the dystrophin gene via automated DHPLC screening and direct sequencing. BMC Genet. 2001;2(1):17. https://doi.org/10.1186/1471-2156-2-17.

Berdugo CA, Zito R, Paulus S, Mahlein A-K. Fusion of sensor data for the detection and differentiation of plant diseases in cucumber. Plant Pathol. 2014;63(6):1344–56. https://doi.org/10.1111/ppa.12219.

Berg G. Plant–microbe interactions promoting plant growth and health: perspectives for controlled use of microorganisms in agriculture. Appl Microbiol Biotechnol. 2009;84(1):11–8. https://doi.org/10.1007/s00253-009-2092-7.

Bolaji Umar O, Khadijat Abdulhamid A, Lukman Bola A, Shehu Abdulbaki A, Amudalat Ranti L, Oluwagbenga Victor K, Ramat Biola M. Stresses in plants: biotic and abiotic. Intechopen; 2022. https://doi.org/10.5772/intechopen.100501.

Boluk G, Alvarez AM, Melzer M, Crockford AB, Arif M, Dobhal S. Genome-informed recombinase polymerase amplification assay coupled with a lateral flow device for in-field detection of Dickeya species. Plant Dis. 2020;104(8):2217–24. https://doi.org/10.1094/pdis-09-19-1988-re.

Bronkhorst AJ, Ungerer V, Holdenrieder S. Early detection of cancer using circulating tumor DNA: biological, physiological and analytical considerations. Crit Rev Clin Lab Sci. 2019;57(4):253–69. https://doi.org/10.1080/10408363.2019.1700902.

Burlakoti S, Kaundal A, Devkota AR, Poudyal S. Beneficial plant–microbe interactions and stress tolerance in maize. Appl Microbiol. 2024;4(3):1000–15. https://doi.org/10.3390/applmicrobiol4030068

Chen H, Song F, Wang B, Huang H, Luo Y, Han X, He H, Lin S, Wan L, Huang Z, Fu Z, Ledesma-Amaro R, Yin D, Mao H, He L, Yang T, Chen Z, Ma Y, Xue EY, et al. Ultrasensitive detection of clinical pathogens through a target-amplification-free collateral-cleavage-enhancing CRISPR-CasΦ tool. Nat Commun. 2025;16(1):3929. https://doi.org/10.1038/s41467-025-59219-x.

Cheng H, Yang Q, Wang R, Luo R, Zhu S, Li M, Li W, Chen C, Zou Y, Huang Z, Xie T, Wang S, Zhang H, Tian Q. Emerging advances of detection strategies for tumor-derived exosomes. Int J Mol Sci. 2022;23(2):868. https://doi.org/10.3390/ijms23020868.

Cheung LS, Orozaliev A, Song Y-A, Sahloul S. Rapid detection and trapping of extracellular vesicles by Electrokinetic concentration for liquid biopsy on Chip. Micromachines. 2018;9(6):306. https://doi.org/10.3390/mi9060306.

Choi N-E, Park E-C, Lee J, Lee J-H, Lee J-Y. Recent advances in organelle-targeted fluorescent probes. Molecules. 2021;26(1):217. https://doi.org/10.3390/molecules26010217.

Cilloni D, Petiti J, Rosso V, Andreani G, Dragani M, Fava C, Saglio G. Digital PCR in myeloid malignancies: ready to replace quantitative PCR? Int J Mol Sci. 2019;20(9):2249. https://doi.org/10.3390/ijms20092249.

Curtis KA, Rudolph DL, Owen SM. Sequence-specific detection method for reverse transcription, loop-mediated isothermal amplification of HIV-1. J Med Virol. 2009;81(6):966–72. https://doi.org/10.1002/jmv.21490.

Das D, Masetty M, Priye A. Paper-based loop mediated isothermal amplification (LAMP) platforms: integrating the versatility of paper microfluidics with accuracy of nucleic acid amplification tests. Chemosensors. 2023;11:163. https://doi.org/10.3390/chemosensors11030163.

Del Río JS, O'Sullivan CK, Mayboroda O, Lobato IM, Katakis I. Enhanced solid-phase recombinase polymerase amplification and electrochemical detection. Anal Bioanal Chem. 2017;409(12):3261–9. https://doi.org/10.1007/s00216-017-0269-y.

Dermody SM, Bhambhani C, Tewari M, Brenner JC, Swiecicki PL. Trans-renal cell-free tumor DNA for urine-based liquid biopsy of cancer. Front Genet. 2022;13(4):879108. https://doi.org/10.3389/fgene.2022.879108.

Devaraj S, Hemarajata P, Versalovic J. The human gut microbiome and body metabolism: implications for obesity and diabetes. Clin Chem. 2013;59(4):617–28. https://doi.org/10.1373/clinchem.2012.187617.

Dindhoria K, Kumar R, Kumar R, Bhargava B. Metagenomic assembled genomes indicated the potential application of hypersaline microbiome for plant growth promotion and stress alleviation in salinized soils. mSystems. 2024;9(3):e0105023. https://doi.org/10.1128/msystems.01050-23.

Ding X, Yin K, Xu Z, Liu C, Sfeir M. Dual-priming isothermal amplification (DAMP) for highly sensitive and specific molecular detection with ultralow nonspecific signals. Anal Chem. 2019;91(20):12852–8. https://doi.org/10.1021/acs.analchem.9b02582.

Dong Z, Zhong V, Lu YY, Jiang Y, Zhou H. Simplifying bioinformatics data analysis through conversation. Cold Spring Harbor Laboratory. 2023; https://doi.org/10.1101/2023.10.29.564479.

Donovan PD, Gonzalez G, Higgins DG, Butler G, Ito K. Identification of fungi in shotgun metagenomics datasets. PLoS One. 2018;13(2):e0192898. https://doi.org/10.1371/journal.pone.0192898.

Du Y, Han X, Tsuda K. Microbiome-mediated plant disease resistance: recent advances and future directions. J Gen Plant Pathol. 2024;91(1):1–17. https://doi.org/10.1007/s10327-024-01204-1.

Elshafei AM, Mansour R. Microbial bioremediation of soils contaminated with petroleum hydrocarbons. Discover Soil. 2024;1(1) https://doi.org/10.1007/s44378-024-00004-5.

Euler M, Wang Y, Hufert FT, Weidmann M, Anda P, Otto P, Escudero R, Tomaso H. Recombinase polymerase amplification assay for rapid detection of Francisella tularensis. J Clin Microbiol. 2012;50(7):2234–8. https://doi.org/10.1128/jcm.06504-11.

Fahey T, Goodwin I, Sabatini R, Stefanelli D, Gardi A, Lamb DW, Pham H. Active and passive electro-optical sensors for health assessment in food crops. Sensors. 2020;21(1):171. https://doi.org/10.3390/s21010171.

Ferrari SR, Grubb J, Bishop DK. The Mei5-Sae3 protein complex mediates Dmc1 activity in Saccharomyces cerevisiae. J Biol Chem. 2009;284(18):11766–70. https://doi.org/10.1074/jbc.c900023200.

Fiala C, Diamandis EP. Utility of circulating tumor DNA in cancer diagnostics with emphasis on early detection. BMC Med. 2018;16(1):166. https://doi.org/10.1186/s12916-018-1157-9.

Fiala C, Diamandis EP. Can a broad molecular screen based on circulating tumor DNA aid in early cancer detection? J Appl Lab Med. 2020;5(6):1372–7. https://doi.org/10.1093/jalm/jfaa138.

Fonjungo PN, Zeh C, Boeras DI, Alexander H, Nkengasong JN, Parekh BS. Access and quality of HIV-related point-of-care diagnostic testing in Global Health programs. Clin Infect Dis. 2015;62(3):369–74. https://doi.org/10.1093/cid/civ866.

Gallei M, Truckenbrodt S, Kreuzinger C, Inumella S, Vistunou V, Sommer C, Tavakoli MR, Agudelo Dueñas N, Vorlaufer J, Jahr W, Randuch M, Johnson A, Benková E, Friml J, Danzl JG. Super-resolution expansion microscopy in plant roots. Plant Cell. 2025;37(4):koaf006. https://doi.org/10.1093/plcell/koaf006.

Gao J, Shen L, Dong B, Rao G, Lu Z, Li B, Li L, Wang H, Zang W, Jiang Z, Yu Y, Li Z. Circulating tumor DNA functions as an alternative for tissue to overcome tumor heterogeneity in advanced gastric cancer. Cancer Sci. 2017;108(9):1881–7. https://doi.org/10.1111/cas.13314.

Ge Q, Zhang Z-Y, Li S-N, Ma J-Q, Zhao Z. Liquid biopsy: comprehensive overview of circulating tumor DNA (Review). Oncol Lett. 2024;28(5):548. https://doi.org/10.3892/ol.2024.14681.

Ghaly TM, Colombi E, Russel D, Tetu SG, Rajabal V. Eco FoldDB: protein structure-guided functional profiling of ecologically relevant microbial traits at the metagenome scale. Environ Microbiol. 2025;27:e70178. https://doi.org/10.1101/2025.04.02.646905.

Ghorbani A, Afsharifar A, Neely GG, Denes CE, Burmistrz M, Niazi A, Eskandari MH, Salari R, Hadifar S, Izadpanah K. A short overview of CRISPR-Cas technology and its application in viral disease control. Transgenic Res. 2021;30(3):221–38. https://doi.org/10.1007/s11248-021-00247-w.

Gong S, Tang B, Pan W, Li J, Li N. Duplex-specific nuclease-assisted CRISPR-Cas12a strategy for MicroRNA detection using a personal glucose meter. Anal Chem. 2021;93(30):10719–26. https://doi.org/10.1021/acs.analchem.1c02478.

Gootenberg JS, Bhattacharyya RP, Abudayyeh OO, Livny J, Dy AJ, Myhrvold C, Collins JJ, Daringer NM, Hung DT, Zhang F, Sabeti PC, Koonin EV, Verdine V, Regev A, Donghia N, Essletzbichler P, Freije CA, Joung J, Lee JW. Nucleic acid detection with CRISPR-Cas13a/C2c2. Science (New York, NY). 2017;356(6336):438–42. https://doi.org/10.1126/science.aam9321.

Gootenberg JS, Collins JJ, Zhang F, Kellner MJ, Joung J, Abudayyeh OO. Multiplexed and portable nucleic acid detection platform with Cas13, Cas12a, and Csm6. Science. 2018;360(6387):439–44. https://doi.org/10.1126/science.aaq0179.

Graham L, Orenstein JM. Processing tissue and cells for transmission electron microscopy in diagnostic pathology and research. Nat Protoc. 2007;2(10):2439–50. https://doi.org/10.1038/nprot.2007.304.

Gray ES, Cooper A, Lo J, Siew K, Millward M, Reid AL, Carlino MS, Pereira MR, Scolyer RA, Long GV, Freeman J, Lee JHJ, Rizos H, Ziman M, Tembe V, Lomma C, Meniawy TM, Khattak MA, Boyd SC. Circulating tumor DNA to monitor treatment response and detect acquired resistance in patients with metastatic melanoma. Oncotarget. 2015;6(39):42008–18. https://doi.org/10.18632/oncotarget.5788.

Gupta R, Kazi TA, Dey D, Ghosh A, Ravichandiran V, Swarnakar S, Roy S, Biswas SR, Ghosh D. CRISPR detectives against SARS-CoV-2: a major setback against COVID-19 blowout. Appl Microbiol Biotechnol. 2021;105(20):7593–605. https://doi.org/10.1007/s00253-021-11583-6.

Hardinge P, Murray JAH. Reduced false positives and improved reporting of loop-mediated isothermal amplification using quenched fluorescent primers. Sci Rep. 2019;9(1):7400. https://doi.org/10.1038/s41598-019-43817-z.

Hassan MM, Poirier AC, La Ragione RM, Grist LF. JMM profile: loop-mediated isothermal amplification (LAMP): for the rapid detection of nucleic acid targets in resource-limited settings. J Med Microbiol. 2022;71(5) https://doi.org/10.1099/jmm.0.001522.

Higgins M, Ravenhall M, Ward D, Phelan J, Ibrahim A, Forrest MS, Clark TG, Campino S. PrimedRPA: primer design for recombinase polymerase amplification assays. Bioinformatics. 2018;35(4):682–4. https://doi.org/10.1093/bioinformatics/bty701.

Hu J, Rampitsch C, Bykova NV. Advances in plant proteomics toward improvement of crop productivity and stress resistancex. Front Plant Sci. 2015;6(764) https://doi.org/10.3389/fpls.2015.00209.

Hu C, Meiners S, Lukas C, Stathopoulos GT, Chen J. Role of exosomal micro RNAs in lung cancer biology and clinical applications. Cell Prolif. 2020;53(6):e12828. https://doi.org/10.1111/cpr.12828.

Huang B, Xu Y. Cellular and molecular mechanisms for elevated CO_2–regulation of plant growth and stress adaptation. Crop Sci. 2015;55(4):1405–24. https://doi.org/10.2135/cropsci2014.07.0508.

Ijaz S. Mitochondrial plasmid: could this be the next hit for genetic engineering? Genet Mol Res. 2011;10(1):391–2.

Ijaz S, Haq IU, Babar M, Nasir B. Disease resistance genes' identification, cloning, and characterization in plants. In: Abd-Elsalam KA, Mohamed HI, editors. Cereal diseases: Nanobiotechnological approaches for diagnosis and management. Singapore: Springer; 2022. https://doi.org/10.1007/978-981-19-3120-8_13.

Jensen K, Thakur S, Patel A, Mendonca-Torres MC, Costello J, Gomes-Lima CJ, Walter M, Wartofsky L, Burman KD, Bikas A, Ylli D, Vasko VV, Klubo-Gwiezdzinska J. Detection of BRAFV600E in liquid biopsy from patients with papillary thyroid cancer is associated with tumor aggressiveness and response to therapy. J Clin Med. 2020;9(8):2481. https://doi.org/10.3390/jcm9082481.

Joseph P, Tawakol A. Imaging atherosclerosis with positron emission tomography. Eur Heart J. 2016;37(39):2974–80. https://doi.org/10.1093/eurheartj/ehw147.

Kashyap PL, Chakdar H, Srivastava AK, Rai P, Pandiyan K, Sharma S, Kumar S. Nanotechnology for the detection and diagnosis of plant pathogens. Springer; 2016. p. 253–76. https://doi.org/10.1007/978-3-319-39306-3_8.

Kaur R, Gupta S, Chauhan A, Mishra V, Sharma MK, Singh J. Harnessing the power of clustered regularly interspaced short palindromic repeats (CRISPR) based microfluidics for next-generation molecular diagnostics. Mol Biol Rep. 2024;51(1):896. https://doi.org/10.1007/s11033-024-09840-8.

Kellner MJ, Koob JG, Gootenberg JS, Zhang F, Abudayyeh OO. SHERLOCK: nucleic acid detection with CRISPR nucleases. Nat Protoc. 2019;14(10):2986–3012. https://doi.org/10.1038/s41596-019-0210-2.

Kreutz JE, Feng Q, Thompson AM, Staheli JP, Chiu DT, Dyen MR, Wang J, Sheen AM. Self-digitization chip for quantitative detection of human papillomavirus gene using digital LAMP. Lab Chip. 2019;19(6):1035–40. https://doi.org/10.1039/c8lc01223g.

Lau HY, Botella JR, Trau M, Wee EJH, Wang Y. Field demonstration of a multiplexed point-of-care diagnostic platform for plant pathogens. Anal Chem. 2016;88(16):8074–81. https://doi.org/10.1021/acs.analchem.6b01551.

Le DT, Vu NT. Progress of loop-mediated isothermal amplification technique in molecular diagnosis of plant diseases. Appl Biol Chem. 2017;60(2):169–80. https://doi.org/10.1007/s13765-017-0267-y.

Lee D, La Mura M, Allnutt TR, Powell W. Detection of genetically modified organisms (GMOs) using isothermal amplification of target DNA sequences. BMC Biotechnol. 2009;9(1):7. https://doi.org/10.1186/1472-6750-9-7.

Leonard S, Reverchon S, Nasser W, Hommais F. Plant-phytopathogen interactions: bacterial responses to environmental and plant stimuli. Environ Microbiol. 2017;19(5):1689–716. https://doi.org/10.1111/1462-2920.13611.

Li J, Von Stetten F, Macdonald J. Review: a comprehensive summary of a decade development of the recombinase polymerase amplification. Analyst. 2019;144(1):31–67. https://doi.org/10.1039/c8an01621f.

Li S, Yin D, Li Q, Yang Y, Wang J, Guan Y, Wei W, Song E, Shi Q, Zhu M, Li Y, Gong Y, Lai H, Jin L, Yi X, Liu J, Cai Z, Liu Y, Liu Q, et al. Circulating tumor DNA predicts the response and prognosis in patients with early breast cancer receiving Neoadjuvant chemotherapy. JCO Precis Oncol. 2020;4(4):244–57. https://doi.org/10.1200/po.19.00292.

Li D, Lai W, Fan D, Fang Q. Protein biomarkers in breast cancer-derived extracellular vesicles for use in liquid biopsies. Am J Phys Cell Phys. 2021;321(5):C779–97. https://doi.org/10.1152/ajpcell.00048.2021.

Li J-Y, Tian Y-Y, Liu J-X, Yang C. Regulation of chloroplast development and function at adverse temperatures in plants. Plant Cell Physiol. 2022;63(5):580–91. https://doi.org/10.1093/pcp/pcac022.

Lin C, Liu X, Zheng B, Ke R, Tzeng C-M. Liquid biopsy, ctDNA diagnosis through NGS. Life. 2021;11(9):890. https://doi.org/10.3390/life11090890.

Manoj P. Droplet digital PCR technology promises new applications and research areas. Mitochondrial DNA Part A. 2014;27(1):742–6. https://doi.org/10.3109/19401736.2014.913168
.

Mesquita SG, Gadd G, Coelho FS, Cieplinski A, Emery A, Lugli EB, Simões TC, Fonseca CT, Caldeira RL, Webster B. Laboratory and field validation of the recombinase polymerase amplification assay targeting the Schistosoma mansoni mitochondrial minisatellite region (SmMIT-RPA) for snail xenomonitoring for schistosomiasis. Int J Parasitol. 2024;54(5):247–56. https://doi.org/10.1016/j.ijpara.2024.01.005.

Mishra M, Das DK, Pramanik DD, Dash SK, Ahmed R, Pramanik A. Recent advancements in the application of circulating tumor DNA as biomarkers for early detection of cancers. ACS Biomater Sci Eng. 2024;10(8):4740–56. https://doi.org/10.1021/acsbiomaterials.4c00606.

Morcillo RJL, Manzanera M. The effects of plant-associated bacterial exopolysaccharides on plant abiotic stress tolerance. Metabolites. 2021;11(6):337. https://doi.org/10.3390/metabo11060337.

Mota DS, Filho JCBS, Gandarilla AMD, Guimarães JM, Brito WR, Mariúba LAM. Recombinase polymerase amplification in the molecular diagnosis of microbiological targets and its applications. Can J Microbiol. 2022;68(6):383–402. https://doi.org/10.1139/cjm-2021-0329.

Mouliere F, Hudecova I, Murphy S, Patel K, Van Der Heijden MS, Massie CE, Ahlborn LB, Pacey S, Parkinson CA, Ros S, Heider K, Watts C, Garcia-Corbacho J, Brenton JD, Piskorz AM, Smith CG, Morris J, Corrie P, Østrup O, et al. Enhanced detection of circulating tumor DNA by fragment size analysis. Sci Transl Med. 2018;10:466. https://doi.org/10.1126/scitranslmed.aat4921.

Moustaka J, Moustakas M. Early-stage detection of biotic and abiotic stress on plants by chlorophyll fluorescence imaging analysis. Biosensors. 2023;13(8):796. https://doi.org/10.3390/bios13080796.

Munawar MA. Critical insight into recombinase polymerase amplification technology. Expert Rev Mol Diagn. 2022;22:725–37. https://doi.org/10.1080/14737159.2022.2109964.

Myint NNM, Verma AM, Fernandez-Garcia D, Sarmah P, Tarpey PS, Al-Aqbi SS, Cai H, Trigg R, West K, Howells LM, Thomas A, Brown K, Guttery DS, Singh B, Pringle HJ, Mcdermott

U, Shaw JA, Rufini A. Circulating tumor DNA in patients with colorectal adenomas: assessment of detectability and genetic heterogeneity. Cell Death Dis. 2018;9(9):894. https://doi.org/10.1038/s41419-018-0934-x.

Nguyen LT, Jain PK, Smith BM. Enhancement of trans-cleavage activity of Cas12a with engineered crRNA enables amplified nucleic acid detection. Nat Commun. 2020;11(1):4096. https://doi.org/10.1038/s41467-020-18615-1.

Nie M, Shen F, Du W, Luo H, Pan J-Z, Song X, Liu M, Zheng M, Lan Y, Li C. Assembled step emulsification device for multiplex droplet digital polymerase chain reaction. Anal Chem. 2019;91(3):1779–84. https://doi.org/10.1021/acs.analchem.8b04313.

Noman M, Song F, Manzoor I, Azizullah A, Li D, Shahid M, Ahmed T, Ijaz U. Plant-microbiome crosstalk: dawning from composition and assembly of microbial community to improvement of disease resilience in plants. Int J Mol Sci. 2021;22(13):6852. https://doi.org/10.3390/ijms22136852.

Novi VT, Meher AK, Abbas A. Visualization methods for loop mediated isothermal amplification (LAMP) assays. Analyst. 2025;150(4):588–99. https://doi.org/10.1039/d4an01287a.

Ohmura H, Hanamura F, Okumura Y, Ando Y, Masuda T, Mimori K, Akashi K, Baba E. Liquid biopsy for breast cancer and other solid tumors: a review of recent advances. Breast Cancer (Tokyo, Japan). 2024;32(1):33–42. https://doi.org/10.1007/s12282-024-01556-8.

Parsons HA, Fleharty M, Rhoades J, Yu F, Love JC, Mayer EL, Neuberg DS, Rotem D, Lin NU, Makrigiorgos GM, Kirkner GJ, Golub TR, Gydush G, Cohen O, Miller KD, Ha G, Li T, Ram P, Leong KW, et al. Sensitive detection of minimal residual disease in patients treated for early-stage breast cancer. Clin Cancer Res. 2020;26(11):2556–64. https://doi.org/10.1158/1078-0432.ccr-19-3005.

Patchsung M, Eiamthong B, Uttamapinant C, Phiwkaow K, Pattama A, Jin X, Jaroensuk J, Homchan A, Kantiwiriyawanitch C, Gootenberg JS, Wongsatit T, Chuthapisith S, Joung J, Ladha A, Phoodokmai T, Tinikul R, Suraritdechachai S, Swangsri J, Zhang F, et al. Clinical validation of a Cas13-based assay for the detection of SARS-CoV-2 RNA. Nat Biomed Eng. 2020;4(12):1140–9. https://doi.org/10.1038/s41551-020-00603-x.

Peng M, Chen C, Hulbert A, Brock MV, Yu F. Non-blood circulating tumor DNA detection in cancer. Oncotarget. 2017;8(40):69162–73. https://doi.org/10.18632/oncotarget.19942.

Reece M, Saluja H, Hollington P, Karapetis CS, Vatandoust S, Young GP, Symonds EL. The use of circulating tumor DNA to monitor and predict response to treatment in colorectal cancer. Front Genet. 2019;10(2):1118. https://doi.org/10.3389/fgene.2019.01118.

Restrepo JC, Martínez Guevara D, Pareja López A, Montenegro Palacios JF, Liscano Y. Identification and application of emerging biomarkers in treatment of non-small-cell lung cancer: systematic review. Cancer. 2024;16(13):2338. https://doi.org/10.3390/cancers16132338

Rich MK, Schorderet M, Reinhardt D. The role of the cell wall compartment in mutualistic symbioses of plants. Front Plant Sci. 2014;5:238. https://doi.org/10.3389/fpls.2014.00238.

Scortichini M. Sustainable Management of Diseases in horticulture: conventional and new options. Horticulturae. 2022;8(6):517. https://doi.org/10.3390/horticulturae8060517.

Semenkovich NP, Szymanski JJ, Earland N, Chauhan PS, Pellini B, Chaudhuri AA. Genomic approaches to cancer and minimal residual disease detection using circulating tumor DNA. J Immunother Cancer. 2023;11(6):e006284. https://doi.org/10.1136/jitc-2022-006284.

Shapiro DM, Eghtesadi SA, Chilkoti A, Ney M. Protein phase separation arising from intrinsic disorder: first-principles to bespoke applications. J Phys Chem B. 2021;125(25):6740–59. https://doi.org/10.1021/acs.jpcb.1c01146.

Shekhawat K, Fröhlich K, García-Ramírez GX, Trapp MA, Hirt H. Ethylene: A Master Regulator of Plant–Microbe Interactions under Abiotic Stresses. Cells. 2022;12(1):31. https://doi.org/10.3390/cells12010031.

Shi X-M, Long F, Suo B. Molecular methods for the detection and characterization of foodborne pathogens. Pure Appl Chem. 2010;82(1):69–79. https://doi.org/10.1351/pac-con-09-02-07.

Skryabin GO, Savelyeva EE, Komelkov AV, Tchevkina EM. Lipid Rafts in Exosome Biogenesis. Biochemistry (Moscow). 2020;85(2):177–91. https://doi.org/10.1134/s0006297920020054.

Stabentheiner E, Zankel A, Pölt P. Environmental scanning electron microscopy (ESEM)—a versatile tool in studying plants. Protoplasma. 2010;246(1–4):89–99. https://doi.org/10.1007/s00709-010-0155-3.

Strayer-Scherer A, Jones JB, Paret ML. Recombinase polymerase amplification assay for field detection of tomato bacterial spot pathogens. Phytopathology. 2019;109(4):690–700. https://doi.org/10.1094/phyto-03-18-0101-r.

Strnad M, Elsterová J, Schrenková J, Vancová M, Rego ROM, Grubhoffer L, Nebesářová J. Correlative cryo-fluorescence and cryo-scanning electron microscopy as a straightforward tool to study host-pathogen interactions. Sci Rep. 2015;5(1):18029. https://doi.org/10.1038/srep18029.

Suenaga H. Targeted metagenomics: a high-resolution metagenomics approach for specific gene clusters in complex microbial communities. Environ Microbiol. 2011;14(1):13–22. https://doi.org/10.1111/j.1462-2920.2011.02438.x.

Sun H, Jiang C, Gao M, Wu C, Wang Q, Jiang S. A review of root exudates and rhizosphere microbiome for crop production. Environ Sci Pollut Res. 2021;28(39):54497–510. https://doi.org/10.1007/s11356-021-15838-7.

Tan C, Fan D, Wang N, Wang F, Wang B, Zhu L, Guo Y. Applications of digital PCR in COVID-19 pandemic. View (Beijing, China). 2021;2(2):20200082. https://doi.org/10.1002/viw.20200082.

Tang H, Cai Q, Hu P, Li H. Comparison of droplet digital PCR to real-time PCR for quantification of hepatitis B virus DNA. Biosci Biotechnol Biochem. 2016;80(11):2159–64. https://doi.org/10.1080/09168451.2016.1196576.

Tenorio Berrío R, Dubois M. Single-cell transcriptomics reveals heterogeneity in plant responses to the environment: a focus on biotic and abiotic interactions. J Exp Bot. 2024;75(17):5188–203. https://doi.org/10.1093/jxb/erae107.

Ter Veer MJT, Nägerl UV, Pfeiffer T. Two-photon STED microscopy for nanoscale imaging of neural morphology in vivo. Methods Mol Biol (Clifton, N.J.). 2017;1663:45–64. https://doi.org/10.1007/978-1-4939-7265-4_5.

Tourlousse DM, Uchino Y, Takahashi S, Kasahara K, Kameyama K, Kawasaki H, Matsuda M, Kuroiwa M, Tanaka Y, Nishiyama M, Koyanagi R, Miura T, Narita K, Ohkuma M, Aoki H, Arioka S, Yamazoe A, Nishiwaki M, Shiina K, et al. Validation and standardization of DNA extraction and library construction methods for metagenomics-based human fecal microbiome measurements. Microbiome. 2021;9(1):95. https://doi.org/10.1186/s40168-021-01048-3.

Varadharajan V, Rajendran R, Runthala A, Madhesh V, Shin H, Swaminathan G, Muthuramalingam P, Park Y, Srinivasan H, Murugan P, Ramesh M. Multi-omics approaches against abiotic and biotic stress-a review. Plants (Basel, Switz). 2025;14(6):865. https://doi.org/10.3390/plants14060865.

Varona M, Anderson JL. Advances in mutation detection using loop-mediated isothermal amplification. ACS Omega. 2021;6(5):3463–9. https://doi.org/10.1021/acsomega.0c06093.

Veldore V, Choughule A, Mandloi N, Gupta R, Routhu T, Dutt A, Noronha V, Vedam R, Prabhash K, Joshi A. Validation of liquid biopsy: plasma cell-free DNA testing in clinical management of advanced non-small cell lung cancer. Lung Cancer: Targets and Therapy. 2018;9(3):1–11. https://doi.org/10.2147/lctt.s147841.

Verma SK, Sharma PC, Singh H. An improved method suitable for isolation of high-quality metagenomic DNA from diverse soils. 3 Biotech. 2017;7(3):171. https://doi.org/10.1007/s13205-017-0847-x.

Walter M, Oecking C, Schütze K, Harter K, Chaban C, Grefen C, Blazevic D, Schumacher K, Kudla J, Näke C, Batistic O, Weckermann K. Visualization of protein interactions in living plant cells using bimolecular fluorescence complementation. Plant J. 2004;40(3):428–38. https://doi.org/10.1111/j.1365-313x.2004.02219.x.

Wandrey M, Jablonska J, Stauber RH, Gül D. Exosomes in cancer progression and therapy resistance: molecular insights and therapeutic opportunities. Life. 2023;13(10):2033. https://doi.org/10.3390/life13102033.

Wang J, Guo X. Adsorption kinetics and isotherm models of heavy metals by various adsorbents: an overview. Crit Rev Environ Sci Technol. 2023;53(21):1837–65. https://doi.org/10.1080/10643389.2023.2221157.

Wang Y, Salazar JK. Culture-independent rapid detection methods for bacterial pathogens and toxins in food matrices. Compr Rev Food Sci Food Saf. 2015;15(1):183–205. https://doi.org/10.1111/1541-4337.12175.
Wang D-G, Brewster JD, Paul M, Tomasula PM. Two methods for increased specificity and sensitivity in loop-mediated isothermal amplification. Molecules. 2015;20(4):6048–59. https://doi.org/10.3390/molecules20046048.
Wang X, Huang J, Chen W, Li G, Li Z, Lei J. The updated role of exosomal proteins in the diagnosis, prognosis, and treatment of cancer. Exp Mol Med. 2022;54(9):1390–400. https://doi.org/10.1038/s12276-022-00855-4.
Wen X, Coradduzza D, Shen J, Scanu AM, Muroni MR, Massidda M, Rallo V, Carru C, Angius A, De Miglio MR. Harnessing minimal residual disease as a predictor for colorectal cancer: promising horizons amidst challenges. Medicina (Kaunas). 2023;59(10):1886. https://doi.org/10.3390/medicina59101886.
Xia X, Wang Y, Huang Y, Zhang H, Lu H, Zheng JC. Exosomal miRNAs in central nervous system diseases: biomarkers, pathological mediators, protective factors and therapeutic agents. Prog Neurobiol. 2019;183:101694. https://doi.org/10.1016/j.pneurobio.2019.101694.
Xu C, Jin W, Wan Y, Li L. Recombinase polymerase amplification (RPA) of CaMV-35S promoter and nos terminator for rapid detection of genetically modified crops. Int J Mol Sci. 2014;15(10):18197–205. https://doi.org/10.3390/ijms151018197.
Yaghoubi Naei V, Bordhan P, Mirakhorli F, Khorrami M, Shrestha J, Nazari H, Kulasinghe A, Ebrahimi Warkiani M. Advances in novel strategies for isolation, characterization, and analysis of CTCs and ctDNA. Therap Adv Med Oncol. 2023;15:17588359231192401. https://doi.org/10.1177/17588359231192401.
Yan L, Zheng Y, Nakayama S, Roembke BT, Sintim HO, Zhou J, Gamson AS. Isothermal amplified detection of DNA and RNA. Mol BioSyst. 2014;10(5):970. https://doi.org/10.1039/c3mb70304e.
Yang N, Zhang H, Han X, Liu Z, Lu Y. Advancements and applications of loop-mediated isothermal amplification technology: a comprehensive overview. Front Microbiol. 2024;15:1406632. https://doi.org/10.3389/fmicb.2024.1406632.
Yazdi ZF, Roshannezhad S, Sharif S, Abbaszadegan MR. Recent progress in prompt molecular detection of liquid biopsy using Cas enzymes: innovative approaches for cancer diagnosis and analysis. J Transl Med. 2024;22(1):1173. https://doi.org/10.1186/s12967-024-05908-y.
Yin K, Qiu J-L. Genome editing for plant disease resistance: applications and perspectives. Philos Trans R Soc B: Biol Sci. 2019;374(1767):20180322. https://doi.org/10.1098/rstb.2018.0322.
Yoshimi K, Takeshita K, Yamayoshi S, Shibumura S, Yamauchi Y, Yamamoto M, Yotsuyanagi H, Kawaoka Y, Mashimo T. CRISPR-Cas3-based diagnostics for SARS-CoV-2 and influenza virus. iScience. 2022;25(2):103830. https://doi.org/10.1016/j.isci.2022.103830.
You Y, Ding C, Wang W, Jiang L, Wang X, Zhao X, Liu X, Zhang D, Wang L. Interspecific plant interaction structures the microbiomes of poplar-soil interface to alter nutrient cycling and utilization. Microbiol Spectrum. 2024;12(2):e0336823. https://doi.org/10.1128/spectrum.03368-23.
Zhang H, Lu B. microRNAs as biomarkers of ovarian cancer. Expert Rev Anticancer Ther. 2020;20(5):373–85. https://doi.org/10.1080/14737140.2020.1760095.
Zhang Y, Liu Y, Liu H, Tang WH. Exosomes: biogenesis, biologic function and clinical potential. Cell Biosci. 2019;9(1):eCollection 2019. https://doi.org/10.1186/s13578-019-0282-2.

Chapter 9
Proteomics: Methods and Applications

Mahrukh Babar, Ali Raza, and Ali Hassan Khan

9.1 Introduction to Proteomics

Proteomics analyzes proteins for their abundance, protein–protein interactions, including both stable and transient interactions, as well as a variety of proteoforms that result from specific Post-Translational Modifications (PTMs) (Yakubu et al. 2019). The word "proteome" refers to all proteins that are either produced or modified by living organisms, including humans, animals, plants, microbes, and even entire organ systems, such as cell cultures and organs from environmental systems. However, the whole proteome of an organ or organ system remains unidentified, as a wide variety of proteins and their different isoforms or species are expressed at various stages in other cells, tissues, and organs (Karlsson et al. 2015).

Proteomics was proposed as a complete atlas of human proteins around 50 years ago. It is a field of rapidly growing interest and research and has become increasingly important since the post-genomic era (Pontén et al. 2011). The current development of involved technology has been a major driving force behind the diversity of methodologies that comprise the field of proteomics. Proteomics research began as a result of developments in four areas: (i) two-dimensional gel electrophoresis (2D-PAGE) developed into a reliable technique for the rapid separation of proteins found in biological complex mixtures; (ii) mass spectrometry techniques for accurate mass and chemical structure analysis continued to advance; (iii) massive

M. Babar (✉)
Faculty of Pharmacy, IBADAT International University, Islamabad, Pakistan
e-mail: mahrukh.babar@pharm.iiui.edu.pk

A. Raza
School of Environmental and Rural Science, University of New England, Armidale, Australia
e-mail: araza3@une.edu.au

A. H. Khan
Horticulture Research Institute, National Agriculture Research Center, Islamabad, Pakistan

S. Ijaz et al. (eds.), *Molecular Biology Methods*,
https://doi.org/10.1007/978-981-92-2642-9_9

genome research and the vast number of peptide/protein sequences catalogued in multiple databases were continuously produced; and (iv) new bioinformatics tools were developed to make it easier to analyze large volumes of MS data to identify intact proteins and their functional or pathological PTMs. All these advances came together in the 1990s, and in 1994, the term "Proteom" was introduced (Latterich et al. 2008).

In the context of certain hypothesis-driven research projects, various reductionist approaches have been employed in the past to identify proteins and protein-based activities or pathways. In reductionist approaches, we move from functions to sequences, whereas in genomic and proteomic approaches, we move from sequences to functions, which provides a detailed understanding of proteins and their sequences; however, this approach is challenging (Treberg et al. 2020). To analyze a large amount of data on proteins over an extensive range of concentrations, science discovery projects must focus on understanding the control and functions of various biological systems.

9.1.1 Role of Proteomics in System Biology

Systems biology is the study and understanding of the physiology of living systems in response to environmental and genetic perturbations (Kitano 2001). It focuses on network-level interactions and emergent behaviors that arise from complex interactions among certain cellular components in the systems. This biological system may be a single cell or group of cells carrying out some biological function, or may be some protein molecules carrying a biological module or a set of proteins and other molecules, like ribosomes, interacting together to either create a protein network or some other important cellular functions (Weston and Hood 2004). The concept of biology is important due to its significant advancements, such as the Human Genome Project, which provides a comprehensive list of all genes and cis-regulatory elements. However, gene expressions do not always correlate with protein abundance or protein activity. A protein can be post-translationally modified, as mentioned earlier, and thus can be inhibited or activated (Chavez et al. 2013).

Furthermore, it is possible to transcribe a gene, but in the process, the mRNA can be degraded. Proteomics helps overcome this situation by providing a complete analysis of functional molecules/proteins.

Furthermore, scientists can gain insights into how protein levels and modifications change over time in relation to stress, apoptosis, signal transduction, and the cell cycle through time-series proteomics (Jiménez et al. 2022). Quantitative data can also be obtained from proteomics using various analytical techniques, such as SILAC, iTRAQ, and TMT, which enable comparisons of different phenotypes, for example, healthy and diseased cells (Li et al. 2020). These techniques provide accurate quantification of protein abundance and dynamics. In addition to these, numerous other proteomic technologies are employed in system biology, including 2-D gel electrophoresis for early-stage protein separation and expression profiling, Liquid Chromatography and Mass spectrometry, label-free quantification, and phosphoproteomics (Dutta et al. 2023).

Proteomics is also integrated with other omics, such as metabolomics (inking protein function to metabolic outcomes), transcriptomics (comparing mRNA and protein levels), genomics (mapping mutations to functional protein changes), and epigenomics (studying how histone-modifying enzymes affect gene regulation via protein expression). These multi-omic approaches provide a comprehensive understanding of regulatory mechanisms and system behaviors (Misra et al. 2019).

9.1.2 Types of Proteomics

Depending on the mode of study, proteomics can be classified into three major categories, including expression proteomics, structural proteomics, and functional proteomics.

Expression proteomics is the field of proteomics that analyzes and compares protein expression levels, both qualitatively and quantitatively, under physiological and pathological conditions. It is used to detect various expressions in different types of cells, including tumor cells, in comparison to healthy or normal cells/tissues, and to compare treated and nontreated cells/tissues. Expression proteomics determines biomarkers for various diseases. It also regulates and determines protein expression, linking it to gene expression. Technologies used in expression proteomics include 2D gel electrophoresis and mass spectrometry, which are employed to detect and quantify proteins (Hanash et al. 2002).

Structural proteomics focuses on determining the three-dimensional structure of proteins. This field provides insight into protein folding and its active sites, as well as the major domains that comprise it. It also aids in computational drug designing by revealing targeted protein structures. It also allows the mapping of protein–protein and protein–ligand interactions. Common technologies involved in structural proteomics are X-ray crystallography and nuclear magnetic resonance (NMR) spectroscopy for the determination of the accurate position of atoms in a protein structure (Renfrey and Featherstone 2002).

Functional proteomics: This field determines the functionality of proteins based on their interactions with other molecules, involving the identification of various protein complexes. These interactions involve protein–protein interactions (PPIs), protein—ligand interactions, and post-translational modifications (PTMs). Functional proteomics also identifies how proteins respond to various environmental changes. Co-immunoprecipitation (Co-IP), yeast two-hybrid screening, and surface plasmon resonance are among the common key technologies used in this field (Monti et al. 2005).

9.2 Technologies Utilized in Proteomics

9.2.1 *Orbitrap Mass Spectroscopy*

9.2.1.1 Principle of Orbitrap

Mass spectroscopy is one of the techniques used in proteomics, and it has been advanced enormously in recent years. One of its kind is Orbitrap mass spectroscopy, which is a combined technique of linear ion trapping and an Orbitrap analyzer. This is a high-resolution mass analyzer used in proteomics for the precise identification and quantification of peptides and proteins. It operates based on the oscillation of ions in an electrostatic field, which oscillate around the z-axis (the central axis) of the electrode. This technique is unique compared to other mass analyzers like quadrupoles or TOF (Time of Flight) because it does not require radiofrequency RF voltages for ion trapping but uses purely an electrostatic field (Michalski et al. 2011).

Orbitraps have a high resolution of around 100,000 at m/z 400; however, advanced models often exceed this value. Resolution is the characteristic that differentiates between two closely spaced masses. Orbitrap achieves high resolution by trapping ions for longer durations, allowing multiple oscillations to be recorded and providing more accurate readings with high resolution. Mass resolution is directly proportional to the number of oscillations; a higher number of oscillations leads to a higher frequency and, consequently, higher resolution (Senyuva et al. 2015).

Orbitrap offers higher accuracy as well. Its accuracy range of mass is around 1–2 ppm (parts per million) (Perry et al. 2008). This accuracy is particularly beneficial for critical techniques such as metabolomics, proteomics, genomics, and environmental testing, which require the precise identification of compounds.

9.2.2 *Instrumentation and Workflow*

The Orbitrap analyzer was manufactured in the early 2000s and consists of a small electrostatic device. Ion packets with high energies are put into this device, and these ions orbit around the central spindle-shaped electrode. A detector picks the image of these ions in motion, and mass spectra with a high resolution are created using a Fourier transformer (FT). The Orbitrap analyzer is one of the key principles of mass spectroscopic analysis, alongside other principles such as quadrupole mass filters, linear ion traps, and time-of-flight (TOF) analyzers. These quadruple mass filters are used to select precursor ions based on their mass-to-charge (m/z) ratio (Makarov et al. 2019).

Furthermore, this instrument features collision cells that facilitate fragmentation through high collision dissociation (HCD) in the MS/MS mode, as explained in this section later. The Orbitrap technology has become widely common in proteomics and relevant fields and is known by the name "LTQ Orbitrap" (Przybylski et al. 2010). There are two modes in which Orbitrap functions. The first one is the MS mode, in which the linear trap collects the ion population and transfers it to an intermediate C-trap for injection and high-resolution analysis in the Orbitrap analyzer. In the other mode, called the MS/MS mode, the linear ion trap records the signal of a mass-dependent scan at low resolution and only keeps a selected mass window, which is triggered by an additional RF field that causes the trapped precursor ions to fragment. These Orbitrap analyzers can achieve mass resolution above 100,000 and accuracy within a few parts per million (ppm) (Kaufmann 2018).

Such proteomics, which are based on this technology, analyze the complex protein structures often derived from living cells, tissues, or even body fluids. These proteins are subjected to proteolytic digestion, and the resulting mixtures are separated using liquid chromatography. After the separation, these mixtures are electrosprayed for mass spectrometric analysis (Slysz and Schriemer 2005).

9.2.3 Sample Preparation

It is a very critical step in proteomics, followed by the Orbitrap technique. When samples are properly prepared, they offer proper identification and quantification with increased specificity, sensitivity, and reproducibility of the results. The primary objective is to convert biological samples, such as cells, tissues, and related living specimens, into suitable mass spectroscopy peptides with minimal sample loss and contamination. The following steps are involved in the process.

1. *Sample collection and lysis*: First and foremost, samples are collected from desired biological sources, including cells, tissues, and body fluids. These samples are subjected to lysis to release proteins, which is achieved using specific lysis buffers. The choice of the lysis buffer depends on the type of sample used and the requirements of downstream analysis. There are various types of lysis methods (Geiger et al. 2010). A few of them are listed below:

    1. *Lysis through detergent-based buffers*: This method utilizes detergent-based buffers. These buffers are used for the solubilization of membrane proteins. It includes SDS and Triton X-100 (Neset et al. 2022).
    2. *Lysis through mechanical disruption*: Cells are physically broken with sonications or bead beating, which act as mechanical disruptors. However, care must be taken to prevent protein denaturation, also known as protein lysis. Protease inhibitors are used for such purposes (Winter and Steen 2011).
    3. *Lysis using chaotropic agents*: These agents disrupt the structure of macromolecules. In proteomics, they can help in breaking protein–protein interactions. Urea and thiourea are examples of such agents (Ashraf Kharaz et al. 2017).

2. *Protein solubilization and cleanup*: For efficient digestion, proteins should be completely solubilized and denatured. This solubilization is performed using certain detergents and chaotropic agents, as mentioned above, which are incompatible with Orbitrap MS. Therefore, these agents must be removed from the sample before LC-MS/MS analysis. There are various methods to remove these compounds, for example, using chemicals such as chloroform, acetone, or with the help of filter-aided sample preparation using molecular weight cut-off filters (Umadevi and Anandaraj 2015).
3. *Protein unfolding*: Reduction is performed to ensure that the intra- and inter-disulfide bonds are reduced. This reduction is achieved via dithiothreitol, also known as DTT or tris(2-carboxyethyl)phosphine (TCEP). Later on, to prevent reformation of these disulfide bonds, proteins are alkylated using compounds like iodoacetoamide (IAA). This process modifies cysteine residues, thereby increasing the digestion efficiency and reproducibility of proteins (Cramer et al. 2016).
4. *Enzymatic cleaving*: Proteins are cleaved using enzymes such as trypsin, which removes the carboxyl end of lysine and arginine (Tian et al. 2018). It is performed under controlled conditions and also includes other proteases, such as chymotrypsin, to increase sequence coverage. Buffer exchange is also done after this step.
5. *Desalting*: These peptides still contain salts and other residues that can suppress ionization. To remove these contaminants, solid-phase extraction is performed. Peptides are eluted with a small volume of organic solvent (e.g., 50–80% acetonitrile with 0.1% formic acid) and dehydrated with the help of vacuum centrifugation before LC–MS/MS analysis (Klatt et al. 2021).
6. *Peptides reconstitution*: Peptides are reconstituted in certain buffers, like 0.1% formic acid. These buffers are MS-compatible. Samples are centrifuged to remove particulates, as these precipitations can clog the MS and decrease the efficiency of the technique (Nepomuceno et al. 2014).

9.2.4 Peptide Separation

Biological samples are often highly complex, comprising tens of thousands of peptides within thousands of proteins. This huge number of peptides can reduce accuracy and create chemical noise; therefore, these peptides need to be separated before proceeding further. It is done via various techniques, including Capillary Electrophoresis (CE), Liquid Chromatography (LC), and Ion Mobility Spectrometry (IMS) (Glazyrin et al. 2022). LC is considered a gold standard. These techniques are explained below one by one.

1. *Liquid chromatography*: This is the standard technique used in peptide separation for proteomics studies, utilizing OrbiTrap. Peptides vary from one another based on their physicochemical properties, such as hydrophobicity, polarity, or

charge, and liquid chromatography separates these peptides according to these properties. There are various types of liquid chromatography. The first one is *Reversed-Phase Liquid Chromatography (RP-LC)*. It is the most commonly used LC technique in proteomics because it is compatible with ESI and allows direct coupling with MS. The stationery phase used in RP-LC is a hydrophobic C18 silica column. The gradient of aqueous solvent (water +0.1% formic acid) and organic solvent (acetonitrile +0.1% formic acid) is used as the mobile phase. The peptides bind to silica, and as the concentration of organic solvent increases, peptides elute gradually. The second type of LC is *High pH reversed-phase fractionation*. This technique is used to separate peptides under high pH conditions using buffers such as ammonium bicarbonate. It is later pooled into various fractions, which are analyzed by using *nLC-MS/MS*. This nLC-MS/MS uses narrow capillaries of around 75 μm ID. As it allows for very low flow rates, it exhibits high sensitivity, thereby increasing proteome coverage.

2. Another type of LC is Strong Cation Exchange Chromatography (SCXC). It separates the peptides based on their charge at a specific pH. Increased salt concentrations are used for elution when cations bind with negatively charged resins. Another technique is *Hydrophilic Interaction Liquid Chromatography (HILC)*. It separates peptides based on their polarity. Polar peptides tend to bind to the stationary phase. After their binding, these peptides are eluted using a high concentration of the aqueous phase. It is commonly used for glycopeptides and phosphorylated peptides (Makarov and Scigelova 2010).
3. *Capillary electrophoresis*: This technique utilizes a thin, fused silica capillary tube through which peptides travel under the influence of an electric field. The separation is based on the mass-to-charge ratio, where peptides with higher charge move faster, while those with lower charge move more slowly. The capillary is 20–100 μm in diameter and ~ 50–100 cm long, and a high voltage power supply of around 10-30 kV is supplied through the capillary (Stolz et al. 2019). To introduce a sample into a capillary, various methods are employed, including electrokinetic injection and siphoning. Capillary electrophoresis is better than LC in terms of resolution and sensitivity, as it can achieve millions of theoretical plate numbers. Moreover, it also provides additional information to hydrophobicity-based RP-LC due to its characteristic feature of separation based on the charge-to-mass ratio (Wenz et al. 2015).
4. *Ion mobility spectrum*: This is another spectrometry technique used in Orbitrap for peptide separation. It includes a chamber filled with an inert gas, such as nitrogen or helium, and peptide ion separation is based on their charge, size, and shape (Bush et al. 2012). A structural descriptor is used to measure separation, as measured by the drift time, which is referred to as the collision cross section (CSS) value. It offers one major benefit over other types: it reduces the chances of false positives and offers improved identification (Gonzales et al. 2016). It also helps in separating co-eluting peptides with the same mass-to-charge ratio that overlap in liquid chromatography. However, additional hardware and calibration are required, along with advanced software, for analysis.

9.2.5 Trapping and Data Output

After protein digestion and peptide separation, they are ionized via electrospray ionization and pushed into the Orbitrap mass spectrometer through an ion-guided capillary. Neutrals and solvent clusters are removed, and radiofrequency only focuses on ions to provide sufficient efficiency and sensitivity. Sometimes, C-trap is used to store ions temporarily before introducing them to the Orbitrap analyzer. This C-trap acts as a buffer, controlling the amount of ions over a controlled injection time until the required amount is achieved. This helps in accurate and sensitive results, as a low number of ions can decrease sensitivity, and a high amount can lead to the space charge effect, causing reduced accuracy. Furthermore, these ions are delivered in the form of small packets to ensure simultaneous oscillations in Orbitrap (Bateman et al. 2009).

After that, the ions are captured in the electrostatic field of the mass analyzer, and the oscillating ions create an image current on the split outer electrodes. This current contains all the information of the ion population, and the signal is recorded as a time-domain transient. Next, the Fourier Transformer converts these time domains into frequency domains. These frequency domains appear as peaks, and each peak corresponds to ions with a specific mass-to-charge ratio. This results in a mass spectrum, where the mass-to-charge (m/z) ratio is identified on the x-axis and the ion intensity is displayed on the y-axis. This y-axis represents intensity based on the abundance of peptides (Hardman and Makarov 2003). However, peptides cannot be identified based on their mass, which is why Orbitrap performs Tandem MS to get information on protein sequencing (Olsen et al. 2009). Typically, the most intense ions are selected from the survey scan, and these precursor ions are fragmented via various methods, such as HCD, CID, or ETD. Orbitrap detects these fragments of ions, and an MS/MS spectrum is generated. This spectrum is compared with theoretical spectra from different protein databases (e.g., UniProt) and the protein sequence is identified (Frese et al. 2011). To decrease false identification, statistical filtering is performed. Lastly, peptide quantification is performed using label-free approaches or isotope labeling methods (SILAC, TMT, iTRAQ). These proteins are then identified for their involvement in various pathways or networks within living processes.

9.2.6 Challenges of Orbitrap in Proteomics

Challenge	Impact on proteomics
Complex sample: proteomics may fail to detect peptides in trace amounts	Reduced sensitivity, incomplete proteomics coverage
Instrument cost and maintenance: high acquisition cost, need for vacuum systems, cryogenic cooling, and regular calibration	Limits accessibility and low-resourced labs

Challenge	Impact on proteomics
Throughput constraints: long duty cycle and slower scanning compared to TOF-based MS	Reduced suitability for high-throughput clinical proteomics
Ion suppression and matrix effects: Co-eluting peptides may suppress ionization efficiency in ESI	Inaccurate quantification and reduced reproducibility
Limited fragmentation efficiency: HCD (higher-energy collisional dissociation) in Orbitrap may lead to incomplete fragmentation for certain peptides	Loss of sequence coverage, difficulty in PTM identification
Data analysis burden: high-resolution spectra generate massive amounts of data, requiring complex algorithms and substantial computational power	Slows down interpretation and increases the likelihood of false positives
Instrument robustness: sensitive to environmental conditions and long-term stability issues	Inconsistent results and reduced reproducibility over long projects

9.3 Matrix-Assisted Laser Desorption/Ionization Mass Spectrometry (MALDI-MS)

MALDI-MS is a spectrometric ionization technique for high-throughput screening of peptides and proteins in proteomics. The mass-to-charge ratio is used for identifying compounds, including their type and concentration. In 1990, the first MALDI-MS instrument was made commercially available. However, before MALDI, the technology of time-of-flight (ToF) was introduced during the 1950s (Vestal 2009). This technique used ionized particles and forced them to fly from the source toward the ion detector. During this forced flight, their time was noted, and the method was termed the technology of time of flight. However, ToF had a very poor resolution and did not apply to MS techniques. After extensive research, it was found that ToF can be incorporated into MALDI to create a combined technique called MALDI-ToF-MS (Darie-Ion et al. 2022).

9.3.1 Principle of MALDI-ToF-MS

The MALDI-ToF technique involves two scientific principles: soft ionization through mass spectrometry and determination of mass-to-charge ratio using the time-of-flight technology (El-Aneed et al. 2009). Previously available techniques, which are used more traditionally, result in the production of extensive fragments. These extensive fragments may be suitable for small compounds, but they make the study of larger compounds, such as proteins and peptides, difficult. MALDI is innovative in that it preserves the bonds of compounds intact, as analytes absorb a small amount of energy from the laser. In contrast, co-crystallized matrix compounds absorb energy and then transfer it to the analyte compound, thereby preventing major bond breakage. This means that the MALDI technique involves an indirect transfer of energy, allowing molecules or analytes with larger masses (e.g., in kilodaltons) to be analyzed effectively without significant fragmentation.

The matrix plays a crucial role in MALDI (Matrix-Assisted Laser Desorption/Ionization) technology. It is an organic acid with high UV absorption properties. It helps in shielding the analyte from destructive laser light by absorbing the laser light itself and passing it on to the analyte indirectly. It transfers its vibrational energy to the analyte compound in a gentle manner. It also surrounds the analyte, which helps prevent the onset of aggregation, allowing them to be individually identified. Thus, it enhances the accuracy and sensitivity of the instrument. The matrix contains a chromophore group. Photons from the laser are absorbed in this chromophore of the matrix. Rapid heating and excitation caused by laser absorption lead to the matrix shifting phases from solid to gas, resulting in the release of a thick cloud of gas into the vacuum chamber.

Neutral analyte molecules and matrix ions collide to form the ultimate product, an ionized analyte. After the generation of these analyte ions, they are detected using the principle of the time-of-flight mass spectrometry mechanism (Wu and Odom 1998). Generated ions are transferred from the smaller acceleration zone toward the larger drift zone. Ions that are lighter in mass reach the drift zone in comparison to the heavy zones. The following formula can be used to properly calculate the mass per charge (*m*/*z*) ratio based on the data the sensor has recorded:

$$m / z = 2eE\left(t / d\right)^2$$

where *m* is the mass of the ionized molecule and *z* is the number of electrons that have been removed from the molecule (Hosseini and Martinez-Chapa 2016).

Previously, the time of flight was measured via linear mode. Depending on masses, the ions take varying amounts of time to reach the detector; lighter ions move more swiftly than larger ones. As a result, heavier ions take up more *m*/*z* than lighter ones. Since all ions gain the same amount of kinetic energy, ions with the same *m*/*z* should theoretically have the same flight duration and reach their destination (ion detector) at the same time. In practice, however, the laser irradiation is not as intense for ions with the same *m*/*z*. This means that every discrete *m*/*z* value has a corresponding kinetic energy distribution (Laštovičková et al. 2010). This process results in a *m*/*z* distribution, which drastically reduces the resolution of the spectrum.

Therefore, a more advanced spectrum mode, called the reflection mode, was developed, which provided higher resolution compared to the linear mode. This advanced method utilizes electric fields and ion mirrors to repel the ions back into the drift zone and propel them toward the detector, which is now positioned on the opposite side of the light path (Liyanage and Lay 2006).

9.3.2 Laser Range of MALDI-ToF

MALDI-ToF is a critical analytical tool in proteomics. This technique is equipped with a laser system used for desorption and ionization of analytes. It includes pulse energies, frequencies, and wavelengths needed for optimal ion generation without causing fragmentation (Vestal 2016).

MALDI analysis uses energies in different ranges:

- In *ultraviolet* lasers, MALDI utilizes wavelengths of 337 nm, also known as the "nitrogen laser" wavelength, or 335 nm, referred to as the YAG laser wavelength. Various organic matrices are used to absorb these wavelengths, such as sinapinic acid, -cyano-4-hydroxycinnamic acid (CHCA), or 2,5-dihydroxybenzoic acid (DHB) (Murray 2021).
- The *Infrared IR* laser uses a 2.9–3.0 μm energy range. This energy corresponds to the absorptional vibration of water and the hydroxyl group. This is why this IR laser is useful for certain biological samples (Fatou et al. 2018).
- The *Pulse energy* falls in the range of microjoules. This energy is adequate for the ionization of samples without causing fragmentation. Lower energy causes poor ionization, whereas higher energy results in fragmentation, so the laser fluence (energy per unit area) must be adjusted carefully (O'Connor and Hillenkamp 2007).

9.3.3 Matrix in MALDI: Selection and Types

A unique feature that sets MALDI-ToF apart from other ionization techniques is the matrices. Matrix is a low molecular weight organic compound co-crystallized with matrix. This matrix is very necessary in absorbing laser energy. It enables soft ionization, thereby preventing fragmentation (Li et al. 2022).

Criteria for selection: Not every compound can be employed as a matrix in the MALDI technology. These matrices should meet some specific requirements, such as:

- The compound must have high absorption at laser wavelengths (UV or IR). Aromatic compounds are useful for this purpose (Robinson et al. 2017).
- Matrices should have functional groups that can donate or accept protons, as MALDI is based on the protonation and deprotonation of compounds. Carboxylic acids and hydroxyl groups are examples of such compounds (Hossain and Limbach 2010).
- To avoid overlapping with the low-mass analyte peaks in the spectrum, these matrices should have a very low molecular weight, typically less than 500 Da (Guo et al. 2002).

- The matrix should be able to form homogeneous crystals. These homogeneous crystallization lead to spot-to-spot variability in ion signals (Önnerfjord et al. 1999).
- The matrix must be compatible with the analyte and should dissolve in solvents. It could prevent side reactions and degradation (Hossain and Limbach 2010).
- As MALDI is performed under vacuum, it must be stable under laser and irradiation without decomposing rapidly (Caricato 2014).

9.3.4 Types of Matrices Used in MALDI-ToF

Matrices are categorized into various types based on the kind of analyte used in the technique.

1. *Matrices for protein compounds*
    - Sinapinic acid is used for analyzing proteins having a large molecular weight (10-300 kDa). It produces a single-charged ion with low fragmentation (Abdelhamid and Wu 2015).
    - α-Cyano-4-hydroxycinnamic Acid (CHCA) is the most widely used matrix for proteins. It provides high energy transfer and gives high resolution (Jaskolla et al. 2009).
    - 2,5-Dihydroxybenzoic Acid (DHB) is a suitable matrix for glycoproteins due to its low ionization (Strupat et al. 1991).
2. *Matrices for Nucleic Acids (DNA/RNA, Oligonucleotides)*

Oligonucleotides have high polarity and are more susceptible to fragmentation than other samples. This is why analyzing oligonucleotides is a very challenging task.

- 3-Hydroxypicolinic acid (3-HPA) is a widely used matrix for such compounds.
- Picolinic acid derivatives (e.g., 2,4,6-trihydroxyacetophenone, THAP) enhance signal intensity and resolution for nucleic acids.

3. *Matrices for Carbohydrates*

2,5-Dihydroxybenzoic acid (DHB) and a-cyano-4-hydroxycinnamic acid (CHCA) are used as suitable matrices for carbohydrates and glycans because carbohydrates are poorly ionized due to their lack of polar groups. The DHB is used for oligosaccharides and glycans. CHCA is used for glycopeptides, but it is relatively less efficient than DHB for free carbohydrates (Vanderschaeghe et al. 2010).

4. *Matrices for Lipids*

Phosphatidylserine and phosphatidylino9-aminocridines are very useful. DHB is used for phospholipid analysis. Occasionally, CHCA or SA can also be used for lipids and sterols (MacIntyre 2011).

5. *Synthetic Polymers*

Special polymers, such as Dithranol (1,8,9-anthracenetriol) and Trans-3-indoleacrylic Acid (IAA), are useful for synthetic polymers like polyethylene glycol and polystyrene (Badia et al. 2011).

9.3.5 Sample Preparation Techniques for MALDI-ToF

Sample preparation is a crucial step in the MALDI-ToF technique, as it enhances ionization and desorption while minimizing the fragmentation of ionized analytes. Because proteins exist in diverse forms in terms of size, charge, and hydrophobicity, certain complexities arise, including conformation and degradation. There are various methods to achieve co-crystallization to ensure soft ionization.

1. *Dried droplet method*: This is one of the first methods for preparing samples. This method involves first preparing a matrix, analyte, and ionizing agent solution, which is then volume-mixed. MALDI spotting can be performed by placing solution mixture droplets onto the MALDI plate and allowing them to dry at room temperature (Weidner et al. 2011).
2. *Fast crystallization method*: Rapid crystallization, which often occurs in a matter of seconds, is the foundation of this approach. This process, typically carried out in a vacuum chamber, produces smaller-scale crystals, which in turn lead to reduced segregation and a notable enhancement in the spectrum's resolution and signal repeatability. Another option for accomplishing the same objective is to use a high-purity nitrogen stream for rapid sample drying (Bouschen et al. 2010).
3. *Spin coating method*: MALDI spotting using the spin coating process is another sample preparation method that yields greater homogeneity. Using this technique, the MALDI plate is inserted into the solution droplets and the spin coating chamber. The mixture is introduced from the top to the spinning plate at a specific rotation speed. When the solvent rapidly evaporates, a thin coating will still be present on the MALDI plate (Bauer et al. 2004).
4. *Layer-by-layer method*: This popular method deposits a droplet of matrix solution as the first layer. If sample preparation requires the use of an ionizing agent, the sample would be deposited as the second layer, as the salt aids the matrix in soft ionization. The polymer solution takes up the top layer in this technique. Between coating steps, the deposited samples are allowed to dry and crystallize at ambient temperature. In its application for investigating peptides and proteins, aromatic polyesters, and organic dendrimers, the layer-by-layer approach is reported to enhance signal detection (Wang et al. 2012).
5. *Electrospray deposition method*: Electrospray deposition is arguably the most promising technology for sample preparation among all the methods, offering numerous benefits over competing approaches. The samples can be deposited in one or two layers using this process, which produces an extremely fine and uniform co-crystallization. More signal intensities and a higher likelihood of repeatability in shot-to-shot and/or spot-to-spot analysis are achieved through electrospray deposition in both scenarios (Wei et al. 2004).

9.3.6 Challenges and Future Directions

Challenge	Future directions
Matrix background interference in low-mass compounds: small peptide molecules often get masked by matrix overlap in spectrometer peaks	Development of universal, low-background matrices with higher ionization efficiency for small molecules
Variable ionization efficiency: inconsistent crystallization leads to spot-to-spot variability in ion crystallization	Advanced microfluidic devices and AI-driven automation for reproducible and uniform crystallization
Limited quantification capability: MALDI struggles with precise quantification, which weakens its role in differential expression studies	Integration with quantitative MS workflows (MALDI-MS/MS, hybrid instruments) and machine-learning-based quantification
Limited dynamic range in complex proteomic samples: biological proteomes contain proteins spanning 10^6–10^9 in concentration differences. MALDI cannot capture this entire range, leading to underrepresentation of low-abundance but functionally significant proteins	Development of multiplexed MALDI imaging with higher resolving power and integration with deep-learning algorithms
Tissue imaging limitations (spatial resolution, ion suppression): in MALDI imaging proteomics, limited spatial resolution and ion suppression prevent detailed mapping of protein distributions at the single-cell level. It restricts its utility in studies of cancer tissue heterogeneity and drug localization analysis	Nanoscale MALDI imaging (sub-cellular resolution), integration with multimodal imaging (MRI, PET)

9.4 Single-Cell Western Blotting Technique

9.4.1 Introduction

Proteins mediate signaling pathways, metabolic reactions, structural integrity, and gene control, making them the essential executors of cellular processes (Miklos and Maleszka 2001). Modern biology and medicine depend on an understanding of protein expression and modification. Since its discovery 40 years ago, Western blotting has been employed for analytical and diagnostic purposes in most fields.

Proteins can be separated using the Western blot approach by type and molecular weight using gel electrophoresis. These findings are then transferred to a membrane, where each unique protein is represented by a band formed. This membrane is used to incubate labeled antibodies that are specific to the protein of interest. When the unbound antibody is removed by washing, only the antibody that is bound to the target protein remains. The associated antibodies are then identified by developing the film (Heda et al. 2020).

There should only be one band visible because the antibodies only attach to the target protein. A standard can be used to determine the amount of protein present because the thickness of the band reflects the amount of protein present. Investigating the regulatory molecular processes underlying energy metabolism, protein turnover, and long-term physiological adaptations is one of the many applications of this method. It can be used, for instance, to investigate cellular localization, protein–protein interactions, kinase activity, protein abundance, or monitoring post-translational modifications (Humphrey et al. 2015).

Western blotting and other conventional protein detection methods have been essential for decades; however, they require large cell populations and provide averaged results that obscure variation within these populations. Recent developments in single-cell analytical tools have significantly enhanced our capacity to examine biological systems at previously unattainable resolution. One of these, Single-Cell Western Blotting (scWestern), is a revolutionary variation of the traditional Western blot that enables the analysis of protein expression at the individual cell level. Through the integration of microfluidics, electrophoresis, and antibody-based detection, scWestern provides valuable insights into rare cell populations, cellular diversity, and dynamic biological processes that are not accessible with bulk techniques (Hughes et al. 2014).

9.4.2 *Principles of Western Blotting and Its Evolution*

9.4.2.1 Traditional Western Blotting

In molecular biology, Western blotting, first introduced in the late 1970s, remains a crucial technique for identifying specific proteins in complex samples. Proteins are extracted from bulk samples, separated using sodium dodecyl sulfate–polyacrylamide gel electrophoresis (SDS-PAGE), and then transferred to a membrane in their native form. After that, primary and secondary antibodies are used to probe the membrane to detect protein expression and changes. Traditional Western blotting is a widely used method; however, it has several drawbacks, including high sample requirements, average results, and reduced sensitivity for rare proteins (Blancher and Jones 2001).

9.4.2.2 Emergence of Single-Cell Approaches

The development of techniques that can probe protein expression at the single-cell level has been fueled by the recognition of cellular heterogeneity in immunology, cancer, and stem cell biology. Although they have limited multiplexing capabilities, flow cytometry and mass cytometry provide valuable insights. This gap is filled by single-cell Western blotting, which offers label-free protein separation and antibody-based detection in individual cells (Nunez and Nomellini 2025).

9.4.3 *Workflow of Single-Cell Western Blotting*

9.4.3.1 Preparation of Cells

This technology utilizes single cells instead of large cell cultures. These cells are extracted from tissues or harvested from cell cultures. These cells should be free from clumps, as aggregation can interfere with the capture of single cells. To ensure this, these cells are filtered through cell strainers. After filtration, the concentration of the suspension is carefully adjusted to ensure that the microwell array contains no more than one cell per well. Cell viability is ensured by using certain viable stains. This step is necessary to ensure that apoptotic or dead cells are identified and removed, as they can add certain protein bodies to the sample and interfere with the results.

9.4.3.2 Cell Capture/Sedimentation

Before cell lysis, cells need to be sedimented. It ensures that cell debris is separated from the protein supernatant and can be processed further. Most proteins remain intact in intracellular structures. In Western blotting, these cells or samples are introduced onto a specialized polyacrylamide gel device containing thousands of microwells. Each microwell can hold a single cell. These cells settle at the bottom of the well and are gently centrifuged to ensure they are captured in the wells. The untrapped cells are washed away. This microwell serves as a specialized chamber for each cell, preventing cross-contamination from neighboring cells (Yamauchi and Herr 2017).

9.4.3.3 In-Well Chemical Solubilization of Proteins and Cell Lysis

The next step involves cell lysis, which is performed in the microwell plates using specific detergents. For instance, sodium dodecyl sulphate is a common mild detergent used for this purpose. Certain salts and reducing agents are also used. These chemicals facilitate the rapid solubilization of proteins after rupturing cell membranes. Buffer is applied across the gel device. These microwells act as electrophoretic media. The lysed cells release proteins, which are captured by electrophoretic separation. Moreover, this lysis should be rapid and efficient to minimize the protein extraction efficiency deficiency between different cells (Hansen et al. 2022).

9.4.3.4 Electrophoretic Separation of Proteins

Polyacrylamide gel electrophoresis (PAGE) is used to separate proteins by applying an electric field. Proteins travel through this gel matrix according to their weight. This gel contains specialized lanes that allow proteins from a single cell to travel in a single lane. This separation is done to prevent mixing with neighboring proteins from other cells (Kim et al. 2018).

9.4.3.5 Photoactivated Protein Immobilization onto the PAGE

Appropriate scWB performance necessitates quick and effective protein immobilization. Following electrophoresis, photoactivated protein immobilization minimizes diffusional losses of protein from the gel into the bulk buffer solution while maintaining protein separation. During this photochemical reaction, the photoactive groups, such as N-(3-((4-benzoylphenyl)formamido)propyl)methacrylamide29 (BPMAC) molecules in the PAG layer, link covalently to proteins via a UV-initiated hydrogen abstraction process, instantly freezing the proteins (Kang et al. 2016).

9.4.3.6 Probing with Antibodies

A blocking solution is used to block the gel and prevent nonspecific binding of antibodies. Specific antibodies targeted for particular proteins are added to the sample and incubated. The proteins are fixed within the gel, allowing the antibodies to diffuse freely and bind to the targeted proteins. Fluorescent-labeled secondary antibodies are introduced to remove unbound primary antibodies. These secondary antibodies bind with primary antibodies and create a fluorescent signal. Multiple proteins can be probed within the same gel using different fluorophore-labeled antibodies (Gomez Martinez et al. 2025).

9.4.3.7 Imaging and Data Acquisition

Using fluorescent microscopes, the gel is imaged. As the protein counts are minimal, specialized sensitive imaging systems are needed for imaging. Protein bands, along with separation lines, are visualized. Then, the data for thousands of cells can be compiled to obtain population-level statistics.

9.4.4 Comparison with Traditional Western Blotting

While single-cell Western blotting enables the examination of cell-to-cell variability, traditional Western blotting yields bulk average protein data. While Western medicine uses individual cells, traditional procedures use millions of cells. Despite its technological complexity, scWestern offers high-resolution information about uncommon populations.

9.4.5 Applications of Single-Cell Western Blotting

Sc Western blotting has various applications.

- It is used in cancer biology for the identification of rare resistant clones and tumor heterogeneity.
- It also has applications in immunology, where it is used to characterize immune cell subsets and signaling pathways.
- In stem cell research, it is used to track differentiation and lineage commitment.
- In addition to biology and precision medicine, it is also used in agriculture to identify resistant disease strains.

9.4.6 Future Perspectives

As single-cell technologies advance, scWestern will integrate with other omics platforms to provide multi-layered cellular insights. Improvements in microfluidics, automation, and computational tools will enhance throughput and accessibility, making scWestern a standard in precision biology and clinical diagnostics.

9.4.7 Conclusion

Single-cell Western blotting is a breakthrough technology that extends the classic Western blot to single-cell analysis. It uncovers protein heterogeneity, enables the characterization of rare cells, and enhances our understanding of complex biological systems. Despite current challenges, scWestern holds great promise for fundamental research and translational applications in medicine.

9.5 Epigenetics in a Proteomic Context

9.5.1 Introduction

Epigenetics is defined as the causal interaction between genes and their products that result in their phenotypes (Meaney 2010). This term was introduced by Waddington in 1939. In epigenetics, gene processes are altered without altering the DNA sequence. Although this term was introduced 85 years ago, it is currently a vibrant and quickly growing field that is questioning and updating conventional theories of possible inheritance. This field of study, which is becoming increasingly significant, examines how phenotypic traits can be influenced by the covalent attachment of chemical groups to DNA and the histone proteins associated with it, without altering the DNA sequence (Jablonka and Lamb 2002).

It includes:

1. DNA methylation
2. DNA hydroxymethylation
3. Histone variants and modifications
4. Nucleosome remodeling
5. Small and large noncoding regulatory RNAs

Several methods are used for mapping these modifications, and one of the most fundamental methods used is chromatin immunoprecipitation (ChIP) (Milne et al. 2009). This method is widely used for studying histone modifications and involves the preparation of extracts. Antibodies that are specific to these histones are used for immunoprecipitation to detect histone modifications. This ChIP technology is now used in combination with other serial analysis of gene expression methods, including ChIP combined with ditag sequencing, genome-wide analysis by direct next-generation sequencing of the immunoprecipitated chromatin (ChIP-seq), or hybridization to a microarray (ChIP-chip). Whole-genome analyses based on ChIP-chips are rather costly, but ChIP-seq is a relatively new technology, and costs are expected to decline as sequence throughput and technology accessibility grow in the future. In addition to H2A.Z, RNA polymerase II, and CTCF in human CD4+ T cells, the first documented application of ChIP-seq was to pinpoint the genome-wide sites of 20 distinct histone lysine and arginine methylations. Thus far, ChIP-seq has been utilized to create genome-wide maps for various histone alterations in cancer or tissue models (Cullum et al. 2011).

9.5.2 Proteomic Approaches for Epigenetic Research

For both large-scale and focused protein interaction mapping projects, affinity purification techniques using epitope tags (also referred to as fusion or affinity tags) and tandem purification techniques are widely employed. These techniques enable researchers to isolate proteins, and their identification is then performed via mass spectroscopy (MS) (Cozzolino et al. 2021).

9.5.2.1 Affinity Purification

Affinity purification is based on identifying proteins that interact with their specific molecules from complex mixtures. Traditionally, affinity purifications have been carried out using antibodies directed against a particular protein. The basic principle involves tagging the protein of interest with the affinity handle, such as the FLAG epitope. An affinity matrix is used to immobilize the proteins. Once the target is immobilized, the protein is purified by removing nonspecific binders via washing. The retained proteins are then eluted and analyzed.

This method requires effective antibody binding to an endogenous protein, which can be a time-consuming and expensive procedure, as well as optimizing the experimental conditions for each antibody–protein pair under study. Numerous methods for protein purification based on the attachment of specific tags to bait proteins have been developed, as this approach facilitates large-scale experiments (Roque et al. 2007).

9.5.2.2 Tandem Purification

A general method for purifying protein complexes is tandem affinity purification (TAP). It entails introducing the construct into the targeted cell lines or species after adding a dual affinity tag to the protein of interest, which consists of two different epitopes separated by a protease cleavage site. It increases the specificity and purity of isolated complexes. The process begins with the tagged protein and associated partners being captured on a resin through the first affinity tag. Then it is washed and eluted by protease cleavage. Then it is subjected to a second affinity tag. The final purified product is subjected to mass spectrometric analysis. The strength of TAP lies in its ability to capture protein complexes with very high specificity and low background, which is particularly important in epigenetics, where protein abundance can be low, and interactions can be highly regulated. However, TAP is not without drawbacks. The introduction of large TAP tags may affect the folding, localization, or activity of the bait protein. The method is also relatively time-consuming and may lead to the loss of very transient interactions that dissociate during the two purification steps. Despite these challenges, TAP remains one of the most reliable approaches for characterizing stable epigenetic complexes (Li et al. 2011).

9.5.3 *Assay for Transposase-Accessible Chromatin Using Sequencing (ATAC-seq)*

It is a recently developed method for mapping chromatin accessibility across the genome. This technique was developed in 2013. This method utilizes the hyperactive Tn5 transposase enzyme activity. This enzyme inserts adaptors into accessible chromatin. Heterochromatin is tightly packed and resistant to the activity of these enzymes. That is why transposases insert adapters into promoters or regulatory elements, which are more accessible (Grandi et al. 2022).

9.5.3.1 Workflow and Instrumentation

First, cells are harvested, and their nuclei are released after the cells undergo lysis. This cell lysis will be performed in a manner that retains chromatin integration. Then these purified nuclei are incubated with the Tn5 transposase enzyme. This enzyme carries adapters. The DNA is cleaved, and adapters are inserted. These DNA-carrying adapters are PCR amplified to generate their sequencing libraries. With the help of Illumina platforms, these libraries are sequenced. This sequencing yields short reads, which are then aligned to the reference genome. Peaks are generated for high-density reads. These peaks indicate accessible regions and can be overlapped with promoters or enhancers (Miskimen et al. 2017).

9.5.3.2 Applications

ATAC-seq has various applications.

- It is employed for chromatin mapping to reveal genome-wide profiles to highlight potential regulatory regions.
- It is also used for transcription factor footprinting as it can reveal certain regions where these factors bind.
- This technology can provide us with mechanistic insights for certain diseases by providing differences in chromatin conditions.
- ATAC-seq also provides insights into single-cell heterogeneity by profiling chromatin accessibility at single-cell resolution (Sun et al. 2019).

9.5.4 *Chromatin Immunoprecipitation Sequencing (ChIP-seq)*

9.5.4.1 Principle

ChIP is a reliable method for DNA purification, as the DNA is in close interaction with the protein; these interactions are highly stable, and a highly specific antibody is available. These ChIP technologies provide information on genome modifications. There are two types of ChIP technology: cross-linked ChIP, also known as XChIP, and native ChIP (NChIP). The latter is appropriate for mapping the DNA target of histone modifiers, whereas XChIP is useful for mapping the target sites of transcription factors and other chromatin-associated proteins (Schulz and Häussler 2014).

9.5.4.2 Workflow

First, a cross-linking agent, typically formaldehyde, is applied to the live cells. Specific DNA sequences associated with a given protein are then precipitated with particular antibodies after the chromatin has been sheared by sonication or enzymatic digestion. The purified DNA products are subsequently examined using a variety of molecular techniques to ascertain the genomic location of DNA following the reversal of the cross-links. A greater signal is the result of more DNA being pulled down and amplified when the target protein interacts with this locus. Without the use of crosslinkers, NChIP employs native chromatin that has been sheared by micrococcal nuclease digestion (Hino et al. 2022).

9.5.5 *ChIP-MS Technology*

The conventional ChIP technology requires a large number of cells for the assay; however, it is not suitable in certain situations, such as cells from rare stem cell populations, small tissue biopsies, or cells from embryos. It only provides information on DNA–protein interaction regions and does not provide insight into modified areas. These problems can be addressed using ChIP in combination with mass spectroscopy. After cell lysis, chromatin is sheared by sonication or enzymatic digestion. Purified DNA products are analyzed to determine the genetic sequence. Moreover, proteins are separated by affinity purification and then analyzed by MS (Han and Garcia 2013).

9.5.6 Future Perspective

By using these ChIP technologies, DNA sequences that are directly or indirectly related to proteins can be sequenced, including their modifications, and their genomes can be identified. Due to advancements in proteomic techniques, subtle changes in protein structures can be identified. However, several difficulties are associated with these techniques. For instance, identifying proteins at single genomic loci, enriching chromatin samples from particular locations in a quantity and purity adequate for MS analysis, and the large number of cells needed for the majority of the combined approaches. However, we think that by combining genomic and proteomic technologies, we can gain a systems biology perspective on epigenetic processes, which will pave the way for the creation of pharmaceutical treatments for illnesses and ailments thought to have an epigenetic basis (Han and Garcia 2013).

9.6 Applications of Proteomics

Proteomics is the cornerstone of biomedical research in diverse scientific and clinical domains. Proteomics can contribute to quantifying proteins, mapping post-translational modifications, and characterizing proteins. In biomedical research, it reveals certain pathways by which kinases, phosphatases, and transcription factors regulate specific signaling cascades. It also helps in finding cell heterogeneity between cells of the same tissue or tumors. It can also predict the complex interaction between proteins to perform cellular functions. Moreover, proteomic technologies such as MALDI-ToF are utilized in clinical diagnostics as prognostic markers for certain diseases. In microbiological laboratories, MALDI-ToF serves as the gold standard for identifying bacteria and fungi. Proteomics also aids in the identification of dysregulated proteins, facilitating the identification and validation of targets during the drug discovery and development process. By mapping protein changes following specific treatments, mechanistic insights can be gained into the therapeutic efficacy of drugs. Proteomics also aids in crop improvement in agriculture by identifying stress proteins and can contribute to the genetic engineering of crops. It can also help stabilize ecosystems by identifying contaminants and toxin exposures. Proteomics complements multi-omics approaches and is useful not only for human health and agriculture but also for the environment (Karr 2008; Latterich et al. 2008).

9.7 Conclusion

Proteomics is one of the most transformative fields in biology. Proteins represent the active executors of functions in cells, tissues, and organisms. Although the genomic system is static, proteins are highly dynamic and continuously adapt to environmental and disease conditions. Certain technologies, such as mass spectrometry, Orbitrap, MALDI-ToF, Western blotting, and ChIP sequencing, aid in the identification of proteins at different stages, both healthy and unhealthy. These technologies help in identification, quantification, characterization, and protein profiling.

References

Abdelhamid HN, Wu HF. Synthesis of a highly dispersive sinapinic acid@ graphene oxide (SA@ GO) and its applications as a novel surface assisted laser desorption/ionization mass spectrometry for proteomics and pathogenic bacteria biosensing. Analyst. 2015;140(5):1555–65.

Ashraf Kharaz Y, Zamboulis D, Sanders K, Comerford E, Clegg P, Peffers M. Comparison between chaotropic and detergent-based sample preparation workflow in tendon for mass spectrometry analysis. Proteomics. 2017;17(13–14):1700018.

Badia JD, Strömberg E, Karlsson S, Ribes-Greus A. Contribution III-E. In: Strategies and analytical procedures for a sustainable plastic waste management, vol. 692; 2011. p. 217.

Bateman KP, Kellmann M, Muenster H, Papp R, Taylor L. Quantitative-qualitative data acquisition using a benchtop Orbitrap mass spectrometer. J Am Soc Mass Spectrom. 2009;20(8):1441–50.

Bauer BJ, Flynn KM, Vogt BD. MALDI of layered polymer films. Polym Mater Sci Eng. 2004;90:457.

Blancher C, Jones A. SDS-PAGE and western blotting techniques. In: Metastasis research protocols: volume I: analysis of cells and tissues. Totowa: Humana Press; 2001. p. 145–62.

Bouschen W, Schulz O, Eikel D, Spengler B. Matrix vapor deposition/recrystallization and dedicated spray preparation for high-resolution scanning microprobe matrix-assisted laser desorption/ionization imaging mass spectrometry (SMALDI-MS) of tissue and single cells. Rapid Commun Mass Spectrom. 2010;24(3):355–64.

Bush MF, Campuzano ID, Robinson CV. Ion mobility mass spectrometry of peptide ions: effects of drift gas and calibration strategies. Anal Chem. 2012;84(16):7124–30.

Caricato AP. MAPLE and MALDI: theory and experiments. In: Lasers in materials science. Cham: Springer; 2014. p. 295–323.

Chavez JD, Weisbrod CR, Zheng C, Eng JK, Bruce JE. Protein interactions, post-translational modifications and topologies in human cells. Mol Cell Proteomics. 2013;12(5):1451–67.

Cozzolino F, Iacobucci I, Monaco V, Monti M. Protein–DNA/RNA interactions: an overview of investigation methods in the-omics era. J Proteome Res. 2021;20(6):3018–30.

Cramer CN, Haselmann KF, Olsen JV, Nielsen PK. Disulfide linkage characterization of disulfide bond-containing proteins and peptides by reducing electrochemistry and mass spectrometry. Anal Chem. 2016;88(3):1585–92.

Cullum R, Alder O, Hoodless PA. The next generation: using new sequencing technologies to analyse gene regulation. Respirology. 2011;16(2):210–22.

Darie-Ion L, Whitham D, Jayathirtha M, Rai Y, Neagu AN, Darie CC, Petre BA. Applications of MALDI-MS/MS-based proteomics in biomedical research. Molecules. 2022;27(19):6196.

Dutta S, Ghosh S, Mishra A, Ghosh R. Oncoproteomics: insight into current proteomic technologies in cancer biomarker discovery and treatment. J Proteins Proteomics. 2023;14(1):1–24.

El-Aneed A, Cohen A, Banoub J. Mass spectrometry, review of the basics: electrospray, MALDI, and commonly used mass analyzers. Appl Spectrosc Rev. 2009;44(3):210–30.

Fatou B, Ziskind M, Saudemont P, Quanico J, Focsa C, Salzet M, Fournier I. Remote atmospheric pressure infrared matrix-assisted laser desorption-ionization mass spectrometry (remote IR-MALDI MS) of proteins. Mol Cell Proteomics. 2018;17(8):1637–49.

Frese CK, Altelaar AM, Hennrich ML, Nolting D, Zeller M, Griep-Raming J, et al. Improved peptide identification by targeted fragmentation using CID, HCD and ETD on an LTQ-Orbitrap Velos. J Proteome Res. 2011;10(5):2377–88.

Geiger T, Cox J, Mann M. Proteomics on an Orbitrap benchtop mass spectrometer using all-ion fragmentation. Mol Cell Proteomics. 2010;9(10):2252–61.

Glazyrin YE, Mironov GG, Kichkailo AS, Berezovski MV. Separation abilities of capillary electrophoresis coupled with ion mobility mass spectrometry for the discrete detection of sequence isomeric peptides. Separations. 2022;9(5):106.

Gomez Martinez AE, Lam T, Herr AE. Paired analyses of nuclear protein targets and genomic DNA by single-cell Western blot and single-cell PCR. Anal Chem. 2025;97(28):15310–9.

Gonzales GB, Smagghe G, Coelus S, Adriaenssens D, De Winter K, Desmet T, Van Camp J. Collision cross section prediction of deprotonated phenolics in a travelling-wave ion mobility spectrometer using molecular descriptors and chemometrics. Anal Chim Acta. 2016;924:68–76.

Grandi FC, Modi H, Kampman L, Corces MR. Chromatin accessibility profiling by ATAC-seq. Nat Protoc. 2022;17(6):1518–52.

Guo Z, Zhang Q, Zou H, Guo B, Ni J. A method for the analysis of low-mass molecules by MALDI-TOF mass spectrometry. Anal Chem. 2002;74(7):1637–41.

Han Y, Garcia BA. Combining genomic and proteomic approaches for epigenetics research. Epigenomics. 2013;5(4):439–52.

Hanash SM, Madoz-Gurpide J, Misek DE. Identification of novel targets for cancer therapy using expression proteomics. Leukemia. 2002;16(4):478–85.

Hansen LL, Lomeli G, Vlassakis J, Herr AE. Single-cell resolution immunoblotting. In: Single cell omics of neuronal cells. New York: Springer; 2022. p. 165–92.

Hardman M, Makarov AA. Interfacing the orbitrap mass analyzer to an electrospray ion source. Anal Chem. 2003;75(7):1699–705.

Heda GD, Shrestha L, Thapa S, Ghimire S, Raut D. Optimization of western blotting for the detection of proteins of different molecular weight. Biotechniques. 2020;68(6):318–24.

Hino S, Sato T, Nakao M. Chromatin immunoprecipitation sequencing (ChIP-seq) for detecting histone modifications and modifiers. In: Epigenomics: methods and protocols. New York: Springer; 2022. p. 55–64.

Hossain M, Limbach PA. A comparison of MALDI matrices. In: Electrospray and MALDI mass spectrometry, fundamentals, instrumentation, practicalities, and biological applications; 2010. p. 215–61.

Hosseini S, Martinez-Chapa SO. Principles and mechanism of MALDI-ToF-MS analysis. In: Fundamentals of MALDI-ToF-MS analysis: applications in bio-diagnosis, tissue engineering and drug delivery; 2016. p. 1–19.

Hughes AJ, Spelke DP, Xu Z, Kang CC, Schaffer DV, Herr AE. Single-cell western blotting. Nat Methods. 2014;11(7):749–55.

Humphrey SJ, James DE, Mann M. Protein phosphorylation: a major switch mechanism for metabolic regulation. Trends Endocrinol Metab. 2015;26(12):676–87.

Jablonka E, Lamb MJ. The changing concept of epigenetics. Ann N Y Acad Sci. 2002;981(1):82–96.

Jaskolla TW, Papasotiriou DG, Karas M. Comparison between the matrices α-Cyano-4-Hydroxycinnamic acid and 4-Chloro-α-Cyanocinnamic acid for trypsin, chymotrypsin, and pepsin digestions by MALDI-TOF mass spectrometry. J Proteome Res. 2009;8(7):3588–97.

Jiménez A, Lu D, Kalocsay M, Berberich MJ, Balbi P, Jambhekar A, Lahav G. Time-series transcriptomics and proteomics reveal alternative modes to decode p53 oscillations. Mol Syst Biol. 2022;18(3):e10588.

Kang CC, Yamauchi KA, Vlassakis J, Sinkala E, Duncombe TA, Herr AE. Single cell–resolution western blotting. Nat Protoc. 2016;11(8):1508–30.
Karlsson R, Gonzales-Siles L, Boulund F, Svensson-Stadler L, Skovbjerg S, Karlsson A, Moore ER. Proteotyping: proteomic characterization, classification and identification of microorganisms–a prospectus. Syst Appl Microbiol. 2015;38(4):246–57.
Karr TL. Application of proteomics to ecology and population biology. Heredity. 2008;100(2):200–6.
Kaufmann A. Analytical performance of the various acquisition modes in Orbitrap MS and MS/MS. J Mass Spectrom. 2018;53(8):725–38.
Kim JJ, Chan PP, Vlassakis J, Geldert A, Herr AE. Microparticle delivery of protein markers for single-cell Western blotting from microwells. Small. 2018;14(48):1802865.
Kitano H. Systems biology: toward system-level understanding of biological systems. In: Foundations of systems biology; 2001. p. 1–36.
Klatt JN, Dinh TJ, Schilling O, Zengerle R, Schmidt F, Hutzenlaub T, Paust N. Automation of peptide desalting for proteomic liquid chromatography–tandem mass spectrometry by centrifugal microfluidics. Lab Chip. 2021;21(11):2255–64.
Laštovičková M, Mazanec K, Benkovská D, Bobál'ová J. Utilization of the linear mode of MALDI-TOF mass spectrometry in the study of glycation during the malting process. J Inst Brew. 2010;116(3):245–50.
Latterich M, Abramovitz M, Leyland-Jones B. Proteomics: new technologies and clinical applications. Eur J Cancer. 2008;44(18):2737–41.
Li K, Chen Z, Zhang Y, Yang X. Applications of iTRAQ and TMT labeling techniques to the study of neurodegenerative diseases. Curr Protein Pept Sci. 2020;21(12):1202–17.
Li D, Yi J, Han G, Qiao L. MALDI-TOF mass spectrometry in clinical analysis and research. ACS Meas Sci Au. 2022;2(5):385–404.
Li Y. The tandem affinity purification technology: an overview. Biotechnol Lett. 2011;33(8):1487–99.
Liyanage R, Lay JO. An introduction to MALDI-ToF MS. In: Identification of microorganisms by mass spectrometry, vol. 169; 2006. p. 39–60.
MacIntyre L. Application of matrix-assisted laser desorption/ionisation (MALDI) imaging mass spectrometry and high resolution Fourier transform mass spectrometry for profiling pharmaceuticvals and biomarkers in a variety of biological samples. Foods. 2011;8:633.
Makarov A, Scigelova M. Coupling liquid chromatography to Orbitrap mass spectrometry. J Chromatogr A. 2010;1217(25):3938–45.
Makarov A, Grinfeld D, Ayzikov K. Fundamentals of Orbitrap analyzer. In: Fundamentals and applications of Fourier transform mass spectrometry. Elsevier; 2019. p. 37–61.
Meaney MJ. Epigenetics and the biological definition of gene× environment interactions. Child Dev. 2010;81(1):41–79.
Michalski A, Damoc E, Hauschild JP, Lange O, Wieghaus A, Makarov A, et al. Mass spectrometry-based proteomics using Q Exactive, a high-performance benchtop quadrupole Orbitrap mass spectrometer. Mol Cell Proteomics. 2011;10(9):M111.011015.
Miklos GLG, Maleszka R. Protein functions and biological contexts. PROTEOMICS: Int Ed. 2001;1(2):169–78.
Milne TA, Zhao K, Hess JL. Chromatin immunoprecipitation (ChIP) for analysis of histone modifications and chromatin-associated proteins. In: Leukemia: methods and protocols. Totowa: Humana Press; 2009. p. 409–23.
Miskimen KL, Chan ER, Haines JL. Assay for transposase-accessible chromatin using sequencing (ATAC-seq) data analysis. Curr Protoc Hum Genet. 2017;92(1):20–4.
Misra BB, Langefeld C, Olivier M, Cox LA. Integrated omics: tools, advances and future approaches. J Mol Endocrinol. 2019;62(1):R21–45.
Monti M, Orru S, Pagnozzi D, Pucci P. Functional proteomics. Clinica Chimica Acta. 2005;357(2):140–50.
Murray KK. Lasers for matrix-assisted laser desorption ionization. J Mass Spectrom. 2021;56(6):e4664.

Nepomuceno AI, Gibson RJ, Randall SM, Muddiman DC. Accurate identification of deamidated peptides in global proteomics using a quadrupole orbitrap mass spectrometer. J Proteome Res. 2014;13(2):777–85.

Neset L, Takayidza G, Berven FS, Hernandez-Valladares M. Comparing efficiency of lysis buffer solutions and sample preparation methods for liquid chromatography–mass spectrometry analysis of human cells and plasma. Molecules. 2022;27(11):3390.

Nunez JH, Nomellini V. Modern techniques for protein assessment. In: Success in academic surgery: basic science; 2025. p. 115–38.

O'Connor PB, Hillenkamp F. MALDI mass spectrometry instrumentation. In: MALDI MS: a practical guide to instrumentation, methods and applications; 2007. p. 29–82.

Olsen JV, Schwartz JC, Griep-Raming J, Nielsen ML, Damoc E, Denisov E, et al. A dual pressure linear ion trap Orbitrap instrument with very high sequencing speed. Mol Cell Proteomics. 2009;8(12):2759–69.

Önnerfjord P, Ekström S, Bergquist J, Nilsson J, Laurell T, Marko-Varga G. Homogeneous sample preparation for automated high throughput analysis with matrix-assisted laser desorption/ionisation time-of-flight mass spectrometry. Rapid Commun Mass Spectrom. 1999;13(5):315–22.

Perry RH, Cooks RG, Noll RJ. Orbitrap mass spectrometry: instrumentation, ion motion and applications. Mass Spectrom Rev. 2008;27(6):661–99.

Pontén F, Schwenk JM, Asplund A, Edqvist PH. The human protein atlas as a proteomic resource for biomarker discovery. J Intern Med. 2011;270(5):428–46.

Przybylski C, Jünger MA, Aubertin J, Radvanyi F, Aebersold R, Pflieger D. Quantitative analysis of protein complex constituents and their phosphorylation states on a LTQ-Orbitrap instrument. J Proteome Res. 2010;9(10):5118–32.

Puig O, Caspary F, Rigaut G, Rutz B, Bouveret E, Bragado-Nilsson E, et al. The tandem affinity purification (TAP) method: a general procedure of protein complex purification. Methods. 2001;24(3):218–29.

Renfrey S, Featherstone J. Structural proteomics. Nat Rev Drug Discov. 2002;1(3):175–6.

Robinson KN, Steven RT, Bunch J. Matrix optical absorption in UV-MALDI MS. J Am Soc Mass Spectrom. 2017;29(3):501–11.

Roque AC, Silva CS, Taipa MÂ. Affinity-based methodologies and ligands for antibody purification: advances and perspectives. J Chromatogr A. 2007;1160(1–2):44–55.

Schulz S, Häussler S. Chromatin immunoprecipitation for ChIP-chip and ChIP-seq. In: Pseudomonas methods and protocols. New York: Springer; 2014. p. 591–605.

Senyuva HZ, Gökmen V, Sarikaya EA. Future perspectives in Orbitrap™-high-resolution mass spectrometry in food analysis: a review. Food Addit Contam Part A. 2015;32(10):1568–606.

Slysz GW, Schriemer DC. Blending protein separation and peptide analysis through real-time proteolytic digestion. Anal Chem. 2005;77(6):1572–9.

Stolz A, Jooß K, Höcker O, Römer J, Schlecht J, Neusüß C. Recent advances in capillary electrophoresis-mass spectrometry: instrumentation, methodology and applications. Electrophoresis. 2019;40(1):79–112.

Strupat K, Karas M, Hillenkamp F. 2, 5-Dihydroxybenzoic acid: a new matrix for laser desorption—ionization mass spectrometry. Int J Mass Spectrom Ion Process. 1991;111:89–102.

Sun Y, Miao N, Sun T. Detect accessible chromatin using ATAC-sequencing, from principle to applications. Hereditas. 2019;156(1):29.

Tian C, Liu K, Sun R, Fu L, Yang J. Chemoproteomics reveals unexpected lysine/arginine-specific cleavage of peptide chains as a potential protein degradation machinery. Anal Chem. 2018;90(1):794–800.

Treberg JR, Martyniuk CJ, Moyes CD. Getting the most out of reductionist approaches in comparative biochemistry and physiology. Comp Biochem Physiol B: Biochem Mol Biol. 2020;250:110483.

Umadevi P, Anandaraj M. An efficient protein extraction method for proteomic analysis of black pepper (Piper nigrum L.) and generation of protein map using nano LC-LTQ Orbitrap mass spectrometry. Plant Omics. 2015;8(6):500–7.

Vanderschaeghe D, Festjens N, Delanghe J, Callewaert N. Glycome profiling using modern glycomics technology: technical aspects and applications. Biol Chem. 2010;2-3:149–61.

Vestal ML. Modern MALDI time-of-flight mass spectrometry. J Mass Spectrom. 2009;44(3):303–17.

Vestal ML. Evolution of quantitative MALDI-TOF mass spectrometry for clinical applications. Clin Chem. 2016;62(1):20–3.

Wang SF, Li X, Agapov RL, Wesdemiotis C, Foster MD. Probing surface concentration of cyclic/linear blend films using surface layer MALDI-TOF mass spectrometry. ACS Macro Lett. 2012;1(8):1024–7.

Wei H, Nolkrantz K, Powell DH, Woods JH, Ko MC, Kennedy RT. Electrospray sample deposition for matrix-assisted laser desorption/ionization (MALDI) and atmospheric pressure MALDI mass spectrometry with attomole detection limits. Rapid Commun Mass Spectrom. 2004;18(11):1193–200.

Weidner S, Knappe P, Panne U. MALDI-TOF imaging mass spectrometry of artifacts in "dried droplet" polymer samples. Anal Bioanal Chem. 2011;401(1):127–34.

Wenz C, Barbas C, López-Gonzálvez Á, Garcia A, Benavente F, Sanz-Nebot V, et al. Interlaboratory study to evaluate the robustness of capillary electrophoresis-mass spectrometry for peptide mapping. J Sep Sci. 2015;38(18):3262–70.

Weston AD, Hood L. Systems biology, proteomics, and the future of health care: toward predictive, preventative, and personalized medicine. J Proteome Res. 2004;3(2):179–96.

Winter D, Steen H. Optimization of cell lysis and protein digestion protocols for the analysis of HeLa S3 cells by LC-MS/MS. Proteomics. 2011;11(24):4726–30.

Wu KJ, Odom RW. Peer reviewed: characterizing synthetic polymers by MALDI MS. Anal Chem. 1998;70:456A–61A.

Yakubu RR, Nieves E, Weiss LM. The methods employed in mass spectrometric analysis of post-translational modifications (PTMs) and protein–protein interactions (PPIs). In: Advancements of mass spectrometry in biomedical research; 2019. p. 169–98.

Yamauchi KA, Herr AE. Subcellular western blotting of single cells. Microsyst Nanoeng. 2017;3(1):1–9.

Chapter 10
Socio-economic Perspectives on Genetically Engineered Crops: From Farm to Market Worldwide

Muhammad Asif Saddique

10.1 Introduction

Evaluating the socio-economic impacts of GE crops is complicated because they are introduced into diverse rural farming systems that differ in crop types, locations, farm sizes, farmer education, government policies, and access to credit. GE crops integrate traditional breeding, molecular biology, and other sciences into a single technology—the seed or vegetative material. They also must adhere to legal frameworks, including national laws as well as international trade and patent agreements. Inventors and regulators decide how these crops are integrated into these systems (Racovita et al. 2015).

Any analysis of genetically engineered (GE) crops must be nuanced, as their social and economic impacts vary across time, regions, and farming communities. This chapter reviews existing evidence on these effects since the introduction of GE crops, drawing on studies, reviews, and meta-analyses to highlight key issues surrounding their adoption. It first examines impacts on income, smallholder farmers, knowledge transfer, gender dynamics, rural communities, and seed decision-making. The discussion then turns to broader dimensions, including consumer acceptance and awareness, international trade, production costs, regulatory frameworks, intellectual property rights, and food security.

This chapter begins by examining the impact of GE crops on farmer incomes. The results can vary due to geographic and temporal differences, as well as among farmers, households, and consumers, with influences such as statistical biases, sampling issues, and survey methods (Smale et al. 2009). Consequently, observed effects encompass a range of benefits, costs, and risks. Following this overview, the committee explores how genetic engineering interacts with other farm-level factors, including gender, community, and farmer knowledge (Klümper and Qaim 2014).

M. A. Saddique (✉)
University of Agriculture, Faisalabad, Pakistan

S. Ijaz et al. (eds.), *Molecular Biology Methods*,
https://doi.org/10.1007/978-981-92-2642-9_10

Insect-resistant (IR) traits help minimize damage from biological threats, while herbicide-resistant (HR) traits reduce crop management time, allowing farmers to focus on off-farm income. These traits are part of a broader range of options for crops. The damage reduction from IR and other GE traits varies by application, affecting their social and economic impacts depending on location and end-users (Sexton and Zilberman 2012).

Agronomic effects include changes in crop yield and the use of insecticides and herbicides for genetically engineered (GE) crops with insect resistance (IR) or herbicide resistance (HR) traits. Beyond their direct agricultural impact, farmers can experience economic effects due to yield variations or costs associated with the use of pesticides and herbicides. Most studies link the impacts of GE crops to gross margins—the difference between total revenue and variable costs—which can influence farm income or household earnings. However, gross margins do not capture the full picture of total farm or household income, as income often comes from both on-farm and off-farm sources (Areal et al. 2013).

Predicting whether biases and uncontrolled confounders will pose problems or estimating their impact is a challenging task. Nonetheless, recognizing that these factors can influence early adoption studies (Smale et al. 2009, 2012) strongly supports making routine testing for such issues a standard part of the research process. The committee concludes that it is impossible to determine whether, or to what extent, early studies on the adoption of GE crops were affected by biases and confounders. Re-examining these studies for a quantitative assessment, if feasible, is essential. More recent research has directly tackled these concerns by employing methods to adjust for biases and uncontrolled confounding variables (Mannion and Morse 2013).

10.2 Assessment of the Economic Impact of Insect-Resistant (IR) Traits

The economic advantages of insect-resistant (IR) crops differ from those of herbicide-resistant (HR) varieties, yet many studies have not clearly distinguished between Bt maize and Bt cotton. On average, IR crops were associated with a 69% increase in profits, largely driven by a 25% rise in yields and a 43% reduction in insecticide expenditures. Most studies reviewed focused on Bt cotton cultivation in India and China (Finger et al. 2011), where significant differences in production costs and yields were observed. For example, while Bt cotton incurred production costs €13 per hectare higher than non-GE varieties, it still generated greater gross margins. Similarly, Bt maize exhibited production costs €14 per hectare higher than conventional maize, yet also achieved superior gross margins (Areal et al. 2013).

Variation in farmers' resources and productivity influenced crop performance outcomes. In Spain, South Africa, and Argentina, gross margins for Bt and non-GE maize showed no statistically significant differences, though insecticide costs were notably lower in Spain and Germany. In Germany, these savings contributed to the adoption of Bt maize as an effective pest control strategy. However, data on management and labor costs were either unavailable or not significant. Importantly, these findings derive from longer-term studies (Fernández-Cornejo et al. 2014). Methodological adjustments were made to isolate the effects of Bt cotton from insecticide use, thereby reducing potential bias and ensuring more reliable estimates.

Over the past decade, the targeted pest (bollworm) has declined in surveyed areas (Huang et al. 2010). In 2008, farmers perceived benefits from planting Bt cotton despite the ongoing use of bollworm sprays (*Helicoverpa zea*). Although less susceptible to Cry 1 Ac and Cry 2 Ab 2, bollworm control was deemed worthwhile, with over 75% of US cotton in Bt varieties that year, despite concerns about insecticide expenses and technological fees (Luttrell and Jackson 2012). However, eggplant was first commercially planted by 20 farmers in Bangladesh in 2014. Ex ante studies in Bangladesh, India, and the Philippines projected the economic impacts of Bt eggplant adoption, emphasizing the indicative nature of the estimates (Islam and Norton 2007). Field trials suggested a 75% reduction in insecticide use against eggplant fruit and shoot borer, resulting in lower costs despite increased expenses for seed and harvesting. These trials are expected to boost marketable fruit yields. Estimated gross margins could increase by 61% in one region (up to 106,351 rupees/ha) and 182% in another (up to 6,868,269 rupees/ha) (Gerpacio and Aquino 2014).

10.3 Economic Evaluation of Herbicide-Resistant Traits

Research on herbicide-resistant (HR) crops is considerably less extensive than studies on insect-resistant (IR) crops. Finger et al. (2011) excluded HR soybean and canola from their meta-analysis due to limited data. Fischer et al. (2015) found that only 20 out of 99 studies on genetically engineered (GE) crops focused on HR varieties. Areal et al. (2013) reported that HR soybeans lowered production costs by €25 per hectare compared to non-GE options; however, this conclusion was based on only six studies. Klümper and Qaim (2014) analyzed data on HR soybean, maize, and cotton, showing a 64% increase in profits for farmers using HR crops, mainly due to a 9% yield increase and a 25% reduction in herbicide costs.

Fernandez-Cornejo et al. (2014) reviewed US research and observed mixed results regarding net returns for HR soybeans—some studies found similar returns, while others found higher returns. Similar patterns were seen for HR maize and cotton, with some studies showing no difference and others indicating gains. Gardner et al. (2009) highlighted labor savings primarily for HR soybean, averaging 14.5%, but found limited evidence for other HR crops.

Fernandez-Cornejo et al. (2005) and earlier reports emphasized that nonmonetary benefits—such as time and effort savings, safety, and environmental advantages—are crucial in the adoption of HR crops. Expenditures on herbicides and insecticides did not vary significantly between GE and non-GE maize, with comparable profit margins. Afidchao et al. (2014) concluded that despite higher seed and fertilizer costs for Bt-HR maize, improved pest control likely offered an overall economic benefit. In 2007, Kniss compared glyphosate-resistant sugar beet fields with non-GE fields in Wyoming, noting similar herbicide use but lower costs and less labor for GE fields, which produced 15% more roots and had higher sucrose content. Despite higher harvest costs, net returns for HR sugar beet farmers were $576 per hectare higher. The study could not be replicated the following year due to widespread adoption of HR sugar beets.

10.4 Real-World Benefits of Existing GMO Crops

The most widespread GE crops—HR soybean, Bt maize, Bt cotton, Bt-HR maize, Bt-HR cotton, and HR canola—were first introduced in the United States, mainly on large farms. However, some, like Bt cotton, are now adopted by smallholders worldwide. Research mainly focuses on countries like India, China, and Pakistan, with large smallholder populations benefiting from GE crops. Evidence shows that leading cotton producers, such as India, China, and Pakistan, use Bt cotton, which benefits small farmers. However, benefits in other regions appear mixed. For instance, a study of 36 farmers in Maharashtra, who grew either Bt or non-Bt cotton, found a positive relationship between input costs, such as insecticides, and total revenue for non-Bt cotton farmers. However, no such link was found for Bt cotton farmers, indicating that many smallholders lack the skills to optimize returns from Bt cotton. Overapplication of insecticides on Bt cotton has been reported in India and China, often due to poor knowledge dissemination; however, improved understanding has reduced overuse. Initial adoption in South Africa's Makhathini Flats was rapid, reaching nearly 3000 farmers by 2001 (a 90% adoption rate), driven by extension services and credit from a monopolizing cotton ginner. However, after these support systems ceased in 2001, adoption sharply declined.

In South Africa, the adoption of GM maize by smallholders is limited mainly due to seed access issues, despite its widespread use among large-scale farmers. Households in Hlabisa, where white maize is a staple, experienced higher yields and labor savings from Bt-HR maize, leading to interest in incorporating the HR trait into existing hybrids, such as PAN 6043. HR soybeans are prevalent in Bolivia, with small farmers comprising 77% of producers and typically operating larger, more educated, and machinery-equipped farms. Adopters reported easier weed management, higher yields, and reduced family labor time, resulting in increased household earnings. However, eggplant has not been widely available or cultivated enough to determine its benefits for smallholders conclusively.

Nevertheless, evidence from India and Bangladesh suggests potential advantages, including reduced insecticide use and higher profits. Critics like Andow raised concerns over the suitability of hybrid Bt eggplants for small farmers, noting that most grow open-pollinated varieties and that hybrids might threaten economic stability; however, these views are now somewhat outdated. In India, regulatory policies now focus on specific host varieties, with simplified approval for other varieties. Economic models suggest that affordable Bt OPVs benefit resource-poor farmers. Surveys in India and other countries indicate that farmers favor hybrids when seeking higher yields, although some might switch to cheaper Bt OPVs (Shi et al. 2009).

Adoption patterns vary, with small eggplant farmers showing potential economic benefits through the adoption of Bt hybrids or OPVs. Similarly, virus-resistant (VR) papaya quickly gained popularity in Hawaii after 1998 due to severe virus damage, with adoption rising from 42% in 2000 to 77% in 2009. China's development of VR papaya in 2007 also saw over 60% cultivation by 2012. Some researchers warn that GE crops might cause "deskilling," reducing farmers' reliance on their skills as new technologies take over tasks traditionally performed manually, a concept dating back to the 1990s. Early studies of hybrid maize and Bt cotton indicated that reliance on advanced technologies can diminish farmers' traditional knowledge and skills, potentially leading to unstable farming environments and increased technological chaos.

10.5 Gender and Rural Communities

Gender-appropriate technologies connect to GE crops. In many regions, farming tools like tractors are seen as masculine (Brandth 2006). GE crops may align more with tools linked to women. In the United States and Europe, female farmers often work in alternative systems (Chiappe and Flora 1998; Peter et al. 2000; Rissing 2012). Many of these systems, especially in developed countries, avoid GE crops due to philosophical reasons or bans, such as USDA organic certification. Women worldwide tend to farm smaller plots (Hoppe and Korb 2013). Though GE crops are seen as gender-neutral or female-friendly, systems managed by women show lower adoption rates. In the United States, social scientists have studied how agricultural changes have impacted rural communities.

The 1948 Goldschmidt thesis posited that industrial agriculture diminishes the quality of life in rural areas (Carolan 2012). More recent studies support this, though debates on specific causes remain (Lyson et al. 2001; Lobao and Stofferahn 2008). Technologies speeding up industrialization reduce family farms, harming rural communities. This chapter examines the impact of GE crops and genetic engineering on communities. While few studies directly assess social impacts, research on farm structures, technologies, and community dynamics provides valuable insights (Lobao and Stofferahn 2008). A previous NRC report (NRC 2010) notes that social impacts are likely to result from the adoption of GE crops. The full effects are uncertain due to limited research, but potential outcomes include changes in labor, farm structure, community viability, and farmer relationships.

10.6 Seed Availability and Pricing

Evidence suggests a correlation between the increased use of land for genetically engineered (GE) crops and a decline in the use of non-GE seeds (Pechlaner 2012). In the United States, the availability of non-GE maize, soybean, and cotton seeds fell by 67%, 51%, and 26%, respectively, from 2005 to 2010 (Heinemann et al. 2014). Several factors could explain these declines. A review of maize hybrid trials from Iowa, Illinois, and Minnesota—major maize-producing states—found that data from Iowa were unavailable for comparison of GE and non-GE seeds. In 2014, these states tested 86 non-GE hybrids against 544 GE hybrids, with non-GE hybrids making up 13.7% and GE hybrids 86.3%. Minnesota's data show that most hybrids, 198 out of 219, contained stacked GE traits, with only 21 having solely herbicide-resistant (HR) traits. In Brazil, from 2008 to 2012, the number of non-GE hybrids decreased from 302 to 263, while the number of GE hybrids increased from 19 to 216 following the approval of GE maize (Parentoni et al. 2013). In both the United States and Brazil, the widespread adoption of GE varieties has limited the availability of non-GE options, although they still exist. The pace of this trend remains uncertain.

Overall, in 2015, nonadopters and partial adopters had fewer hybrid choices than before the introduction of GE crops, as shown by Krishna et al. (2016) in their study on cotton varietal diversity in India. For farmers interested in growing GE crops, cost can be a barrier, especially for smallholders with limited resources. Although GE seeds tend to be more expensive than other seed types, this impacts adoption mainly if benefits such as lower insecticide costs, reduced crop damage, or labor savings do not offset the higher price. The critical factors are the proportion of total costs that seed purchases represent and how farmers can recover that investment.

Typically, seed costs account for a small portion of overall production expenses, but they can still pose a significant financial challenge. A policy change in India in 2006 reduced this gap to 68%. In Spain, Germany, and Argentina, the costs of Bt maize seed were 9.9%, 17%, and 36% higher, respectively (Finger et al. 2011). Seed prices vary regionally and are influenced by pest infestation levels; costs tend to be lower where pest pressure is light and yield gaps are small. A 2010 survey in the Philippines found that seed costs for all GE maize varieties were 60% higher than for non-GE ones. Efforts are ongoing to lower these costs through humanitarian licenses, which enable researchers to develop GE crops without paying biotech royalty fees (Frewer et al. 2011).

10.7 Consumers' Acceptance and Awareness of the Market

Over the past 20 years, numerous scholarly investigations have focused on understanding consumers' acceptance and willingness to purchase food derived from genetically engineered (GE) crops. These investigations have utilized methodologies such as surveys, choice experiments, and hedonic analyses (Costa-Font et al. 2008; Rollin et al. 2011). In excess of 100 studies conducted across more than 20 countries have examined consumers' willingness to pay (WTP) for genetically modified (GM) foods, assessing both the significance and magnitude of premiums consumers might be willing to pay to select or avoid GE crops (Colson and Rousu 2013). This discount is contingent upon factors such as the type of genetic modification, the food product involved, and the manner in which the modification influences the final product. They further observed that consumers in the United States generally exhibit greater acceptance of GE foods compared to European consumers. Similarly, Colson and Huffman (2011) discovered that WTP is influenced by the information available at the point of purchase; emphasizing benefits tends to enhance WTP, whereas perceived risks can diminish it.

Phillips and Hallman (2013) posited that consumers' perceptions are heavily influenced by the manner in which GE foods are presented—either emphasizing benefits or highlighting risks—and that these perceptions are shaped by prior knowledge and other factors. Finally, Colson and Rousu (2013) questioned the overall validity of WTP studies, noting that they may provide limited insights into actual consumer behavior, as many consumers remain unaware that GE foods are available in the market. They also highlighted the variation in findings across different studies and questioned whether these results genuinely reflect real-world purchasing behavior.

10.8 Seed-Market Dominance

Concerns about utility or plant patents for GE crops include their potential to increase seed market concentration, which could diminish the benefits of a competitive market, such as fair pricing (Glenna and Cahoy 2009). Market concentration is assessed through various methods, including monitoring changes in market share of top firms, patent holdings, or the use of the Herfindahl-Hirschman Index (HHI). Fuglie and Toole (2014) noted that when four or fewer firms hold over 40% of the market, it indicates potential concentration, with the global seed share controlled by four firms rising from 21.1% in 1994 to 53.9% in 2009, coinciding with the spread of GE crops. Whether GE crops accelerated this trend remains debated. Patent control, another measure, reveals that since the 1980s, 37 companies have held patents on GE maize, and 118 on other GE crops. However, in 2008, three firms controlled 85% of GE maize patents and 70% of non-maize GE crop patents, indicating significant market concentration.

The HHI index, employed by the US DOJ, measures industry concentration. Scores above 1800 suggest high concentration. The world's crop-seed sector's HHI increased from 171 in 1994 to 991 in 2009, with the US maize, soybean, and cotton sectors surpassing 1800 by 1999. Despite this, market power does not always lead to consumer harm; fair sharing of benefits can occur, as seen with Bt cotton (Falck-Zepeda et al. 2000). US research from 1997 to 2008 suggests moderate market power and efficiency, but further research is needed to understand the impacts on seed prices. Interestingly, seed prices for stacked traits are often lower than the combined cost of individual traits, possibly due to economies of scope or market segmentation strategies (Stiegert et al. 2010; Shi et al. 2010).

Trait stacking may cause farmers to purchase more expensive seeds than necessary, such as paying for Bt traits when only herbicide resistance is required, raising concerns about market dominance. Increased seed market concentration could lead to higher prices for farmers if competition decreases, although some evidence suggests that costs have been offset by yield gains and input savings (NRC 2010). Public trust varies: scientists, professionals, and advocacy groups tend to trust research on GE crops, while skepticism remains toward government and industry sources (Lang and Hallman 2005; Lang 2013). The global presence of major firms and stringent international patent protections raise concerns about market dominance and its broader social and economic implications, which could impact international R&D, particularly if US patent policies hinder innovation (Strauss 2009; Jefferson et al. 2015).

10.9 Food Security

Many scholars argue that genetic engineering could be pivotal in addressing global hunger (e.g., Borlaug 2000; Juma 2011). Although genetically engineered (GE) crops possess the potential to enhance yields (Anthony and Ferroni 2012), their effectiveness in alleviating hunger depends on the development of appropriate crop varieties and the establishment of supportive political, social, and cultural environments. For subsistence farmers, safeguarding yields against pests, diseases, drought, and extreme weather conditions can stabilize food supplies and mitigate hunger. Existing GE crops may help protect yields where access to such technology is limited, and new varieties—such as disease-resistant cassava and climate-adaptive rice—are being developed for a broader range of crops.

Nevertheless, as acknowledged, genetically engineered (GE) crops alone are insufficient to address all challenges faced by smallholders. Even if a GE crop increases yield or nutritional value, its benefits are heavily dependent on the social and economic conditions surrounding its production and distribution (Tripp 2009). An increasing number of GE crop developers originate from developing countries, particularly India and China.

African countries (Parisi et al. 2016) offer promising prospects for future genetically engineered (GE) crops that can be customized for specific regions, governments, or farmers, thereby enhancing productivity or nutritional content at the local level. Addressing the complex challenges faced by smallholder farmers and food-insecure populations is crucial for advancing efforts to reduce hunger. Although global food production currently meets demand, about one in nine people still lack adequate access to food (FAO 2015). While genetically engineered (GE) crops can contribute to a comprehensive food security strategy, solving food insecurity requires "multi-pronged solutions" (Qaim and Kouser 2013).

References

Afidchao MM, Musters CJM, Wossink A, Balderama OF, de Snoo GR. Analysing the farm level economic impact of GM corn in The Philippines. NJAS Wagening J Life Sci. 2014;70–71:113–21.

Areal FJ, Riesgo L, Rodriguez-Cerezo E. Economic and agronomic impact of commercialized GM crops: a meta-analysis. J Agric Sci. 2013;151:7–33.

Borlaug NE. Ending world hunger: the promise of biotechnology and the threat of antiscience zealotry. Plant Physiol. 2000;124:487–90.

Carolan M. The sociology of food and agriculture. Abingdon: Routledge; 2012.

Chiappe MB, Flora CB. Gendered elements of the alternative agriculture paradigm. Rural Sociol. 1998;63:372–93.

Colson G, Huffman WE. Consumers' willingness to pay for genetically modified foods with product-enhancing nutritional attributes. Am J Agric Econ. 2011;93:358–63.

Colson G, Rousu MC. What do consumer surveys and experiments reveal and conceal about consumer preferences for genetically modified foods? GM Crops & Food. 2013;3:158–65.

Costa-Font M, Gil JM, Traill WB. Consumer acceptance, valuation of and attitudes towards genetically modified food: review and implications for food policy. Food Policy. 2008;33:99–111.

Falck-Zepeda JB, Traxler G, Nelson RG. Rent creation and distribution from biotechnology innovations: the case of Bt cotton and herbicide-tolerant soybeans in 1997. Agribusiness. 2000;16:21–32.

FAO (Food and Agriculture Organization). The State of Food Insecurity in the World 2015. Rome: FAO; 2015.

Fernandez-Cornejo J, Hendricks C, Mishra A. Technology adoption and off-farm household income: the case of herbicide-tolerant soybeans. J Agric Appl Econ. 2005;37:549–63.

Fernandez-Cornejo J, Wechsler SJ, Livingston M, Mitchell L. Genetically engineered crops in the United States. Washington, DC: United States Department of Agriculture–Economic Research Service; 2014.

Finger R, El Benni N, Kaphengst T, Evans C, Herbert S, Lehmann B, Morse S, Stypak N. A meta-analysis on farm-level costs and benefits of GM crops. Sustainability. 2011;3:743–62.

Frewer LJ, Bergmann K, Brennan M, Lion R, Meertens R, Rowe G, Siegrist M, Vereijken C. Consumer response to novel agri-food technologies: implications for predicting consumer acceptance of emerging food technologies. Trends Food Sci Technol. 2011;22:442–56.

Fuglie KO, Toole AA. The evolving institutional structure of public and private agricultural research. Am J Agric Econ. 2014;96:862–83.

Gardner JG, Nehring RF, Nelson CH. Genetically modified crops and household labor savings in US crop production. AgBioforum. 2009;12:303–12.

Gerpacio RV, Aquino AP, editors. Socioeconomic impacts of Bt eggplant: ex-ante case studies in The Philippines. Ithaca/Los Baños: International Service for the Acquisition of Agri-biotech Applications and SEAMEO Southeast Asian Regional Center for Graduate Study and Research in Agriculture; 2014.

Glenna LL, Cahoy DR. Agribusiness concentration, intellectual property, and the prospects for rural economic benefits from the emerging biofuel economy. South Rural Sociol. 2009;24:111–29.

Heinemann JA, Massaro M, Coray DS, Agapito-Tenfen SZ, Wen JD. Sustainability and innovation in staple crop production in the US Midwest. Int J Agric Sustain. 2014;12:71–88.

Hoppe RA, Korb P. Characteristics of women farm operators and their farms. Washington, DC: U.S Department of Agriculture–Economic Research Service; 2013.

Huang J, Mi J, Hai L, Wang Z, Chen R, Hu R, Rozelle S, Pray C. A decade of Bt cotton in Chinese fields: assessing the direct effects and indirect externalities of Bt cotton adoption in China. Sci China Life Sci. 2010;53:981–91.

Islam SMF, Norton GW. Bt eggplant for fruit and shoot borer resistant in Bangladesh. In: Ramasamy C, Selvaraj KN, Norton GW, Vijayaraghavan K, editors. Economic and environmental benefits and costs of transgenic crops: ex-ante assessment. Coimbatore: Tamil Nadu Agricultural University; 2007. p. 91–106.

Jefferson DJ, Graff GD, Chi-Ham CL, Bennett AB. The emergence of agbiogenerics. Nat Biotechnol. 2015;33:819–23.

Juma C. Science meets farming in Africa. Science. 2011;334:1323.

Klümper W, Qaim M. A meta-analysis of the impacts of genetically modified crops. PLoS One. 2014;9:e111629.

Krishna VV, Qaim M, Zilberman D. Transgenic crops, production risk, and agrobiodiversity. Eur Rev Agric Econ. 2016;43:137–64.

Lang JT. Elements of public trust in the American food system: experts, organizations, and genetically modified food. Food Policy. 2013;41:145–54.

Lang JT, Hallman WK. Who does the public trust? The case of genetically modified food in the United States. Risk Anal. 2005;25:1241–52.

Lobao L, Stofferahn C. The community effects of industrialized farming: social science research and challenges to corporate farming laws. Agric Hum Values. 2008;25:219–40.

Luttrell RG, Jackson RE. Helicoverpa zea and Bt cotton in the United States. GM Crops Food. 2012;3:213–27.

Lyson TAT, Robert J, Welsh R. Scale of agriculture production, civic engagement and community welfare. Soc Forces. 2001;80:311–27.

Mannion AM, Morse S. GM crops 1996–2012: a review of agronomic, environmental and socio-economic impacts, Working Paper 04/13. Center for Environmental Strategy, University of Surrey; 2013. [May 9, 2016]. https://www.surrey.ac.uk/ces/files/pdf/04-13%20Morse_Mannion_GM%20Crops.pdf.

NRC (National Research Council). The impact of genetically engineered crops on farm sustainability in the United States. Washington, DC: National Academies Press; 2010.

Parentoni SN, Augusto de Miranda R, Garcia JC. Implications on the introduction of transgenics in Brazilian maize breeding programs. Crop Breed Appl Biotechnol. 2013;13:9–22.

Parisi C, Tillie P, Rodríguez-Cerezo E. The global pipeline of GM crops out to 2020. Nat Biotechnol. 2016;34:31–6.

Pechlaner G. Corporate crops: biotechnology, agriculture, and the struggle for control. Austin: University of Texas Press; 2012.

Peter G, Bell MM, Jarnagin S, Bauer D. Coming back across the fence: masculinity and the transition to sustainable agriculture. Rural Sociol. 2000;65:215–33.

Phillips DM, Hallman WK. Consumer risk perceptions and marketing strategy: the case of genetically modified food. Psychol Mark. 2013;30:739–48.

Qaim M, Kouser S. Genetically modified crops and food security. PLoS One. 2013;8:e64879.

Racovita M, Oboryo DN, Craig W, Ripandelli R. What are the non-food impacts of GM crop cultivation on farmers' health. Environ Evid. 2015;3:1.

Rissing AL. Iowan women farmers' perspectives on alternative agriculture and gender. Journal of Agriculture, Food Systems & Community Development. 2012;3:127–36.

Rollin F, Kennedy J, Wills J. Consumers and new food technologies. Trends Food Sci Technol. 2011;22:99–111.

Sexton S, Zilberman D. Land for food and fuel production: the role of agricultural biotechnology. In: Graff Zivin JS, Perloff JM, editors. The intended and unintended effects of U.S. agricultural and biotechnology policies. Chicago: University of Chicago Press; 2012. p. 269–88.

Shi G, Chavas J-P, Stiegert KW. Pricing of herbicide-tolerant seeds: a market structure approach. AgBioforum. 2009;12:326–33.

Smale M, Zambrano P, Gruère G, Falck-Zepeda J, Matuschke I, Horna D, Nagarajan L, Yerramareddy I, Jones H. Measuring the economic impacts of transgenic crops in developing agriculture during the first decade: approaches, findings, and future directions. Washington, DC: International Food Policy Research Institute; 2009.

Smale M, Zambrano P, Paz-Ybarnegaray R, Fernández-Montaño W. A case of resistance: herbicide-tolerant soybeans in Bolivia. AgBioforum. 2012;15:191–205.

Stiegert K, Shi G, Chavas JP. Innovation, integration, and biotech revolution: the case of U.S. seed markets. Choices. 2010;25

Strauss DM. The application of TRIPS to GMOs: international intellectual property rights and biotechnology. Stanford J Int Law. 2009;45:287–320.

MIX
Papier aus verantwortungsvollen Quellen
Paper from responsible sources
FSC® C105338

If you have any concerns about our products,
you can contact us on
ProductSafety@springernature.com

In case Publisher is established outside the EU,
the EU authorized representative is:
Springer Nature Customer Service Center GmbH
Europaplatz 3, 69115 Heidelberg, Germany

Printed by Libri Plureos GmbH
in Hamburg, Germany